2008
Advanced
Biology A2
Student Workbook

Advanced *Biology A2* 2008

Student Workbook

Previous annual editions 1999-2007
Tenth edition 2008 (second printing)

ISBN 978-1-877329-97-5

Copyright © **2007** Richard Allan
Published by **BIOZONE International Ltd**

Printed by REPLIKA PRESS PVT LTD using paper produced from renewable and waste materials

About the Writing Team

Tracey Greenwood joined the staff of Biozone at the beginning of 1993. She has a Ph.D in biology, specialising in lake ecology, and taught undergraduate and graduate biology at the University of Waikato for four years.

Lyn Shepherd joined as an author in 2003, bringing her 20 years experience as a secondary school biology teacher to Biozone.

Lissa Bainbridge-Smith joined Biozone as an author this year after working in industry for 8 years. Lissa has an MSc. from Waikato University, majoring in aquatic toxicology.

Richard Allan has had 11 years experience teaching senior biology at Hillcrest High School in Hamilton, New Zealand. He attained a Masters degree in biology at Waikato University, New Zealand.

As always, the authors acknowledge and thank our graphic artist Dan Butler for his continued dedication to the job at hand.

Purchases of this workbook may be made direct from the publisher:

BIOZONE

www.biozone.co.uk

UNITED KINGDOM:

BIOZONE Learning Media (UK) Ltd.
P.O. Box 23698, Edinburgh EH5 2WX, **Scotland**
Telephone: (131) 557 5060
FAX: (131) 557 5030
E-mail: sales@biozone.co.uk

AUSTRALIA:

BIOZONE Learning Media Australia
P.O. Box 7523, GCMC 4217 QLD, **Australia**
Telephone: +61 (7) 5575 4615
FAX: +61 (7) 5572 0161
E-mail: info@biozone.com.au

NEW ZEALAND:

BIOZONE International Ltd.
P.O. Box 13-034, Hamilton, **New Zealand**
Telephone: +64 (7) 856 8104
FAX: +64 (7) 856 9243
E-mail: sales@biozone.co.nz

CONTENTS (continued)

CODES: Δ **Upgraded** this edition   ☆ **New** this edition   Activity is marked: • to be done; ✓ when completed

CONTENTS *(continued)*

Supplementary Material for A2 on the
Teacher Resource CD-ROM

AQA-A: Biology & Human Biology

△ Ecosystems and Energy
Biology: Module 5: 14.5, 14.7, 14.9

△ Physiology and the Environment
Biology: Module 6: 15.1-15.2, 15.4-15.7

△ The Human Life Span
Human Biology: Module 7: 16.1-16.4, 16.6, 16.9, 16.11

AQA-B: Biology

△ Energy Flow and Nutrient Cycles
Module 5a: 14.1-14.2

△ Applied Ecology .... Option Module 6

△ Microbes and Disease .... Option Module 7

△ Behaviour and Populations .... Option Module 8

CIE: Applications in Biology

△ Biodiversity and Conservation .... Q

△ Gene Technology .... R

△ Biotechnology .... S

△ Crop Plants .... T

△ Aspects of Human Reproduction .... U

OCR

△ Growth, Development & Reproduction .. Option 01

△ Applications of Genetics .... Option 02

△ Environmental Biology .... Option 03

△ Microbiology and Biotechnology .... Option 04

△ Mammalian Physiology & Behaviour ...... Option 05

Edexel: Biology & Human Biology

Farming and Conservation
Biology: Unit 5B3 (in part)
Human Biology: Unit 5H1 (in part)

Chromosome Mutations & Aneuploidy
Biology: Unit 5B.4 (in part)
Human Biology: Unit 5H.1 (in part)

Gene Technology
Biology: Unit 5B.4 (in part)
Human Biology: Unit 5H.1 (in part)

Human Evolution
Human Biology: Unit 5H.2

△ Microbiology and Biotechnology .... Option A

△ Food Science .... Option B

△ Human Health and Fitness .... Option C

SNAB

△ Conservation Issues
Unit 4, Topic 5: 4.5.17-4.5.19

△ Bacteria and Viruses
Unit 4, Topic 6: 4.6.4

△ Muscles and Performance
Unit 5, Topic 7: 5.7.10-5.7.11

△ Scans and Diagnosis
Unit 5, Topic 8: 5.8.8

WJEC: Biology

Digestion and Absorption
Module BI4: 4.4

Microbiology
Module BI4: 4.5, 4.8

Chromosome Mutations & Aneuploidy
Module BI5: 5.1

Fruits and Seeds
Module BI5: 5.3(b)

Biology Higher (AS & A2)

Viruses
Unit 1 (e): part i

Chromosome Mutations & Aneuploidy
Unit 2 (a): part iii-v

Adaptation
Unit 2 (c): part 2 (i-v)

Environmental Influences
Unit 3 (a): part 4

Monitoring Populations
Unit 3 (c): part iii

CCEA: Biology (AS & A2)

The Digestive System
Module 2: 2.1, part c (iii)

Biological Pest Control
Module 2: 2.3 (in part)

Human Impact
Module 4: 4.8

Reproductive Strategies
Module 5: 5.2

Chromosome Mutation & Aneuploidy
Module 5: 5.4 (in part)

Aspects of Diversity
Module 5: 5.6, 5.8-5.10 (in part)

CODES: △ **Upgraded** this edition ☆ **New** this edition **Activity** is marked: ⚬ to be done; ☑ when completed

Preface to the 2008 Edition

This tenth edition of the workbook is designed to meet the needs of students enrolled in biology courses for: **AQA specifications A** and **B**, **EDEXCEL**, **OCR**, Cambridge International Examinations (**CIE**), and Salters-Nuffield Advanced Biology (**SNAB**). It is also suitable for senior level biology courses in **Wales**, **Northern Ireland**, and **Scotland**. Previous editions of this workbook have received very favourable reviews (see our web site: **www.biozone.co.uk** for details). A new feature this year is the page tab identifying "**Related activities**" in the workbook and "**Web links**". These will help students to locate related material within the workbook and provides access to web links and activities (including animations) that will enhance their understanding of the topic. See page 11 to find out more about these. A further enhancement is the placement of two page activities back-to-back facilitating easier removal of discrete activities without upsetting the flow of information. Supplementary material and extension activities continue to be available with a limited photocopy licence on Biozone's **Teacher Resource CD-ROM**. A guide to using the Teacher Resource CD-ROM (TRC) to best effect has been included in the introductory section of the workbook and can be used in conjunction with the revised course guides. As in previous years, we have updated the listed resources and include crosswords on the TRC as a student-friendly way in which to test vocabulary. Revised and new activities are indicated in the contents pages. These annual upgrades are in keeping with our ongoing commitment to providing up-to-date, relevant, interesting, and accurate information.

A Note to the Teacher

This workbook has been produced as a student-centred resource, and benefits students by facilitating independent learning and critical thinking. Biozone's workbooks motivate and challenge a wide range of students by providing a highly visual format, a clear map through the course, and a synopsis of supplementary resources. In modern biology, a single textbook may no longer provide all the information a student needs to grasp a topic. This workbook is a generic resource and **not a textbook**, and we make a point of referencing texts from other publishers. Above all, we are committed to continually revising and improving this resource **every year**. The price remains below £11 for students, as a reflection of our commitment to providing high-quality, cost effective resources for biology. Please **do not photocopy** from this workbook. We cannot afford to supply single copies to schools and still provide annual updates as we intend. If you think it is worth using, then we recommend that the students themselves own this resource and keep it for their own use. A free model answer book is supplied with your **first order** of 5 or more workbooks.

How Teachers May Use This Workbook

This workbook may be used in the classroom to guide students through each topic. Some activities may be used to introduce topics while others may be used to consolidate and test concepts already covered by other means. The workbook may be used as the primary tool in teaching some topics, but it should not be at the expense of good, 'hands-on' biology. Students may attempt the activities on their own or in groups. The latter provides opportunities for healthy discussion and peer-to-peer learning. Many of the activities may be set as homework exercises. Each page is perforated, allowing for easy removal of pages to be submitted for marking. This has been facilitated by the back-to-back format of two page activities. Teachers may prescribe the activities to be attempted by the students (using the check boxes next to the objectives for each topic), or they may allow students a degree of freedom with respect to the activities they attempt. The objectives for each topic will allow students to keep up to date even if they miss lessons and teachers who are away from class may set work easily in their absence. I thank you for your support.

Richard Allan

Acknowledgements

We would like to thank those who have contributed towards this edition: • Campus Photography at the University of Waikato for photographs of monitoring instruments • Dr Douglas Cooper and the University of California San Francisco, for the use of the SEM of a podocyte (http://www.sacs.ucsf.edu/home/cooper/Anat118/urinary/urinary98.htm) • Corel Corporation, for use of their eps clipart of plants and animals from the Corel MEGAGALLERY collection • Genesis Research and Development Corp. Auckland, for the photo used on the HGP activity • Human Genome Sciences, Inc. for photos of large scale DNA sequencing • Kurchatov Inst., for the photographs of Chornobyl • Photo of a swollen thorn *Acacia* courtesy of Alex Wild • Photo of Paul Thomas courtesy of Wellington Harrier Athletic Club • Photo of Richard Hall courtesy of Helen Hall • Cereal Research Centre, Agriculture and Agri-Food Canada, for their use of the photograph of a pheromone trap • Missouri Botanical Gardens for their photograph of egg mimicry in *Passiflora* • Jan Morrison for her line diagrams • Sirtrack NZ Ltd, for the photo of penguin radiotracking • TechPool Studios, for their clipart collection of human anatomy: Copyright ©1994, TechPool Studios Corp. USA (some of these images were modified by Richard Allan and Tracey Greenwood) • Totem Graphics, for their clipart collection of plants and animals • Janice Windsor, for her photograph of a baboon in East Africa The 3D modelling by Dan Butler using Poser IV, Curious Labs and Bryce.

Photo Credits

Royalty free images, purchased by Biozone International Ltd, are used throughout this workbook and have been obtained from the following sources: istockphotos (www.istockphoto.com) • Corel Corporation from various titles in their Professional Photos CD-ROM collection; IMSI (International Microcomputer Software Inc.) images from IMSI's MasterClips® and MasterPhotosTM Collection, 1895 Francisco Blvd. East, San Rafael, CA 94901-5506, USA; ©1996 Digital Stock, Medicine and Health Care collection; ©Hemera Technologies Inc, 1997-2001; © 2005 JupiterImages Corporation www.clipart.com; ©Click Art, ©T/Maker Company; ©1994., ©Digital Vision; Gazelle Technologies Inc.; PhotoDisc®, Inc. USA, www.photodisc.com • Human 3D models were created using Poser IV, Curious Labs. Photos kindly provided by individuals or corporations have been indentified by way of coded credits as follows: **BH**: Brendan Hicks (University of Waikato), **BOB**: Barry O'Brien (Uni. of Waikato), **CDC**: Centers for Disease Control and prevention, Atlanta, USA, **COD**: Colin O'Donnell, **DoC**: Dept of Conservation NZ, **EII**: Education Interactive Imaging, **EW**: Environment Waikato, **HGSI**: Dena Borchardt, at Human Genome Sciences Inc., **JDG**: John Green (University of Waikato), **JR-PE**: Jane Roskruge, **NASA**: National Aeronautics and Space Administration, **RA**: Richard Allan, **RCN**: Ralph Cocklin, **TG**: Tracey Greenwood, **WBS**: Warwick Silvester (Uni. of Waikato), **UCSF**: University of California San Francisco, **WMU**: Waikato Microscope Unit.

Special thanks to all the partners of the Biozone team for their support.

Cover Photographs

Main photograph: Cheetahs (*Acinonyx jubatus*) are the smallest and most specialised of the so-called big cats. Well known as the fastest land mammal, cheetahs are relatively non-aggressive, relying on speed for hunting success. Cheetahs are the sole member of their genus and their conservation has been hampered by the lack of genetic diversity and inbred nature of the population, a legacy of a genetic bottleneck about 10,000 years ago. PHOTO: photos.com ©JUPITERIMAGES.

Background photograph: Autumn leaves, Image ©2005 JupiterImages Corporation www.clipart.com

Contents

CODES: Δ **Upgraded** this edition   ☆ **New** this edition   **Activity** is marked: ▪ to be done;   ✓ when completed

How to Use this Workbook

This workbook is designed to provide you with a resource that will make the subject of biology more enjoyable and fun to study. The workbook addresses the second year requirements (**A2 schemes**) for the following biology courses: **AQA specifications A and B, Edexcel, OCR,** Cambridge International Examinations (**CIE**), and the Salters-Nuffield Advanced Biology (**SNAB**). This workbook is also appropriate for senior level biology courses in **Northern Ireland, Wales,** and **Scotland (Biology Higher)**. The Syllabus Guides on pages 13-19 (or on the Teacher Resource CD-ROM) indicate where the material required by your syllabus is covered. This workbook will reinforce and extend the ideas developed by your teacher. It is **not a textbook**; its aim is to complement the texts written for your course. The workbook provides the following useful resources for each topic:

Guidance Provided for Each Topic

Topic outcomes:
This panel identifies the learning objectives relevant to the topic for each designated course. Attempt only those objectives that relate to your course. See pages 13-19 for a listing of your syllabus requirements.

Learning objectives:
These provide a map of the topic content. Completing the relevant learning objectives will help you to satisfy the knowledge requirements of your course. Your teacher may add to or omit points from this list.

Key words:
Key words are displayed in **bold** type in the learning objectives and should be used to create a glossary as you study each topic. From your own reading and your teacher's descriptions, write your own definition for each word. Only the terms relevant to your learning objectives should be used to create your glossary. Free glossary worksheets are also available from our web site.

Use the check boxes to mark objectives to be completed.
Use a **dot** to be done (•).
Use a **tick** when completed (✓).

Comprehensive textbooks
The *Textbook Reference Grid* on pages 8-9 lists the major comprehensive textbooks available for your course (texts providing coverage of all or most of the course). The grid provides the page or chapter numbers from each text relevant to each topic in the workbook.

Supplementary texts:
References to supplementary texts, which have only a restricted topic coverage, are provided as appropriate in each topic.

Supplementary resources from Biozone Supporting Presentation MEDIA are noted where appropriate. Computer software and videos relevant to every topic in the workbook are provided on the **Teacher Resource CD-ROM** (which may be purchased separately). See page 12 for details.

Periodical articles:
Ideal for those seeking more depth or the latest research on a specific topic. Articles are sorted according to their suitability for student or teacher reference. Visit your school, public, or university library for these articles.

Internet addresses:
Access our database of links to more than **800** web sites (updated regularly) relevant to topics covered in the manual. Go to Biozone's own web site: **www.biozone.co.uk** and link directly to listed sites using the *BioLinks* button.

Activity Pages

The activities and exercises make up most of the content of this workbook and they are designed to reinforce the concepts you have covered. Your teacher may use the activity pages to introduce a topic for the first time, or you may use them to revise ideas already covered. They are excellent for use in the classroom, and as homework exercises and revision. In most cases, the activities should not be attempted until you have carried out the necessary background reading from your textbook. Your teacher should have a **model answers book** with the answers to each activity (see the Index for details). Because this workbook caters for more than one exam board, you will find some activities that may not be relevant to your course. Although you will miss out these pages, our workbooks still represent exceptional value.

Introductory paragraph:
The introductory paragraph sets the 'scene' for the focus of the page. Note any words that appear in **bold**, as they are 'key words' worthy of including in a glossary of biological terms for the topic.

Easy to understand diagrams:
The main ideas of the topic are represented and explained by clear, informative diagrams.

Tear-out pages:
Each page of the book has a perforation that allows easy removal. Your teacher may ask you to remove activity pages from this workbook for marking. Your teacher may also ask you to tear out the pages and place them in a folder with other work on the topic.

Write-on format:
Your understanding of the main ideas of the topic is tested by asking questions and providing spaces for your answers. Where indicated by the space available, your answers should be concise. Questions requiring more explanation or discussion are spaced accordingly. Answer the questions adequately according to the questioning term used (see the opposite page).

Activity code and links:
Activity codes (explained right) help to identify the type of activities and the skills they require. Most activities require knowledge recall as well as the application of knowledge to explain observations or predict outcomes.

Use the **Related activities** indicated to visit pages that may help you with understanding the material or answering the questions.

Web links indicate additional material of assistance or interest (either web pages or pdf activities). You can access these from: *www.biozone.co.uk/weblink/UK2-9975.html*

Activity Level

1	=	Simple questions not requiring complex reasoning
2	=	Some complex reasoning may be required
3	=	More challenging, requiring integration of concepts

PRDA 3

Related activities: Gene Pool Exercise, Genetic Drift
Web links: Genetic Drift Simulation, Oceanic Island Colonisers

Type of Activity

D	=	Includes some data handling and/or interpretation
P	=	includes a paper practical
R	=	May require research outside the page
A	=	Includes application of knowledge to solve a problem
E	=	Extension material

Explanation of Terms

Questions come in a variety of forms. Whether you are studying for an exam or writing an essay, it is important to understand exactly what the question is asking. A question has two parts to it: one part of the question will provide you with information, the second part of the question will provide you with instructions as to how to answer the question. Following these instructions is most important. Often students in examinations know the material but fail to follow instructions and do not answer the question appropriately. Examiners often use certain key words to introduce questions. Look out for them and be clear as to what they mean. Below is a description of terms commonly used when asking questions in biology.

Commonly used Terms in Biology

The following terms are frequently used when asking questions in examinations and assessments. Students should have a clear understanding of each of the following terms and use this understanding to answer questions appropriately.

Account for: Provide a satisfactory explanation or reason for an observation.

Analyse: Interpret data to reach stated conclusions.

Annotate: Add **brief** notes to a diagram, drawing or graph.

Apply: Use an idea, equation, principle, theory, or law in a new situation.

Appreciate: To understand the meaning or relevance of a particular situation.

Calculate: Find an answer using mathematical methods. Show the working unless instructed not to.

Compare: Give an account of similarities and differences between two or more items, referring to both (or all) of them throughout. Comparisons can be given using a table. Comparisons generally ask for similarities more than differences (see contrast).

Construct: Represent or develop in graphical form.

Contrast: Show differences. Set in opposition.

Deduce: Reach a conclusion from information given.

Define: Give the precise meaning of a word or phrase as concisely as possible.

Derive: Manipulate a mathematical equation to give a new equation or result.

Describe: Give a detailed account, including all the relevant information.

Design: Produce a plan, object, simulation or model.

Determine: Find the only possible answer.

Discuss: Give an account including, where possible, a range of arguments, assessments of the relative importance of various factors, or comparison of alternative hypotheses.

Distinguish: Give the difference(s) between two or more different items.

Draw: Represent by means of pencil lines. Add labels unless told not to do so.

Estimate: Find an approximate value for an unknown quantity, based on the information provided and application of scientific knowledge.

Evaluate: Assess the implications and limitations.

Explain: Give a clear account including causes, reasons, or mechanisms.

Identify: Find an answer from a number of possibilities.

Illustrate: Give concrete examples. Explain clearly by using comparisons or examples.

Interpret: Comment upon, give examples, describe relationships. Describe, then evaluate.

List: Give a sequence of names or other brief answers with no elaboration. Each one should be clearly distinguishable from the others.

Measure: Find a value for a quantity.

Outline: Give a brief account or summary. Include essential information only.

Predict: Give an expected result.

Solve: Obtain an answer using algebraic and/or numerical methods.

State: Give a specific name, value, or other answer. No supporting argument or calculation is necessary.

Suggest: Propose a hypothesis or other possible explanation.

Summarise: Give a brief, condensed account. Include conclusions and avoid unnecessary details.

In Conclusion

Students should familiarise themselves with this list of terms and, where necessary throughout the course, they should refer back to them when answering questions. The list of terms mentioned above is not exhaustive and students should compare this list with past examination papers / essays etc. and add any new terms (and their meaning) to the list above. The aim is to become familiar with interpreting the question and answering it appropriately.

Resources Information

Your set textbook should always be a starting point for information. There are also many other resources available, including scientific journals, magazine and newspaper articles, supplementary texts covering restricted topic areas, dictionaries, computer software and videos, and the internet.

A synopsis of currently available resources is provided below. Access to the publishers of these resources can be made directly from Biozone's web site through our resources hub: **www.biozone.co.uk/resource-hub.html**, or by typing in the relevant addresses provided below. Most titles are also available through amazon.co.uk. Please note that our listing any product in this workbook does not, in any way, denote Biozone's endorsement of that product.

Comprehensive Biology Texts Referenced

Appropriate texts for this course are referenced in this workbook. Page or chapter references for each text are provided in the text reference grid on page 8. These will enable you to identify the relevant reading as you progress through the activities in this workbook. Publication details of texts referenced in the grid are provided below and opposite. For further details of text content, or to make purchases, link to the relevant publisher via Biozone's resources hub or by typing: **www.biozone.co.uk/resources/uk-comprehensive-pg1.html**

Bailey, M. and K. Hirst, 2001
Collins Advanced Modular - Biology A2, 2/edn
Publisher: HarperCollins Publishers Ltd.
Pages: 250
ISBN: 0-00-327752-6
Comments: *Appropriate to AQA A2-level Biology (B) specification. Student support packs for AQA B and Edexcel specifications are available at extra cost.*

Baker, M., B. Indge, and M. Rowland, 2001
AQA Biology Specification A: Further Studies in Biology (A2)
Publisher: Hodder & Stoughton
Pages: 276
ISBN: 0-340-80244-8
Comments: *Written specifically for the AQA biology specification A. The text covers, in full, the first two modules of the A level course. Extension material is provided.*

Boyle, M. and K. Senior, 2002
Collins Advanced Science: Biology, 2/e
Publisher: HarperCollins Publishers Ltd.
Pages: 675
ISBN: 0-00-713600-5
Comments: *Fully revised to reflect changes in the specifications. Includes a new section on plant biology and a new chapter on biotechnology.*

Bradfield, P., J. Dodds, J. Dodds, and N. Taylor, 2002
A Level Biology
Publisher: Longman
Pages: 464
ISBN: 0-582-42945-5
Comments: *Written to match A2 level specifications, providing synoptic questions and core content.*

Clegg, C.J., 2000
Introduction to Advanced Biology
Publisher: John Murray
Pages: 528
ISBN: 0-7195-7671-7
Comments: *Matches new AS/A2 specifications, but aims to provide a bridge from GCSE (grade C pass) to A level. Detailed specification matches are available on CD-ROM, FREE on demand.*

Clegg, C.J. and D.G. McKean, 2000
Advanced Biology: Principles and Applications 2/e
Publisher: John Murray
Pages: 712
ISBN: 0-7195-7670-9
Comments: *Student study guide also available. A general text with specific references for use by biology students of the AS and A2 curricula in the UK.*

Fullick, A., 2000
Heinemann Advanced Science: Biology, 2 edn
Publisher: Heinemann Educational Publishers
Pages: 760
ISBN: 0-435-57095-1
Comments: *Suitable as a general text for use with all specifications. The earlier edition (1994) is still available.*

Jones, M., R. Fosbery, D. Taylor, and J. Gregory, 2007
CIE Biology AS and A Level, 2 ed.
Publisher: Cambridge University Press
Pages: 424
ISBN: 0-521-53674-X
Comments: *This text meets the new CIE requirements and covers the complete AS level syllabus, the core A level syllabus, and the new Applications of Biology section.*

Jones, M., and J. Gregory, 2001
Biology 2
Publisher: Cambridge University Press
Pages: 152
ISBN: 0-521-79714-4
Comments: *Provides coverage of the core material for the second year of Advanced Level Biology (OCR), and covers the Advanced Level module, 'Central Concepts'.*

Jones, M. and G. Jones, 1997
Advanced Biology
Publisher: Cambridge University Press
Pages: 560
ISBN: 0-521-48473-1
Comments: *Provides full coverage of the core material included in all advanced biology syllabuses. Suitable for students studying for a range of qualifications.*

Kent, N. A. 2000
Advanced Biology
Publisher: Oxford University Press
Pages: 624
ISBN: 0-19-914195-9
Comments: *Each book comes with a free CD-ROM to help with specification planning. Book is formatted as a series of two page concept spreads.*

Lea, C., P. Lowrie, and S. McGuigan, (S. McGuigan, ed), 2001
Advanced Level Biology for AQA: A2
Publisher: Heinemann Educational Publishers
Pages: 315
ISBN: 0-435-58081-7
Comments: *Appropriate for AQA specification B. Accompanying Resource Pack, providing support material, and extra questions and tasks, is also available.*

Roberts, M., G. Monger, M. Reiss, 2000
Advanced Biology
Publisher: NelsonThornes
Pages: 780
ISBN: 0-17-438732-6
Comments: *Provides complete coverage of the new Advanced Level specifications. Additional support provided from a complementary website.*

SNAB Project Team, 2005
Salters-Nuffield Advanced Biology AS Student Book
Publisher: Heinemann Educational Publishers
Pages: 229
ISBN: 978-0-435628-57-4
Comments: *Developed for the SNAB specification using an innovative contextual approach. Although an AS text, this volume contains reading for A2 coverage in Biozone's A2 workbook.*

SNAB Project Team, May 2006
Salters-Nuffield Advanced Biology A2 Student Book
Publisher: Heinemann Educational Publishers
Pages: information not available
ISBN: 978-0-435628-58-1
Comments: *Developed for the SNAB specification using an innovative contextual approach. Students explore basic biological concepts within relevant and high interest contexts, e.g. biodiversity and forensics.*

Taylor D.J., N.P.O. Green, & G.W. Stout, 1997
Biological Science 1 & 2, 3 ed. (hardback)
Publisher: Cambridge University Press
Pages: 984
ISBN: 0-521-56178-7
Comments: *An older but comprehensive and accurate text, with clear line diagrams and photographs and methodical treatment of a wide range of topic areas.*

Toole G. and S. Toole, 2000
New Understanding Biology for Advanced Level
Publisher: NelsonThornes
Pages: 672
ISBN: 0-7487-3964-5
Comments: *Fully revised to match AS/A2 specifications. Also available with study guide, which provides support material.*

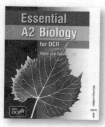

Toole G. and S. Toole, 2004
Essential A2 Biology for OCR
Publisher: NelsonThornes
Pages: approx 170
ISBN: 0-7487-8518-3
Comments: *An appealing and student-friendly text written to meet the core A2 specifications for OCR. A comprehensive glossary is included.*

Williams, G., 2000
Advanced Biology for You
Publisher: NelsonThornes
Pages: 464
ISBN: 0-7487-5298-6
Comments: *Covers the current AS level specifications and the core topics of all A2 specifications.*

Supplementary Texts

For further details of text content, or to make purchases, link to the relevant publisher via Biozone's resources hub or by typing: www.biozone.co.uk/resources/uk-supplementary-pg1.html

Barnard, C., F. Gilbert, F., and P. McGregor, 2007
Asking Questions in Biology: Key Skills for Practical Assessments & Project Work, 256 pp.
Publisher: Benjamin Cummings
ISBN: 978-0132224352
Comments: *Covers many aspects of design, analysis and presentation of practical work.*

Cadogan, A. and Ingram, M., 2002
Maths for Advanced Biology
Publisher: NelsonThornes
ISBN: 0-7487-6506-9
Comments: *Fully covers the new maths requirements of the AS/A2 biology specifications. Includes worked examples.*

Freeland, P., 1999
Hodder Advanced Science: Microbes, Medicine, and Commerce, 160 pp.
Publisher: Hodder and Stoughton
ISBN: 0340731036
Comments: *Combines biotechnology, pathology, microbiology & immunity in a comprehensive text.*

Fullick, A., 1998
Human Health and Disease, 162 pp.
Publisher: Heinemann Educational Publishers
ISBN: 0435570919
Comments: *An accompanying text for courses with modules in human health and disease. Covers both infectious and non-infectious disease.*

Indge, B., 2004
Data and Data Handling for A2 Level Biology, 96 pp.
Publisher: Hodder Arnold H&S
ISBN: 0340812753
Comments: *Develops the data handling skills required at A2 level. Suitable for all specifications.*

Jones, A., R. Reed, and J. Weyers, 4th ed. 2007
Practical Skills in Biology, approx. 300 pp.
Publisher: Pearson
ISBN: 978-0-131775-09-3
Comments: *Provides information on all aspects of experimental and field design, implementation, and data analysis. Contact www.amazon.co.uk*

Morgan, S., 2002
Advanced Level Practical Work for Biology
Publisher: Hodder and Stoughton
ISBN: 0-340-84712-3
Comments: *128 pages catering for the practical and investigative requirement of A level. Covers experimental planning, technique, observations and measurement, and interpretation and analysis.*

Rowett, H.G.Q, 1999
Basic Anatomy & Physiology, 4 ed. 132 pp.
Publisher: John Murray
ISBN: 0-7195-8592-9
Comments: *A revision of a well established reference book for the basics of human anatomy and physiology. Accurate coverage of required AS/A2 content with clear, informative diagrams.*

6

Advanced Biology Readers (John Murray Publishers)
Two supplementary texts supporting Advanced Biology options. Useful as teacher reference and student extension for core topics.

Chenn, P., 1997.
Microorganisms and Biotechnology, 176 pp.
ISBN: 0-71957-509-5
Thorough coverage of the nature of microorganisms, their culture and growth, and their various roles in biotechnology. It includes chapters on the genetic engineering of microbes and enzyme technology.

Jones, N., A. Karp., & G. Giddings, 2001.
The Essentials of Genetics, 224 pp.
ISBN: 0-7195-8611-9
Thorough supplemental for genetics and evolution. Comprehensive coverage of cell division, molecular genetics, and genetic engineering is provided, and the application of new gene technologies to humans is discussed in a concluding chapter.

Bath Advanced Sciences Series (NelsonThornes)
The Bath Advanced Sciences series provides a number of titles (including a full text) to cover Advanced Level material for biology.

Taylor, J., 2001.
Microorganisms and Biotechnology, 192 pp.
ISBN: 0-17-448255-8
Comments: *Provides information for microbiology and biotechnology topics at A2 level. Other older titles in this series include Applied Genetics, 1990 (ISBN: 0174385110) and Biochemistry and Molecular Biology, 1994 (ISBN: 0174482078).*

Cambridge Advanced Sciences / Advanced Biology Topics (Cambridge University Press)
Modular titles. Those in the Advanced Sciences Series cover the A2 options for OCR, but are suitable as teacher reference and student extension reading for core topics in other AS/A2 biology courses.

Dockery, M. and M. Reiss, 1999. **Advanced Biology Topics: Behaviour**, 120 pp.
ISBN: 0521597544
An introductory text covering the basics of human and animal behaviour including environmental influence, behavioural development, courtship, and social interaction.

Gregory, J. 2000.
Applications of Genetics, 82 pp.
ISBN: 0521787254
Topics covered include the basis of variation, selective breeding and heritability (including AI in humans and livestock), genetic diversity and cloning, genetic engineering (techniques and ethics), and a chapter on human genetics.

Jones, M. and G. Jones, 2002.
Mammalian Physiology and Behaviour, 104 pp.
ISBN: 0521797497
Covers mammalian nutrition, the structure and function of the liver, support and locomotion, the nervous system, and senses and behaviour. Questions and exercises are provided and each chapter includes an introduction and summary.

Lowrie, P. & S. Wells, 2000.
Microbiology and Biotechnology, 112 pp.
ISBN: 0521787238
This text covers the microbial groups important in biotechnology, basic microbiological techniques, and the various applications of microbes, including the industrial-scale production and use of microbial enzymes.

Reiss, M. & J. Chapman, 2000.
Environmental Biology, 104 pp.
ISBN: 0521787270
An introduction to environmental biology covering agriculture, pollution, resource conservation and conservation issues, and practical work in ecology. Questions and exercises are provided and each chapter includes an introduction and summary.

Taylor, D. 2001.
Growth, Development & Reproduction, 120 pp.
ISBN: 0521787211
Includes coverage of asexual reproduction, sexual reproduction in flowering plants and mammals, and regulation of growth and development. Questions and exercises are provided and each chapter includes an introduction and summary.

Collins Advanced Modular Sciences (HarperCollins)
Modular-style texts covering material for the A2 options for AQA specification B, but suitable as teacher reference and student extension reading for core topics in other A2 biology courses.

Allen, D, M. Jones, and G. Williams, 2001.
Applied Ecology, 104 pp.
ISBN: 0-00-327741-0
Includes coverage of methods in practical ecology, the effects of pollution on diversity, adaptations, agricultural ecosystems and harvesting (including fisheries), and conservation issues. Local examples are emphasised.

Hudson, T. and K. Mannion, 2001.
Microbes and Disease, 104 pp.
ISBN: 0-00-327742-9
Coverage of selected aspects of microbiology including the culture and applications of bacteria, and the role of bacteria and viruses in disease. Immunity, vaccination, and antimicrobial drug use are covered in the concluding chapter.

Murray, P. & N. Owens, 2001.
Behaviour and Populations, 82 pp.
ISBN: 0-00-327743-7
This text covers an eclectic range of topics including patterns of behaviour, reproduction and its control, human growth and development, human populations, infectious disease, and issues related to health and lifestyle.

Nelson Advanced Sciences (NelsonThornes)
Modular-style texts covering material for the A2 specifications for Edexcel, but suitable as teacher reference and student extension reading for core topics in other AS/A2 biology courses.

Adds, J., E. Larkcom & R. Miller, 2004.
Exchange and Transport, Energy and Ecosystems, revised edition 240 pp.
ISBN: 0-7487-7487-4
Includes exchange processes (gas exchanges, digestion, absorption), transport systems, adaptation, sexual reproduction, energy and the environment, and human impact. Practical activities are included in several of the chapters.

Adds, J., E. Larkcom & R. Miller, 2003.
Molecules and Cells, revised edition 112 pp.
ISBN: 0-7487-7484-X
Includes coverage of the basic types of biological molecules, with extra detail on the structure and function of nucleic acids and enzymes, cellular organisation, and cell division. Practical activities are also provided.

Adds, J., E. Larkcom & R. Miller, 2004.
Genetics, Evolution, and Biodiversity, revised edition, 200 pp.
ISBN: 0-7487-7492-0
A range of topics including photosynthesis and the control of growth in plants, classification and quantitative field ecology, populations and pest control, conservation, Mendelian genetics and evolution, gene technology and human evolution.

Adds, J., E. Larkcom & R. Miller, 2003.
Respiration and Coordination, revised edition 208 pp.
ISBN: 0-7487-7489-0
Covers Unit 4 and options for Edexcel: metabolic pathways, internal regulation and nervous coordination, microbiology and biotechnology, food science, and health & exercise physiology.

Adds, J., E. Larkcom, R. Miller, & R. Sutton, 1999. **Tools, Techniques and Assessment in Biology**, 160 pp.
ISBN: 0-17-448273-6
A course guide covering basic lab protocols, microscopy, quantitative techniques in the lab and field, advanced DNA techniques and tissue culture, data handling and statistical tests, and exam preparation. Includes useful appendices.

Illustrated Advanced Biology (John Murray Publishers)
Modular-style texts aimed as supplements to students of AS and A2 level biology courses.

Clegg, C.J., 2002.
Microbes in Action, 92 pp.
ISBN: 0-71957-554-0
Microbes and their roles in disease and biotechnology. It includes material on the diversity of the microbial world, microbiological techniques, and a short, but useful, account of enzyme technology.

Clegg, C.J., 1999.
Genetics and Evolution, 96 pp.
ISBN: 0-7195-7552-4
Concise but thorough coverage of molecular genetics, genetic engineering, inheritance, and evolution. An historical perspective is included by way of introduction, and a glossary and a list of abbreviations used are included.

Clegg, C.J., 1998.
Mammals: Structure & Function, 96 pp.
ISBN: 0-7195-7551-6
An excellent, clearly written supplementary text covering classification and basic mammalian anatomy and physiology. Note: This text is now out of print, but many schools will have copies and it is still available from amazon books.

Clegg, C.J., 2003
Green Plants: The Inside Story, approx. 96 pp.
ISBN: 0 7195 7553 2
The emphasis in this text is on flowering plants. Topics include leaf, stem, and root structure in relation to function, reproduction, economic botany, and sensitivity and adaptation.

Biology Dictionaries
Access to a good biology dictionary is valuable when dealing with biological terms. Some selected titles are listed below. All are available from www.amazon.co.uk. For further details, link to the relevant publisher via Biozone's resources hub or by typing: **www.biozone.co.uk/resources/dictionaries-pg1.html**

Clamp, A.
AS/A-Level Biology. Essential Word Dictionary, 2000, 161 pp. Philip Allan Updates.
ISBN: 0-86003-372-4.
Carefully selected essential words for AS and A2. Concise definitions are supported by further explanation and illustrations where required.

Collin, P.H.
A Dictionary of Ecology and Environment 5 ed., 2001, 560 pp. Peter Collin Publishers Ltd
ISBN: 0747572011
A revised edition, with over 9000 definitions from all areas of ecology and environmental science, including climate and energy conservation.

Hale, W.G. **Collins: Dictionary of Biology** 4 ed., 2005, 528 pp. Collins.
ISBN: 0-00-720734-4.
Updated to take in the latest developments in biology and now internet-linked. This dictionary is specifically designed for advanced school students, and undergraduates in the life sciences.

Henderson, I.F, W.D. Henderson, and E. Lawrence. **Henderson's Dictionary of Biological Terms**, 1999, 736 pp. Prentice Hall.
ISBN: 0582414989
This edition has been updated, rewritten for clarity, and reorganised for ease of use. An essential reference and the dictionary of choice for many.

McGraw-Hill (ed). **McGraw-Hill Dictionary of Bioscience**, 2 ed., 2002, 662 pp. McGraw-Hill.
ISBN: 0-07-141043-0
22 000 entries encompassing more than 20 areas of the life sciences. It includes synonyms, acronyms, abbreviations, and pronunciations for all terms. Accessible, yet comprehensive.

Thain, M. **Penguin Dictionary of Biology**, 2004, 750 pp. Penguin.
ISBN: 0-14-101396-6
Concise reference with definitions to more than 6000 terms. It covers fundamental concepts, core vocabulary, and new advances in the subject.

Periodicals, Magazines and Journals
Articles in journals and magazines can be of great value in providing current information on specific topics. They can be accessed in your school, local, public, and university libraries. Periodicals referenced in this workbook are listed below. For further details, link to the relevant publisher via Biozone's resources hub (**www.biozone.co.uk/resources/resource-journal.html**)

Biological Sciences Review: *An informative, readable quarterly publication for teachers and students of biology.* Enquiries: Philip Allan Publishers, Market Place, Deddington, Oxfordshire OX 15 OSE.
Tel: 01869 338652 **Fax**: 01869 338803
E-mail: sales@philipallan.co.uk *or subscribe from their web site.*

New Scientist: *Widely available weekly magazine providing cuurent research summaries.* Subscription enquiries: Reed Business Information Ltd. 151 Wardour St. London WIV 4BN **Tel**: (UK and intl):+44 (0) 1444 475636 **E-mail**: ns.subs@qss-uk.com *or subscribe from their web site.*

Scientific American: *A monthly magazine containing mostly specialist feature articles. Articles range in level of reading difficulty and assumed knowledge.* Subscription enquiries: 415 Madison Ave. New York. NY10017-1111 **Tel**: (outside North America): 515-247-7631 *or subscribe from their web site.*

School Science Review: *A quarterly journal published by the ASE for science teachers in 11-19 education. SSR includes articles, reviews, and news on current research and curriculum development. Free to Ordinary Members of the ASE or available on subscription.* Subscription enquiries:
Tel: 01707 28300
Email: info@ase.org.uk *or visit their web site.*

Biologist: *Published five times a year, this journal from the IOB includes articles relevant to teachers of biology in the UK. Articles referenced in the workbook can be identified by title and volume number. The IOB also publish the Journal of Biological Education, which provides articles and reviews relevant to those in the teaching profession. Visit their web site for more information.*

Textbook Reference Grid

TOPIC IN WORKBOOK	Bailey & Hirst 2001	Baker et al. 2001	Boyle & Senior 2002	Bradfield et al. 2002	
Cellular Metabolism	2-17	93-103	133-186	Chpt 1-2	
Populations and Interactions	168-188	79-87, 108 (symbiosis)	663-689, 695-696	379-402, 409-416	
Practical Ecology	156-167	75-79	635-645 (in part)	388, 402-408	
Sources of Variation	118-124	7-13, 31-40	Chpt. 24, 27 (in part)	105-109	
Inheritance	96-117	13-27	435-444, 503-520	57-99	
Population Genetics and Speciation	124-134	41-51, 54-59, 68	521-546	110-150	
Patterns of Evolution	133-134	Not covered	Not covered	140-144	
Biodiversity and Classification	134-137	59-69	N/A	155-182	
Homeostasis	18-33	174-192	269-312	293-341	
Muscles and Movement	86-93	N/A	388-394	269-285	
Responses and Coordination	48-85	249-270	345-382	Chpt. 6-7 & 11	

Figures refer to page numbers unless indicated otherwise

TOPIC IN WORKBOOK	Kent 2000	Lea et al. 2001	Roberts et al. 2000	SNAB AS 2005	
Cellular Metabolism	86-110	10-13, 156-183	138-155, 197-213	121-124, 129, 186-187	
Populations and Interactions	512-525, 546-547	216-231	663-667, 690-717	183-185, 208	
Practical Ecology	570-572	232-242	Parts of chpt 37-39 as required	N/A	
Sources of Variation	78-79, 414-415	14-19, 38-45	461-465, 585-586, 588-594	50-1, 66-8, 82-3, 86-7, 127-8, 135, 172, 188	
Inheritance	416-431	6-9, 14-35	571-583, 587-590	84-85, 127-28, 135-43	
Population Genetics and Speciation	436-437, 442-454	46-71	Chpt. 36, 42-44 as required	172-174, 180	
Patterns of Evolution	460-461	Not covered	727-734	N/A	
Biodiversity and Classification	464-489, 494-501	See companion AS text	79-102	N/A	
Homeostasis	136-158, 184-187	79-100	267-315 as required	70	
Muscles and Movement	214-221	144-153	401-421	N/A	
Responses and Coordination	188-208, 226-33, 308-13	74-78, 101-143	345-399	N/A	

Clegg 2000	Clegg & McKean 2000	Fullick 2000	Jones & Gregory 2001	Jones *et al.* 2007	Jones & Jones 1997
169-204	211, 246-269, 314-319	177-196, 224-235	1-27, 63-64	Chpt. 15-16	169-204
374-391	48-55, 64-71, 80-83	510-513, 524-557, 570-572	28-36, 38-44	N/A	433-443, 463-468
376-377	68-69	NA	36-38	N/A	428-432
103-106, 115-116, 436-438	196-199, 616-617, 624-627	368-371, 440-42, 452-455, 461-464	47-52, 61-64	Chpt. 17 as reqd	88-94, 107-12, 147-50
439-456	604-625	429-449	51-60	Chpt. 17 as reqd	127-131, 133-143
458-471	656-663	495-505	67-79	Chpt. 18	132-133, 147-168
Not covered	664-667	Not covered	Not covered	Not covered	162
344-368	18-43	479-492	80-85	Chpt. 20	497-524
252-272	372-391 (in part)	287-294, 314-323	87-105	Chpt. 19 as reqd	323-348
304-313	484-495	156-165	N/A	N/A	362-374 as reqd
274-303	418-467, 516-522	241-285, 298-312	106-123	Chpt. 19 as reqd	297-321, 349-356

SNAB A2 2006	Taylor *et al.* 1997	Toole & Toole 2000	Toole & Toole 2004	Williams 2000
N/A	197-226, 264-277, 806	43, 269-284, 316-329	6-31	284-309
Chpt. 5	312-326, 332-333	347-348, 354-361, 366-368, 378-383	32-49	406-415, 424, 431
N/A	349-374	348-354	N/A	416-422
N/A	783-90, 814-17, 828-33	139-145, 161-162, 179-181, 186-195	50-53, 68-71, 82-83	119-122, 361, 370-373
NA	808-828	151-172	54-67	354-369
N/A	884-879, 881-885	199-210	84-97	377-384, 388
N/A	879-880	Not covered	94-95	Not covered
Chpt. 5	3-76	74-97	76-81	385-387
Chpt. 6-8 as reqd	598-608, 665-698	476-486, 497-511, 514-516	98-109, 126-131	310-325
Chpt. 7	623-639	565-574, 644-646	N/A	340-341
Chpt. 8	532-597, 609-622	526-561, 577-589	110-125, 132-135	326-348

Using the Internet

The internet is a powerful resource for locating information. There are several key areas of Biozone's web site that may be of interest to you. Go to the **BioLinks** area to browse through the hundreds of web sites hosted by other organisations. These sites provide a supplement to the activities provided in our workbooks and have been selected on the basis of their accurate, current, and relevant content. We have also provided links to biology-related **podcasts** and **RSS newsfeeds**. These provide regularly updated information about new discoveries in biology; perfect for those wanting to keep abreast of changes in this dynamic field.

The BIOZONE website: www.biozone.co.uk

The current internet address (URL) for the web site is displayed here. You can type a new address directly into this space.

Use Google to search for web sites of interest. The more precise your search words are, the better the list of results. EXAMPLE: If you type in "biotechnology", your search will return an overwhelmingly large number of sites, many of which will not be useful to you. Be more specific, e.g. "biotechnology medicine DNA uses".

Find out about our superb **Presentation Media**. These slide shows are designed to provide in-depth, highly accessible illustrative material and notes on specific areas of biology.

Podcasts: Access the latest news as audio files (mp3) that may be downloaded to your ipod (mp3 player) or played directly off your computer.

RSS Newsfeeds: See breaking news and major new discoveries in biology directly from our web site.

The **Resource Hub** provides links to the supporting resources referenced in the workbook. These resources include comprehensive and supplementary texts, biology dictionaries, computer software, videos, and science supplies.

Access the **BioLinks** database of web sites related to each major area of biology.

News: Find out about product announcements, shipping dates, and workshops and trade displays by Biozone at teachers' conferences around the world.

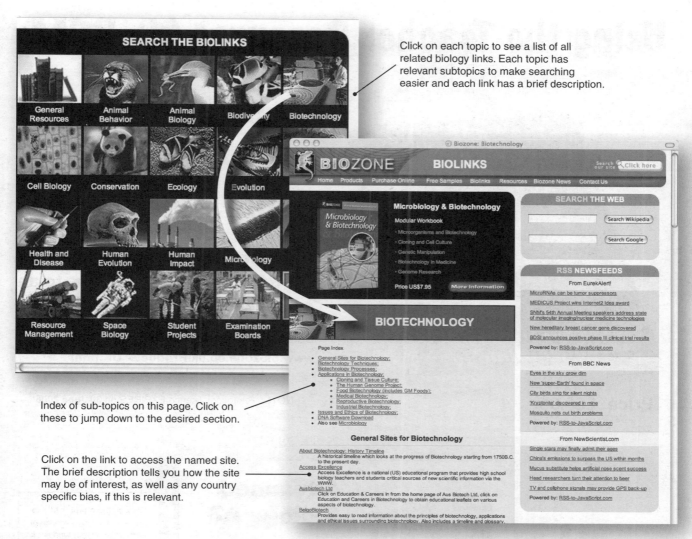

Click on each topic to see a list of all related biology links. Each topic has relevant subtopics to make searching easier and each link has a brief description.

Index of sub-topics on this page. Click on these to jump down to the desired section.

Click on the link to access the named site. The brief description tells you how the site may be of interest, as well as any country specific bias, if this is relevant.

Weblinks:

Go to: **www.biozone.co.uk/weblink/UK2-9975.html**

Throughout this workbook, some pages make reference to additional or alternative activities, as well as web sites that have particular relevance to the activity. See example of page reference below:

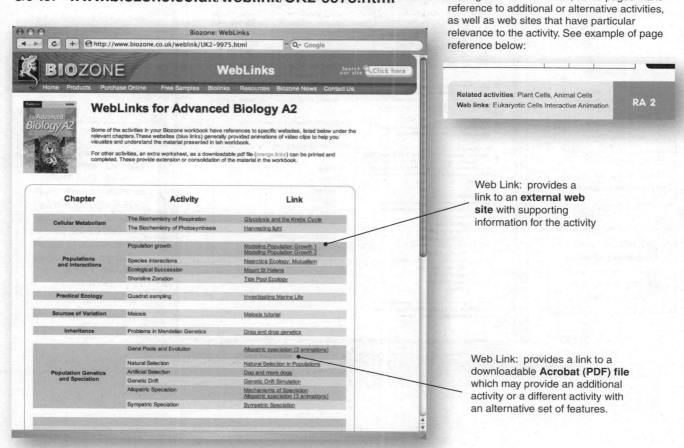

Web Link: provides a link to an **external web site** with supporting information for the activity

Web Link: provides a link to a downloadable **Acrobat (PDF) file** which may provide an additional activity or a different activity with an alternative set of features.

Using the Teacher Resource CD-ROM

Supporting resources for this workbook

Acrobat PDF files supplied on CD-ROM provide hyperlinks to an extensive collection of resources.

Details of this product can be viewed at:
www.biozone.co.uk/Products_UK.html

Price: £59.95

NOTE: Photocopy licence EXPIRES on **30 June 2008**

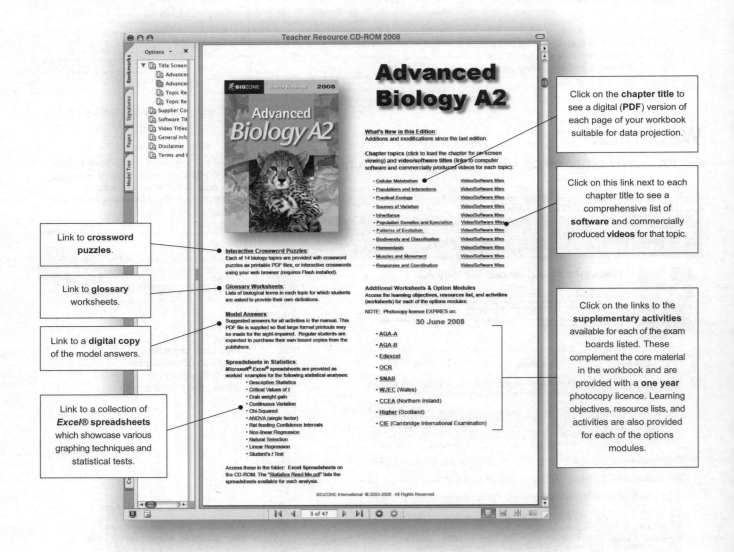

Link to **crossword puzzles**.

Link to **glossary** worksheets.

Link to a **digital copy** of the model answers.

Link to a collection of *Excel*® **spreadsheets** which showcase various graphing techniques and statistical tests.

Click on the **chapter title** to see a digital (**PDF**) version of each page of your workbook suitable for data projection.

Click on this link next to each chapter title to see a comprehensive list of **software** and commercially produced **videos** for that topic.

Click on the links to the **supplementary activities** available for each of the exam boards listed. These complement the core material in the workbook and are provided with a **one year** photocopy licence. Learning objectives, resource lists, and activities are also provided for each of the options modules.

AQA Specification A

AQA-A offers a **Biology** or a **Human Biology** route. **AS** candidates must complete **three** assessment units. For Biology, modules 1, 2, and 4 and for Human Biology, modules 1, 3, and 4. Advanced Level candidates must complete the same 3 assessment units from the AS scheme and three units from the A2 scheme. For Biology, modules 5, 6, 8a and 8b and for Human Biology, 5, 6, 9a and 9b.

AS Module

Topics in AS workbook (unless indicated)

Module 1: Molecules, Cells and Systems

10.1	Prokaryotic and eukaryotic cell structure.	
10.2	Electron microscopy. Cell organelles. Centrifugation.	Cell Structure
10.3	Cell membranes and cellular transport.	Cell Membranes and Transport
10.4	Biological molecules. Biochemical tests.	Biological Molecules
10.5	Enzyme structure and function.	
10.6	Cell specialisation and organisation. The structure and function of epithelial tissues and blood.	Cell Structure Transport Systems
10.7	Circulation and capillaries. Gas exchange and lung function.	Transport Systems Gas Exchange
10.8	Heart structure and function. The effects of exercise.	Transport Systems

Module 2: Making Use of Biology   *Biology only*

11.1	Isolation and applications of enzymes.	Biological Molecules
11.2	Mitosis and the cell cycle. The significance of meiosis.	Cell Division and Cloning
11.3	DNA structure and replication. The genetic code and protein synthesis.	The Genetic Code
11.4	Production and use of recombinant DNA. Production of human proteins.	Gene Technology
11.5	Principles of immunology (ABO blood grouping). Genetic fingerprinting and PCR.	Defence and the Immune System Gene Technology
11.6	Control of cereal crop plants. Fertilisers, pesticides, and pest management.	● TRC: Crop production
11.7	Reproductive hormones in mammals. Control of reproduction.	Sexual Reproduction ● TRC: Biotechnology and Reproduction

Module 3: Pathogens and Disease   *Human Biology only*

12.1	Pathogenic bacteria and viruses (HIV, *Salmonella*, and *M. tuberculosis*).	Infectious Disease ● TRC: Bacterial Growth
12.2	Adaptations and life cycle of parasites (includes malarial parasite).	Energy and Ecosystems Infectious Disease
12.3	The body's defence systems against disease. The immune system. Vaccinations and immunisation.	Transport Systems Defence and the Immune System
12.4	Mitosis, cell cycle., significance of meiosis.	Cell Division and Cloning
12.5	DNA structure and replication. The genetic code. Protein synthesis, gene expression.	The Genetic Code
12.6	Production and use of recombinant DNA. Production of human proteins.	Gene Technology
12.7	Non-communicable disease: basis and risk factors of heart disease and cancer.	Aspects of Human Health Cell Division and Cloning
12.8	Diagnosis of disease using DNA probes and enzymes as analytical reagents and biosensors.	Aspects of Human Health Biological Molecules ● TRC: Pre- and Post-Natal Testing
12.9	Control and treatment of diseases using drugs. Antibiotics. Monoclonal antibodies.	Infectious Disease Defence and the Immune System

Module 4: Centre Assessed Coursework

Assessment of planning and implementation of practical work, collection and presentation of raw data, data analysis and evaluation, ability to select and retrieve information and communication skills.	Skills in Biology

A2 Module

Topics in A2 workbook (unless indicated)

Module 5: Inheritance, Evolution and Ecosystems

14.1	Meiosis, Mendelian inheritance, application of the chi-squared test.	Sources of Variation Inheritance
14.2	Sample variation. Causes of variation: meiosis, mutation, environment, polygeny.	
14.3	Hardy-Weinberg principle. Allele frequencies and selection. Speciation.	Population Genetics and Speciation
14.4	Species concept. The five kingdom classification.	Population Genetics and Speciation Biodiversity and Classification
14.5	Ecosystem numbers and distribution. Diversity indices. Succession.	Populations and Interactions Practical Ecology ● TRC: Ecosystems & Energy
14.6	Biochemistry of photosynthesis.	Cellular Metabolism
14.7	Ecological pyramids. Energy transfer between trophic levels.	Energy Flow and Nutrient Cycles ● TRC: Ecosystems & Energy
14.8	Respiration, respiratory quotient, ATP.	Cellular Metabolism
14.9	Carbon and nitrogen cycling and the role of microorganisms in these.	Energy Flow and Nutrient Cycles
14.10	Causes and effects of deforestation. Conservation and forest resources.	Populations and Interactions

Module 6: Physiology and the Environment   *Biology only*

15.1	Dicotyledon water uptake and transport. Potometers, Xeromorphic adaptations.	● TRC: Physiology & the Environment
15.2	Homeostasis, blood glucose control, thermoregulation	
15.3	Liver and kidney function. Nitrogenous waste and water balance.	Homeostasis
15.4	Fick's law and adaptation for gas exchange.	
15.5	Transport of respiratory gases.	
15.6	Digestion and absorption in mammals (including control of digestive secretion).	● TRC: Physiology & the Environment
15.7	Metamorphosis and insect nutrition.	
15.8	Components of the nervous system.	
15.9	Receptors and stimuli.	
15.10	Role of the nervous system in behaviour: reflexes, ANS, taxes, kineses.	Responses & Coordination

Module 7: The Human Life Span   *Human Biology only*

16.1	Sexual reproduction: gametogenesis & fertilisation.	
16.2	Foetal development. The placenta. The role of hormones in pregnancy.	● TRC: The Human Life Span
16.3	Growth and sexual development.	
16.4	Digestion and absorption in mammals (including control of digestive secretion).	
16.5	Nutrients and their role. Diet and BMR.	AS: Aspects of Human Health ● TRC: The Human Life Span
16.6	Transport of respiratory gases	● TRC: The Human Life Span
16.7	Components of the nervous system.	Responses & Coordination
16.8	Receptors and stimuli.	
16.9	Muscle structure and function	Muscles and Movement
16.10	Spinal reflexes and the ANS.	Responses & Coordination
16.11	Homeostasis: thermoregulation, control of blood glucose	Homeostasis ● TRC: The Human Life Span

Modules 8a/9a: Written synoptic papers
Modules 8b/9b: Centre assessed coursework

Synoptic paper. Assessment of planning and implementation of practical work, statistical analysis, interpretation and evaluation of results, and communication skills.	Skills in Biology

AQA Specification B

AS candidates taking the AQA-B course must complete **three** compulsory assessment units. **Advanced Level (As+A2) candidates** must complete the same three compulsory assessment units from the AS scheme and **three** compulsory units from the A2 scheme, **plus one option unit** from the three offered (Option 6, or Option 7, or Option 8).

AS Module	Topics in AS workbook *(unless indicated)*
Module 1: Core Principles	
10.1 Biological molecules. Biochemical analyses.	Biological Molecules
10.2 Cell structure and organelles, microscopy, centrifugation, cell differentiation	Cell Structure
10.3 Cell membranes and cellular transport.	Cell Membranes & Transport
10.4 Surface area: volume ratios. Gas exchange and ventilation	Cell Membranes & Transport Gas Exchange
10.5 Enzyme structure and function	Biological Molecules
10.6 Human digestion and absorption.	● TRC: Digestion
Module 2: Genes and Genetic Engineering	
11.1 DNA structure and replication. The genetic code, transcription, and translation. Mutations and their consequences.	The Genetic Code ● TRC: Mutations
11.2 Mitosis and the cell cycle. Cloning.	Cell Division and Cloning
11.3 Gametogenesis and fertilisation. Meiosis	Sexual Reproduction
11.4 Principles and applications of gene technology. GMOs and gene therapy.	Gene Technology
Module 3a: Physiology and Transport	
12.1 Transport systems: heart structure and function. Role of blood and tissue fluid.	Transport Systems
12.2 Control of breathing and heart rate in humans. Physiological responses to exercise.	Gas Exchange Transport Systems
12.3 Energy and exercise: energy for muscle contraction. Muscle fatigue and blood lactate.	Gas Exchange A2: Muscles & Movement
12.4 Transport in plants (including root structure) Structural adaptations of xerophytes.	Transport Systems
Module 3b: Coursework	
Candidates undertake experiments and investigations. Skills include: planning and implementation of practical work, analysing evidence and drawing conclusions, and evaluating evidence and procedures.	Skills in Biology

A2 Module	Topics in A2 workbook *(unless indicated)*
Module 4: Energy, Control, and Continuity	
13.1 Photosynthesis, respiration, and ATP. 13.2 Biochemistry of photosynthesis. 13.3 Biochemistry of respiration.	Cellular Metabolism
13.4 Detecting stimuli: receptors and effectors. Reflex arc. Effects of hormones.	Homeostasis Responses & Coordination
13.5 Homeostasis.: thermoregulation, control of blood glucose, excretion, osmoregulation.	Homeostasis
13.6 The mammalian eye. Transmission of nerve impulses. Synapses. Effects of drugs. 13.7 Structure and function of the brain & ANS.	Responses & Coordination
13.8 Ultrastructure of skeletal muscle. Sliding filament theory of muscle contraction.	Muscles & Movement
13.9 Inheritance: genotype, meiosis & fertilisation, sex determination, mono- and dihybrid cross	Sources of Variation Inheritance
13.10 Continuous and discontinuous variation. Causes of genetic variation.	Sources of Variation
13.11 Natural selection and speciation	Population Genetics and Speciation
13.12 Principles and importance of taxonomy. The five kingdom classification.	Population Genetics and Speciation Principles of Classification
Module 5a: Environment	
14.1 Energy transfer and trophic levels. Ecological pyramids. 14.2 Carbon and nitrogen cycles, including the role of microorganisms in these.	AS: Energy & Ecosystems ... or ... ● TRC: Energy Flow and Nutrient cycles
14.3 Biological and physical sampling methods.	Practical Ecology
14.4 Features of populations. Population stability. Ecological succession.	Populations & Interactions
14.5 Human environmental impact: hedgerows, monocultures, pesticides, pollution.	Populations & Interactions ● TRC: Human Impact
Module 5b: Coursework	
Candidates will be assessed on planning, implementing, analysis, evaluation, and synthesis of principles and concepts.	AS: Skills in Biology A2: Practical Ecology
Option Module 6: Applied Ecology	
Ecosystem diversity, pollution, adaptations of organisms, agricultural ecosystems, harvesting, conservation.	● TRC: Option Module 6 *Applied Ecology*
Option Module 7: Microbes and Disease	
Features of bacteria, culture of microorganisms, commercial use of microorganisms, microbial pathogens.	● TRC: Option Module 7 *Microbes and Disease*
Option Module 8: Behaviour and Populations	
Innate and learned behaviours, reproductive behaviours, pregnancy, human growth and development, human population growth, effects of lifestyle on health.	● TRC: Option Module 8 *Behaviour and Populations*

Salters-Nuffield

The **Salters-Nuffield Advanced Biology** (SNAB) GCE specification is divided into six units. The Advanced Subsidiary (AS) specification comprises Units 1, 2, and 3, with the practical assessment of Unit 3 consisting of two parts. The Advanced Level Specification comprises Units 1-6 and includes synoptic assessment. Practical skills are assessed in Unit 5 with submission of a written report of an experimental investigation that they have devised and carried out. Because of the contextual nature of the SNAB specification, it is recommended that students purchase both the AS and A2 volumes together and use them over the two-year duration of the course. This will enable the necessary scope of content to be covered as indicated below.

AS Unit

Topics in AS workbook
(unless indicated)

Unit 1: Topic 1 (1.1): Lifestyle, Health and Risk

Internal transport in animals: heart & blood	Transport Systems
Blood clotting	Defence & Immune system
Atherosclerosis, CVD, energy and diet, BMI	Aspects of Human Health
Simple carbohydrates and lipids	Biological Molecules

Unit 1: Topic 2 (1.2): Genes and Health

Gas exchange surfaces and lung structure	Gas Exchange
Membrane structure, cell transport processes	Cell Membranes & Transport
Nucleic acids, genetic code, protein synthesis	The Genetic Code
Amino acids and proteins, enzymes	Biological Molecules
DNA replication	The Genetic Code
Mutations: causes and consequences of CF	A2: Sources of Variation
Monohybrid inheritance, pedigree analysis	A2: Inheritance
Principles and applications of gene therapy	Gene Technology
Genetic profiling and screening.	Gene Technology
Prenatal testing: amniocentesis, CVS	Aspects of Human Health

Unit 2: Topic 3 (2.3): The Voice of the Genome

Ultrastructure of eukaryotic cells. Function of rER and Golgi. Prokaryotic v eukaryotic cells.	Cell Structure
Mitosis in the cell cycle.	Cell Division and Cloning
Mammalian gametes and fertilisation	Sexual Reproduction
Meiosis as a source of variation	Sources of Variation
Stem cell research	Cell Division and Cloning
Cell specialisation through gene switching	Cell Structure
Control of gene expression	A2: Cellular Metabolism
Effects of environment on genotype expression	A2: Sources of Variation
Polygenic inheritance	A2: Inheritance
Cancers as a result of uncontrolled cell division	Cell Division and Cloning
Genetic and environmental causes of cancer	Aspects of Human Health
HGP: principal outcomes and ethical issues	The Genetic Code

Unit 2: Topic 4 (2.4): Plants and Climate Change

Importance of water and inorganic ions to plants	Biological Molecules
Structure and function of polysaccharides	A2: Cellular Metabolism
Ultrastructure of plant cells v animal cells	Cell Structure
Structure and function of xylem. Transpiration.	Transport Systems
Seed structure and adaptations for dispersal	Sexual Reproduction
Uses of starch and plant oils by humans	Energy and Ecosystems
Genetic modification of plants (incl ethical issues)	Gene Technology
Role of the carbon cycle in regulating CO2	Energy and Ecosystems
Causes and consequences of global warming	TRC: Global warming
Reducing carbon emission: biofuels	TRC: Biofuels
Increased temperature: effect on enzyme activity	Biological Molecules

Unit 3: Report and Practical Review

Areas of practical work are: use of organisms, microscopy, lab techniques, and ecological methodology. Candidates are assessed on knowledge and experimental and investigative skills developed through the AS course.	Skills in Biology Cell Structure A2: Practical Ecology

A2 Unit

Topics in A2 workbook
(unless indicated)

Unit 4: Topic 5 (4.5): On the Wild Side

Taxonomy, biodiversity, and classification keys	Biodiversity & Classification
Genetic diversity: mutation, recombination	Sources of Variation
Dihybrid inheritance	Inheritance
Practical investigation of species ecology	Practical Ecology
Habitat, distribution, and adaptation	AS: Energy and Ecosystems
Photosynthesis in C3 plants	Cellular Metabolism
Primary productivity and ecological efficiency	AS: Energy and Ecosystems
Natural selection, gene mutation, and evolution	Popn Genetics & Speciation
Development of evolutionary theory	Popn Genetics & Speciation
Ecological succession	Populations and Interactions
Conservation of endangered species	TRC: Conservation Issues

Unit 4: Topic 6 (4.6): Infection, Immunity, and Forensics

Forensic pathology. Bacteria and viruses.	AS: Infectious Disease
Bacterial infection: TB. Viral infection: HIV.	AS: Infectious Disease
Non specific defence mechanisms. B cells,	AS Defence and the
T cells and the immune response. Immunity.	Immune System
Negative feedback and homeostasis	Homeostasis
Routes of infection. Immune system evasion.	AS: Infectious Disease
Antibiotics and their mode of action	AS: Infectious Disease
Development of antibiotic resistance	Sources of Variation
Evolution of pathogens	Popn Genetics & Speciation

Unit 5: Topic 7 (5.7): Run For Your Life

Movement and muscle physiology	Muscles and Movement
Role of ATP. Glycolysis and cellular respiration	Cellular Metabolism
Effects of exercise: variations in and control of ventilation rate and cardiac output.	AS: Gas Exchange AS: Transport Systems
Fast twitch v slow twitch muscle fibers	Muscles and Movement
Homeostasis: thermoregulation during exercise	Homeostasis
Exercise: too much v too little	AS: Aspects of Human Health

Unit 5: Topic 8 (5.8): Grey Matter

Nervous system structure and function. Vision.	Responses and Coordination
Nervous v hormonal regulatory systems	Homeostasis
Structure of the human brain	Responses and Coordination
Imaging: MRI, fMRI, CT scans	AS: Aspects of Human Health
Learning: habituation, conditioning, insight	Responses and Coordination
The effects of drugs at synaptic connections	Responses and Coordination
Continuous variation as a result of polygeny	Inheritance

Practical Assessment in Unit 5

Planning and implementing practicals: observations and recording, data interpretation and analysis, evaluation and communication of results.	AS: Skills in Biology A2: Practical Ecology

Unit 6: Synoptic Paper

Compulsory questions and an essay. This unit provides the chance to make connections between topics and to use skills and ideas in new contexts.	Integrated throughout

Edexcel

Edexcel offers a **Biology** or **Biology (Human)** route for AS and A2 candidates. AS candidates must complete three compulsory assessment units: for 8040 Biology, units 6101, 6102, and 6103, and for 8042 Biology (Human) 6101, 6112, and 6103. A2 candidates must complete the same three units from the AS scheme, plus three compulsory units (including one option) from the A2 scheme.

AS Unit	Topics in AS workbook *(unless indicated)*
6101 Unit 1: Molecules and Cells	
1.1 Biological molecules and biochemical tests. DNA replication. Transcription & translation. Human Genome Project.	Biological Molecules The Genetic Code
1.2 Structure and function of enzymes (incl. immobilisation and commercial uses)	Biological Molecules
1.3 Cell structure and organisation. Microscopy. Cell membranes and transport. Tissues.	Cell Structure Cell Membranes & Transport
1.4 Chromosome structure and mitosis. Cloning in plants and animals.	The Genetic Code Cell Division and Cloning
6102 Unit 2: Exchange, Transport & Reproduction *Biology only*	
2B.1 Surface area to volume ratios. Gas exchanges. Use of a respirometer. Digestion and absorption in humans	Cell Membranes & Transport Gas Exchange ● TRC: Digestion & Absorption
2B.2 Transport in flowering plants (incl. root structure). Transport in mammals.	Transport Systems
2B.3 Structural and physiological adaptations in plants and animals.	Gas Exchange Transport Systems
2B.4 Sexual reproduction, meiosis, and variation. Reproductive systems.	Sexual Reproduction
6112 Unit 2H: Exchange, Transport & Reproduction in Humans *Human Biology only*	
2H.1 Structure & function of the breathing system. Effects of smoking. Digestion and absorption	Gas Exchange Aspects of Human Health ● TRC: Digestion & Absorption
2H.2 The circulatory system, blood, heart structure and function. Artificial pacemakers.	Transport Systems Aspects of Human Health
2H.3 Human adaptation to extreme environments. Homeostasis and thermoregulation.	Transport Systems A2: Homeostasis
2H.4 Human reproduction and development. Effects of ageing.	Sexual Reproduction Aspects of Human Health
6103 Unit 3: Energy and the Environment	
3.1 Modes of nutrition. Adaptations to diet. 3.2 Ecosystem structure and function. 3.3 Trophic relationships, energy transfers, ecological pyramids, productivity. 3.4 Carbon, nitrogen, and water cycles.	Energy and Ecosystems
3.5 Sustainable management of energy. 3.6 Human impact on the environment.	● TRC: Energy and the Environment
Paper 01: T1 Practical Assessment of Coursework	
Practical work is integrated into each unit as appropriate. Students carry out an investigation and present a report which will be teacher assessed.	Skills in Biology

A2 Unit	Topics in A2 workbook *(unless indicated)*
6104 Unit 4: Respiration and Coordination, and Options	
4.1 Enzymes and metabolic pathways. The biochemistry of cellular respiration.	Cellular Metabolism
4.2 Principles of homeostasis: kidney structure and function, blood glucose regulation. Photosensitivity: the eye, phytochrome. nervous and hormonal coordination.	Homeostasis Responses and Coordination
Option A: Microbiology and Biotechnology Diversity and classification of microbes, culture of bacteria, commercial use of microorganisms in biotechnology.	● TRC: Option A *Microbiology & Biotechnology*
Option B: Food Science Food and diet, additives and preservatives, food storage and spoilage, fermentation i in food biotechnology.	● TRC: Option B *Food Science*
Option C: Human Health and Fitness Physiology of cardiovascular, pulmonary, musculoskeletal, and lymphatic systems (including disorders). Effects of exercise.	● TRC: Option C *Human Health and Fitness*
6105 Unit 5B: Genetics, Evolution, & Biodiversity *Biology only*	
5B.1 Photosynthesis and plant mineral nutrition.	Cellular Metabolism
5B.2 Control of plant growth: plant hormones.	Responses and Coordination
5B.3 Taxonomy: the five kingdom classification. Population distribution and dynamics. Ecological succession. Predator-prey. Sampling to investigate distribution. Pest control, bioaccumulation, IPM, habitat management, farming and biodiversity.	Biodiversity and Classification Populations and Interactions Practical Ecology ● TRC: Farming and Conservation
5B.4 Continuous and discontinuous variation. Effect of environment on gene expression. Mendelian genetics, multiple alleles, sex determination, mutation. Natural selection, gene pools, reproductive isolating mechanisms, speciation. Gene technology, including applications of GMOs, PCR, genetic fingerprinting.	Sources of Variation Inheritance ● TRC: Chromosome mutations and aneuploidy Population Genetics and Speciation ● AS: Gene Technology ...or... TRC: Gene Technology
6115 Unit 5H: Genetics, Human Evolution & Biodiversity *Human Biology only*	
5H.1 Continuous and discontinuous variation. Effect of environment on gene expression. Mendelian genetics, multiple alleles, sex determination, mutation. Natural selection, gene pools, reproductive isolating mechanisms, speciation. Gene technology, including applications of GMOs, PCR, genetic fingerprinting.	Sources of Variation Inheritance TRC: Chromosome mutations and aneuploidy Population Genetics and Speciation AS: Gene Technology ...or ... ● TRC: Gene Technology
5H.2 Physical and cultural evolution of humans.	● TRC: Human Evolution
5H.3 Human population growth.	Populations and Interactions
5H.4 Biodiversity and conservation: population distribution. Ecological succession. Sampling to investigate distribution. Pest control, bioaccumulation, IPM, habitat management, farming and biodiversity.	Populations and Interactions ● Practical Ecology TRC: Farming and Conservation
Paper 01: T1 or Paper 02: W2. Paper 03 (Synoptic Paper)	
Practical work is integrated into each unit as appropriate. Students carry out an investigation and present a report which will be teacher assessed.	Skills in Biology

Candidates taking the OCR AS course are required to complete the compulsory units 2801, 2802, plus 2803 (either components 01 and 02, or components 01 and 03). Candidates taking the OCR A2 course need to complete the requirements for the AS course, as well as 2804, 2805 (one of the five offered options), and 2806 (either components 01 and 02, or components 01 and 03).

AS Module	Topics in AS workbook (unless indicated)
5.1 Module 2801: Biology Foundation	
5.1.1 Cell structure: cell organisation, and characterisation, eukaryote and prokaryote cells, organelles, microscopy.	Cell Structure
5.1.2 Biological molecules, water and inorganic ions. Hydrolysis and condensation. 5.1.3 Enzyme structure and function.	Biological Molecules
5.1.4 Membrane structure and cell transport. Gas exchange in mammalian lungs. Root uptake systems.	Cell Membranes & Transport Gas Exchange Transport Systems
5.1.5 DNA structure & replication, protein synthesis. Gene manipulation: human insulin and factor VIII as examples.	The Genetic Code Gene Technology
5.1.6 Nuclei replication & division. Uncontrolled cell division (cancer). Chromosomes and mitosis. Meiosis in sexual reproduction.	Cell Division & Cloning Sexual Reproduction
5.1.7 Ecosystem structure and function. Energy transfer, nitrogen cycle.	Energy and Ecosystems
5.2 Module 2802: Human Health and Disease	
5.2.1 Health versus disease. Patterns of disease. Examples of diseases and disorders. The HGP and health and disease.	Aspects of Human Health Genetic Code
5.2.2 Balanced diet, energy and nutrient requirements. Malnutrition. Diet and CVD.	Aspects of Human Health
5.2.3 Human gas exchange system. The short and long term effects of exercise (incl. pulse and blood pressure).	Gas Exchange Transport Systems Aspects of Human Health
5.2.4 Health effects of smoking. Preventing & treating smoking-related diseases.	Aspects of Human Health
5.2.5 Causes, transmission, control and prevention of infectious disease: HIV/AIDS, TB, cholera and malaria.	Infectious Disease
5.2.6 The immune response, immunity and vaccination. Allergies.	Defence & Immune System
5.3 Module 2803: Transport/Experimental Skills 1	
5.3.1 Structure and function of the mammalian transport system. Adaptation to altitude. 5.3.2 Structure and function of the mammalian heart (incl. control of heart activity. 5.3.3 Transport in multicellular plants: xylem and phloem structure and distribution. Transpiration. Translocation. Xerophytes.	Transport Systems
Experimental Skills 1: *Option (candidates take either 02 or 03)*	
Component 02: Coursework 1: Internal assessment of experimental and investigative work. Planning and implementing a study, recording observations, measurements, data analysis and evaluation, drawing conclusions.	Skills in Biology
Component 03: Practical Examination 1: External assessment of experimental and investigative skills. Planning, implementing a study, recording observations, measurements, data analysis and evaluation, drawing conclusions. Components consist of a planning task, experimentation, and microscopy.	Skills in Biology Cell Structure

A2 Module	Topics in A2 workbook (unless indicated)
5.4 Module 2804: Central Concepts	
5.4.1 Structure and role of ATP. Biochemistry of respiration. Respiratory quotients. 5.4.2 Dicot leaf structure. Chloroplasts. Biochemistry of photosynthesis. Limiting factors.	Cellular Metabolism
5.4.3 Population dynamics and species interactions. Ecological succession. Sampling distribution and abundance. Conservation v production. Sustainable management of forest ecosystems.	Populations & Interactions Practical Ecology Populations & Interactions
5.4.4 Meiosis. Genetic crosses, sex linkage, codominance, multiple alleles. Chi-squared. Mutation. Genes and environment Gene regulation in bacteria (*Lac* operon). Implications of the HGP.	Inheritance Sources of Variation Cellular metabolism Sources of Variation
5.4.5 Five kingdom classification system. Species & phylogeny. Natural selection. Factors affecting gene pools. Speciation. Adaptation to environment.	Biodiversity & Classification Popn Genetics & Speciation
5.4.6 Homeostasis. Kidney structure and function. Nervous system structure and function. Endocrine regulation (blood glucose). Role of hormones in flowering plants.	Homeostasis Responses & Coordination Homeostasis Responses & Coordination
5.5 Module 2805: Options in Biology	
01: Growth, Development & Reproduction Growth and development, asexual reproduction, sexual reproduction in humans and angiosperms. Role of hormones in growth and rreproduction.	TRC: Option 01 *Growth, Development, and Reproduction*
02: Applications of genetics Genetic variation, selective breeding, GE techniques, genetic screening and gene therapy, genetic fingerprinting, transplant compatibility. Ethical issues.	TRC: Option 02 *Applications of Genetics*
03: Environmental Biology Practical ecology, pollution, agricultural ecosystems, resource conservation (fisheries, land reclamation, recycling), nation and international conservation.	TRC: Option 03 *Environmental Biology*
04: Microbiology and Biotechnology Nature of microorganisms, culturing techniques, applications of biotechnology.	TRC: Option 04 *Microbiology and Biotechnology*
05: Mammalian Physiology & Behaviour Mammalian nutrition, the liver, locomotion, nervous system and sensory reception, principles of behaviour (innate and learned).	TRC: Option 05 *Mammalian Physiology and Behaviour*
5.10 Module 2806: Unifying Concepts in Biology Experimental Skills 2	
Component 01: Unifying concepts in Biology (compulsory) Exam questions based on concepts covered in the AS scheme, and module 2804 of A2.	
Experimental Skills 2: *Option (candidates take either 02 or 03)*	
Component 02: Coursework 2 Internal assessment of experimental and investigative work. Skills include planning and implementing a study, recording, analysing, and evaluating data, drawing conclusions.	AS Skills in Biology Practical Ecology
Component 03: Coursework 3 External assessment of experimental and investigative skills including planning and implementing a study, recording, analysing, and evaluating data, drawing conclusions. Components consist of planning a task, experimentation, and microscopy.	AS Skills in Biology AS Cell Structure Practical Ecology

The subject content of the **Cambridge International Examination (CIE)** programme is divided into AS and A2. The A2 includes a core and an Applications of Biology section, which is studied in its entirety, by all A2 candidates. Candidates taking the CIE AS will be assessed on the Learning Outcomes A-K. A level candidates will be assessed on the Learning Outcomes L-U. The *Applications of Biology* section accounts for about 12% of the A level course. The acquisition of practical skills (and their assessment) underpins the course.

AS Module	Topics in AS workbook
Core Syllabus (*sections L-U are in Advanced biology A2*)	
A Light microscopy, electron microscopy, and cell structure. Eukaryote and prokaryote cells. Functions of organelles.	Cell Structure
B Structure and role of carbohydrates, lipids, and proteins. Water and inorganic ions. Hydrolysis and condensation reactions.	Biological Molecules
C Enzyme action, enzyme activity and enzyme inhibitors.	
D Fluid mosaic model of membrane structure. Transport across the membrane.	Cell Membranes and Transport
E Replication and division of nuclei and cells. Role of meiosis in sexual reproduction. Chromosome behaviour during mitosis. Uncontrolled cell division (cancer).	Cell Division and Cloning
F DNA structure and replication. The role of DNA in protein synthesis. Nucleotide base pairing.	The Genetic Code
G Transport in multicellular plants: structure and distribution of xylem and phloem in dicots. Transpiration. Translocation. Xerophytes. Structure and function of mammalian transport systems (including the heart). Haemoglobin and gas transport, gas exchange and altitude.	Transport Systems
H The structure and function of the human respiratory system. The effects of smoking on gas exchange. Smoking related diseases.	Gas Exchange / Aspects of Human Health
I Causes and transmission of infectious diseases, their control and prevention. HIV/AIDS, TB, cholera and malaria. The use of antibiotics.	Infectious Disease
J Structure and function of the immune system. Types of immunity. Vaccinations.	Aspects of Human Health
K The ecosystem concept (habitat, niche, populations, communities). Energy transfer, ecological efficiency, the nitrogen cycle.	Energy and Ecosystems
Meeting Assessment Objectives	
A Knowledge with understanding, including the use of scientific vocabulary, understanding scientific ideas and concepts, using instruments and scientific measurements, applying of scientific and technological techniques.	Skills in Biology
B Handling information and solving problems: organising and extracting information. Manipulating and presenting data, and drawing conclusions. Applying knowledge to solve problems.	
C Experimental skills and investigations. Following detailed instructions, using techniques and apparatus correctly, making accurate observations, and interpreting the data to form predictions. Designing, planning and carrying out a scientific experiment or investigation.	

A2 Module	Topics in A2 workbook (*unless indicated*)
Core Syllabus (*sections A-K are in Advanced biology AS*)	
L Energy requirements in living organisms. The structure and function of ATP. Cellular respiration, and energy transfer. Aerobic and anaerobic respiration. Respiratory quotient and the energy value of substrates. Respirometers.	Cellular Metabolism
M Photosynthesis and energy transfer. The biochemistry of photosynthesis. Limiting factors in photosynthesis. The structure of a dicot leaf.	Cellular Metabolism
N Homeostasis principles. Kidney function and structure. Nervous system: sensory receptors, neurones, action potential, and synapses. Endocrine glands (pancreas). Control of blood glucose (diabetes treatment). Role of hormones in flowering plants.	Homeostasis / Responses & Coordination
O Genetic transfer. Meiosis. Genes and alleles. Monohybrid and dihybrid crosses, sex linkage, codominance, multiple alleles. Chi-squared. Mutations and environmental effects on phenotype.	Sources of Variation / Inheritance
P Natural and artificial selection. The role of natural selection in evolution. The role of environmental factors and isolating mechanisms. The role of artificial selection on livestock improvement. Factors affecting allele frequencies (malaria and sickle cell anaemia).	Population Genetics and Speciation
Q The five kingdom classification system. The importance of biodiversity. Conservation issues: endangered species and strategies to protect them.	Biodiversity & Classification / ● TRC: Conservation Issues
R Gene technology, techniques in gene technology. Uses of gene technology (insulin production, DNA sequencing, genetic fingerprinting, and genetic screening). Benefits and hazards of gene technology. Ethical issues.	AS Gene Technology
S Biotechnology. Industrial use of microorganisms. Large scale production of microorganisms. Enzyme technology. The use of monoclonal antibodies.	● TRC: Biotechnology
T Crop plant reproduction and adaptations. Methods of improving crop production.	● TRC: Crop Plants
U Human reproduction. Gametogenesis (mitosis, growth, meiosis and maturation). The role of hormones in the menstrual cycle. Contraception. In-vitro fertilisation.	● TRC: Aspects of Human Reproduction
Meeting Assessment Objectives	
A Knowledge with understanding. Including the use of scientific vocabulary, understanding scientific ideas and concepts, using instruments and scientific measurements, the application of scientific and technological techniques.	Skills in Biology / Practical Ecology
B Handling information and solving problems. Understand how to organise and extract information. Know how to manipulate and present data, and draw conclusions. Apply knowledge to problem solve.	
C Experimental skills and investigations. Following detailed instructions, using techniques and apparatus correctly, making accurate observations, and interpreting the data to form predictions. Designing, planning and carrying out a scientific experiment or investigation.	

WJEC

Welsh Joint Education Committee (WJEC) candidates in the AS course are required to complete the three compulsory assessment units BI1, BI2, and BI3. A2 candidates are required to complete the same three units from the AS scheme, as well as three compulsory units (BI4, BI5, BI6) from the A2 scheme.

Introduction

AS Module	Topics in AS workbook *(unless indicated)*
Module BI1: Fundamental Concepts and Organisation	
1.1 Biological molecules: (a-b) Carbohydrates, proteins, lipids, water. Condensation and hydrolysis reactions.	Biological Molecules
1.2 Cell organisation: (a-b) Eukaryotic cell structure & organelles. Prokaryotic cell structure. Viruses. Levels of organisation, tissues.	Cell Structure
1.3 Cell membrane and transport: (a-b) Cellular transport mechanisms.	Cell Membranes & Transport
1.4 Enzymes: (a-c) Nature of enzymes and activity. Enzyme inhibitors. **1.5 Application of enzymes:** (a-b) Enzymes as biosensors. Immobilised enzymes.	Biological Molecules
1.6 Nature of the genetic code: (a-c) DNA and RNA structure. DNA replication. Transcription and translation. (d) DNA technology, HGP. Gene therapy for CF, genetic fingerprinting. Genetic engineering techniques, genetic fingerprinting. Ethical issues of GMOs.	The Genetic Code The Genetic Code Gene Technology
1.7 Genetic information: (a-b) Mitosis and meiosis. Cytokinesis. Meiosis and fertilisation. Asexual and sexual reproduction.	Cell Division & Cloning Sexual Reproduction ● TRC: Sexual Reproduction
Module BI2: Adaptations and Ecology	
2.1 The need for transport systems: (a) Unicellular organisms. Surface area to volume ratios. (b) Multicellular organisms. Transport systems and gas exchange.	Cell Membranes & Transport Gas Exchange Transport Systems
2.2 Gas exchange: (a-b) Gas exchange in plants and animals with respect to environment.	Gas Exchange
2.3 Transport systems: (a) Structure and function of a dicot root. Transpiration and translocation. Xerophyte adaptations. (b) Human circulatory system: blood, tissue fluid, lymph, gas transport.	Transport Systems
2.4 Transfer of energy: (a-b) Ecosystems & energy transfer. Heterotrophic nutrition. Food chains & webs, ecological pyramids, nutrient cycles.	Energy and Ecosystems
2.5 Nutrition and population growth: (a-b) Population growth and regulation. Ecological succession.	● TRC: Nutrition and Population Growth A2: Popns & Interactions
2.6 Human Impact: (a-b) Deforestation, overfishing. Pest control. Human impact on nutrient cycles.	● TRC: Human Impact
Module BI3: Practical Work	
Skills include: planning and implementing practical work (including microscopy), observations and recording, interpretation and analysis of results, evaluation and communication of results.	Skills in Biology

A2 Module	Topics in A2 workbook *(unless indicated)*
Module BI4: Biochemistry and Health	
4.1 (a-c) The role and synthesis of ATP. **4.2** (a-d) The biochemistry of respiration. **4.3** (a-d) Pigments and light absorption. Biochemistry of photosynthesis. Factors influencing photosynthesis.	Cellular Metabolism
4.4 (a-b) Gut structure & function. Histology of the ileum. Digestion & absorption.	● TRC: Digestion & Absorption
4.5 (a-b) Bacterial classification. Microbiological techniques.	● TRC: Microbiology
4.6 (a-b) Causes and transmission of disease. Control and prevention of disease.	AS Infectious Disease
4.7 (a-b) Barriers to infection. The immune response and immunity, vaccination.	AS Defence & Immune System
4.8 (a) Industrial fermentation (penicillin). Monoclonal antibodies. Antibiotic resistance. (b) Drug use and antibiotic resistance. Vaccination programmes.	● TRC: Microbiology AS: Defence & Immune System Sources of Variation Popn Genetics & Speciation AS: Defence & Immune System
Module BI5: Variety and Control	
5.1 (a-b) Alleles, Mendelian inheritance, sex linkage, mutagens and mutation. Random assortment, fertilisation, and variation.	Sources of Variation Inheritance ● TRC: Chromosome Mutations & Aneuploidy
5.2 (a-b) Continuous variation (t-test). Gene pool, genetic drift, natural selection, reproductive isolation and speciation. (c) Evolutionary consequences of human impact on the environment. Artifical selection. Conservation.	● Inheritance Popn Genetics & Speciation TRC: Conservation of Biodiversity
5.3 (a-b) Angiosperm reproduction.	AS Sexual Reproduction ● TRC: Fruits and Seeds
5.4 (a-b) Human reproduction.	AS Sexual Reproduction
5.5 Plant and animal cloning.	AS Cell Division & Cloning
5.6 (a-b) Five kingdom classification. Binomial nomenclature. Species and phylogeny.	Biodiversity & Classification Popn Genetics & Speciation
5.7 (a-b) Homeostasis, the mammalian kidney, endocrine control.	Homeostasis
5.8 Response to stimuli. The ear.	Responses & Coordination
5.9 (a-c) The nervous system. Nerve impulses, the brain, reflexes. (d) Effectors, skeletal muscle structure, sliding filament hypothesis.	Responses & Coordination Muscles & Movement
Module BI6: Practical Work	
Practical work is integrated into each unit.	Practical Ecology AS Skills in Biology

Cellular Metabolism

AQA-A	AQA-B	SNAB	Edexcel	OCR
Complete:	Complete:	Complete:	Complete:	Complete:
1-3, 5-6, 10-18, 21, 24	1-3, 5-6, 10-17, 21, 24	1-3, 5-6, 10-17, 21-24	1-6, 10-17, 20-27	1-3, 5-21, 24-25

Learning Objectives

☐ 1. Compile your own glossary from the **KEY WORDS** displayed in **bold type** in the learning objectives below.

ATP and Metabolic Pathways *(pages 22-23, 29-30)*

☐ 2. Explain the need for energy in living things and the universal role of ATP in metabolism, as illustrated by specific examples e.g. **glycolysis**, **active transport**, **anabolic reactions**, movement, and **thermoregulation**.

☐ 3. Describe the structure of **ATP** as a phosphorylated nucleotide. Describe its synthesis from ADP and inorganic phosphate and explain how it stores and releases its energy.

☐ 4. Explain what is meant by a **metabolic pathway**. Describe the role of enzymes in the control of metabolic pathways as illustrated by specific examples, e.g. **oxidoreductases** and **hydrolases**, **anabolism** and **catabolism**, control of phenylalanine metabolism.

☐ 5. Outline the principles involved in **photosynthesis** and **cellular respiration**, explaining in which way the two processes can be considered opposites.

☐ 6. Appreciate that both photosynthesis and cellular respiration involve the molecule **ATP** and a hydrogen carrier molecule (**NADPH$_2$** in photosynthesis, and **NADH$_2$** in respiration).

The operon hypothesis *(pages 31-32 and see the TRC: Gene Control in Eukaryotes)*

☐ 7. Understand the term **operon** as being a unit of genes that function in a coordinated way under the control of an operator gene.

☐ 8. Explain how simple metabolic pathways are regulated in bacteria, as illustrated by the *lac* operon in *E. coli*. Outline the principles involved in **gene induction** in the *lac* operon, identifying how lactose activates transcription and how metabolism of the substrate is achieved. Explain the adaptive value of gene induction.

☐ 9. Appreciate that the end-product of a metabolic pathway can activate a repressor and switch genes off (**gene repression**). Appreciate the adaptive value of gene repression for the control of a metabolic end-product.

Cellular Respiration

Introduction to respiration *(pages 22-24)*

☐ 10. Describe the role of mitochondria in cellular respiration. Draw and label the structure of a **mitochondrion**, including the **matrix**, the outer and inner membrane and the **cristae**. Briefly state the function of each.

☐ 11. Identify the main steps in cellular respiration: **glycolysis**, **Krebs cycle** (*tricarboxylic acid cycle*) **electron transport system** (*electron transport chain, ETS, or respiratory chain*). On a diagram of a mitochondrion, indicate where each stage occurs. Recognise glycolysis as first stage in cellular respiration and the major anaerobic pathway in cells.

☐ 12. Identify **glucose** as the main respiratory substrate. Appreciate that other substrates can, through conversion, act as substrates for cellular respiration.

☐ 13. Outline **glycolysis** as the phosphorylation of glucose and the subsequent splitting of a 6C sugar into two triose phosphate molecules (2 X **pyruvate**). State the net yield of ATP and NADH$_2$ from glycolysis and appreciate that the subsequent metabolism of pyruvate depends on the availability of oxygen.

Aerobic respiration *(pages 25-26)*

☐ 14. Describe the complete oxidation of glucose to CO_2, with reference to:
- The conversion of pyruvate to **acetyl-coenzyme A**.
- The entry of acetyl CoA into the Krebs cycle by combination with **oxaloacetate**.
- The **Krebs cycle** (as a series of oxidation reactions involving release of CO_2, the production of NAD.H$_2$ or FAD.H$_2$, and the regeneration of oxaloacetate).
- The *role* of the coenzymes NAD and FAD.
- Synthesis of **ATP** by **oxidative phosphorylation** in the electron transport chain.
- The role of oxygen as the terminal electron acceptor and the formation of water.
- The net yield of ATP from aerobic respiration compared to the yield from glycolysis.

☐ 15. Understand the terms **decarboxylation** and **dehydrogenation** as they relate to the Krebs cycle.

Fermentation *(page 27)*

☐ 16. Understand the situations in which the pyruvate formed in glycolysis may not undergo complete oxidation. Describe the following examples of **fermentation**, identifying the H$^+$ acceptor to each case:

  (a) Formation of **lactic acid** in muscle.
  (b) Formation of **ethanol** in yeast.

  NOTE: Appreciate that, although fermentation is often used synonymously with anaerobic respiration, they are not the same. Respiration always involves hydrogen ions passing down a chain of carriers to a terminal acceptor, and this does not occur in fermentation. In anaerobic respiration, the terminal H$^+$ acceptor is a molecule other than oxygen, e.g. Fe^{2+} or nitrate.

☐ 17. Compare and explain the differences in the yields of ATP from aerobic respiration and from fermentation.

Respiratory quotients *(page 28)*

☐ 18. Explain the relative energy values of carbohydrate, lipid, and protein as respiratory substrates. Explain the term **respiratory quotient** (RQ). Explain what RQ reveals about the substrate being respired.

□ 19. Use a simple respirometer to measure RQ and the effect of temperature on respiration rate.

Photosynthesis

The structure of the dicot leaf *(pages 33, 39-40)*

□ 20. Recognise the leaf as the main photosynthetic organ in plants. Describe the internal and external structure of a dicotyledonous leaf with reference to: the location of the palisade tissue and the distribution of chloroplasts. If required, compare dicot leaf structure with monocot leaf structure and describe adaptations of the leaf to maximising photosynthesis in different environments.

Chloroplasts *(pages 34-35)*

□ 21. Describe the structure and role of **chloroplasts**, identifying the **stroma**, **grana**, lamellae (**thylakoids**), and location of the chlorophylls and other pigments.

□ 22. Describe the role of **chlorophyll a** and **b**, and **accessory pigments** (e.g. carotenoids) in light capture. Explain what is meant by the terms **absorption spectrum** and **action spectrum** with respect to the light absorbing pigments.

□ 23. Investigate chloroplast pigments using chromatography.

Photosynthesis in C3 plants *(pages 37-38)*

□ 24. Describe, using diagrams, the reactions of photosynthesis in a C3 plant with reference to:

The light dependent phase (LDP) with reference to:
- Where in the chloroplast the LDP occurs and the location and role of the photosystems.
- The **photoactivation** of chlorophyll.

- The splitting of water (**photolysis**) to produce protons and electrons.
- The production of O_2 as a result of photolysis.
- The transfer of energy to ATP (photophosphorylation) and the formation of NADPH$_2$ (reduced NADP).

The light independent phase (LIP) with reference to:
- Where in the chloroplast the LIP occurs.
  The **Calvin cycle** including:
- The fixation of carbon dioxide into a 5C compound, **ribulose bisphosphate** (RuBP).
- The reduction of **glycerate-3-phosphate** (PGA) to **carbohydrate** and the role of **ATP** and **NADPH$_2$** (formed in the light dependent phase) in this.
- The regeneration of the ribulose bisphosphate.

Photosynthetic rate *(pages 36, 39)*

□ 25. Understand the term **photosynthetic rate**. With reference to **limiting factors**, describe the effect of each of the following on photosynthetic rate:
- Light intensity and wavelength.
- Temperature.
- Carbon dioxide concentration.

□ 26. Explain what is meant by the (light) **compensation point**. Interpret the compensation point for plants in different environments.

Plant mineral nutrition *(page 41)*

□ 27. Describe the uptake of mineral ions in plant roots. Describe the function of some important plant mineral ions, including: nitrate (NO_3^-), phosphate (PO_4^{3-}), and magnesium (Mg^{2+}) ions.

 See the 'Textbook Reference Grid' on pages 8-9 for textbook page references relating to material in this topic.

Supplementary Texts
See page 6 for additional details of this text:

■ Adds, J. *et al.*, 2003. **Respiration and Coordination** (NelsonThornes), pp. 1-8 (respiration).

See page 7 for details of publishers of periodicals:

STUDENT'S REFERENCE

Metabolic pathways and their control

■ **Tyrosine** Biol. Sci. Rev., 12 (4) March 2000, pp. 29-30. *The central metabolic role of the amino acid tyrosine (includes errors in tyrosine metabolism).*

■ **Leptin** Biol. Sci. Rev., 15(3) February 2003, pp. 30-32. *The role of the hormone leptin in the regulation of body mass and the control of obesity.*

Presentation MEDIA to support this topic:
GENES & INHERITANCE
• Nature of Genes

Presentation MEDIA to support this topic:
CELL BIO & BIOCHEM:
• Cellular Energetics

The negative feedback mechanisms regulating gain and loss in mass are illustrated.

■ **Gene Structure and Expression** Biol. Sci. Rev., 12 (5) May 2000, pp. 22-25. *The nature of genes and the control of gene expression. Includes gene induction in the Lac operon in E. coli.*

■ **Genes that Control Genes** New Scientist, 3 November 1990 (Inside Science). *The control of gene expression in prokaryotes by gene induction and repression. The operon model is explained.*

Chlorophyll and photosynthesis

■ **Growing Plants in the Perfect Environment** Biol. Sci. Rev., 15(2) November 2002, pp. 12-16. *To manipulate the growth of plants in controlled environments, one must understand how plants grow and what influences photosynthetic rate.*

■ **Chemistry that Comes Naturally** New Scientist, 31 July 1993, pp. 24-28. *The processes of photosynthesis and respiration.*

■ **Green Miracle** New Scientist, 14 August 1999, pp. 26-30. *The mechanism by which plants split water to make oxygen remains a mystery.*

■ **Chlorophyll** Biol. Sci. Rev., 8(3) January 1996, pp. 28-30. *The chlorophyll molecule: how it absorbs light and its role in photosynthesis.*

■ **Why Don't Plants Wear Sunhats?** Biol. Sci. Rev., 9(3) Jan. 1997, pp. 32-35. *Plants need light, but too much is damaging - how do they cope?*

ATP and cellular respiration

■ **The Role of ATP in Cells** Biol. Sci. Rev., 19(2) Nov. 2006, pp. 30-33. *Synthesis and uses of ATP.*

■ **Fat Burns in the Flame of Carbohydrate** Biol. Sci. Rev., 15(3) February 2003, pp. 37-41. *A thorough account of both carbohydrate metabolism and how fatty acid oxidation feeds into the Krebs cycle. Starvation and ketosis are described and several points for discussion are included.*

■ **Fuelled for Life** New Scientist, 13 January 1996 (Inside Science). *Energy and metabolism: ATP, glycolysis, electron transport system, Krebs cycle, and enzymes and cofactors.*

■ **Glucose Catabolism** Biol. Sci. Rev., 10(3)

Jan. 1998, pp. 22-24. *Glucose in cells: oxidative phosphorylation and the role of mitochondria.*

TEACHER'S REFERENCE

■ **Molecular Machines that Control Genes** Scientific American, February 1995, pp. 38-45. *Regulation of gene action by protein complexes that assemble on DNA.*

■ **The Bigger Picture** New Scientist, June 2000, pp. 54-61. *The complexity of biochemical pathways and the effects of altering genetic constitution.*

■ **Measuring the Metabolism of Small Organisms** Scientific American, December 1995, pp. 84-85. *Methods of measuring and monitoring respiration and metabolic rate in small organisms.*

See pages 10-11 for details of how to access **Bio Links** from our web site: **www.biozone.co.uk**. From Bio Links, access sites under the topics:

GENERAL BIOLOGY ONLINE RESOURCES > Online Textbooks and Lecture Notes: • S-Cool! A level biology revision guide • Learn.co.uk • Mark Rothery's biology web site ... *and others*

CELL BIOLOGY AND BIOCHEMISTRY: • Cell and molecular biology online ... *and others* > **Biochemistry and Metabolic Pathways:** • Calvin cycle (C3 cycle) • Cellular energy references • Cellular respiration • Cycle (Krebs cycle, Citric Acid Cycle) • Electron transport chain • Energy, enzymes, and catalysis problem set • Glycolysis • Learning about photosynthesis • Chapter 7: Metabolism and biochemistry ... *and others*

GENETICS > Molecular Genetics (DNA): • Prokaryotic genetics and gene expression chapter • Model of Lac operon (animation) • Induction of the Lac operon • Molecular genetics of prokaryotes

Energy in Cells

A summary of the flow of energy within a plant cell is illustrated below. Animal cells have a similar flow except the glucose is supplied by ingestion rather than by photosynthesis. The energy not immediately stored in chemical bonds is lost as heat. Note the role of ATP; it is made in cellular respiration and provides the energy for metabolic reactions, including photosynthesis.

Energy Transformations in a Photosynthetic Plant Cell

Photosynthesis is a chemical process that captures light energy and stores it as potential chemical energy.

Light energy

Oxygen

***Note:** Heterotrophic organisms (with the exception of photoheterotrophs) are dependent on organic molecules ('food') to provide the ultimate energy source for cellular respiration.

Photosynthesis → **Glucose** *

Oxygen

Other uses of glucose

Fuel

Carbon dioxide + water

ADP

Respiration

ATP provides energy for metabolic reactions. While some energy is stored in chemical bonds, some is lost as heat

ATP

Water

Heat energy

Carbon dioxide

Cellular respiration is a chemical process that releases energy from glucose to make the energy available (in the form of ATP) to power metabolic reactions.

1. Distinguish between **heterotrophs**, **photosynthetic autotrophs**, and **chemosynthetic autotrophs** with respect to how these organisms derive their source of energy for metabolism:

_____

_____

_____

_____

_____

_____

2. In 1977, scientists working near the Galapagos Islands in the equatorial eastern Pacific found warm water spewing from cracks in the mid-oceanic ridges 2600 metres below the surface. Clustered around these hydrothermal vents were strange and beautiful creatures new to science. The entire community depends on sulfur-oxidising bacteria that use hydrogen sulfide dissolved in the venting water as an energy source to manufacture carbohydrates. This process is similar to photosynthesis, but does not rely on sunlight to provide the energy for generating ATP and fixing carbon:

   (a) Explain why a community based on photosynthetic organisms is not found at this site: _____

   _____

   (b) Name the ultimate energy source for the bacteria: _____

   (c) This same chemical that provides the bacteria with energy is also toxic to the process of cellular respiration; a problem that the animals living in the habitat have resolved by evolving various adaptations. Explain what would happen if these animals did not possess adaptations to reduce the toxic effect on cellular respiration:

   _____

   (d) Name the energy source classification for these sulfur-oxidising bacteria: _____

Related activities: The Role of ATP in Cells, Cellular Respiration, Photosynthesis

The Role of ATP in Cells

The molecule ATP (adenosine triphosphate) is the universal energy carrier for the cell. ATP can release its energy quickly; only one chemical reaction (hydrolysis of the terminal phosphate) is required. This reaction is catalysed by the enzyme ATPase. Once ATP has released its energy, it becomes ADP (adenosine diphosphate), a low energy molecule that can be recharged by adding a phosphate. This requires energy, which is supplied by the controlled breakdown of respiratory substrates in cellular respiration. The most common respiratory substrate is glucose, but other molecules (e.g. fats or proteins) may also be used.

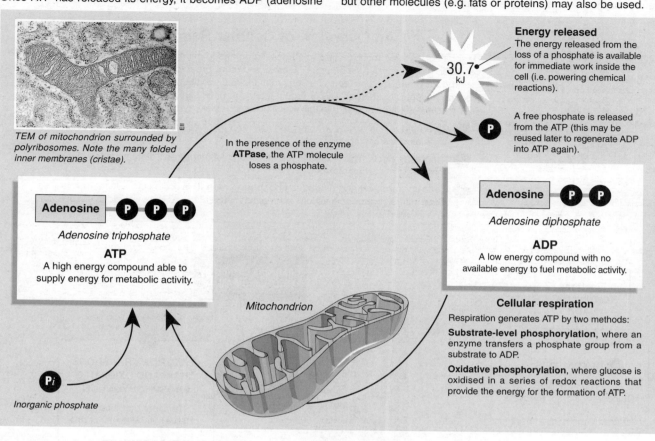

TEM of mitochondrion surrounded by polyribosomes. Note the many folded inner membranes (cristae).

In the presence of the enzyme **ATPase**, the ATP molecule loses a phosphate.

Energy released
The energy released from the loss of a phosphate is available for immediate work inside the cell (i.e. powering chemical reactions).

30.7 kJ

A free phosphate is released from the ATP (this may be reused later to regenerate ADP into ATP again).

Adenosine P P P
Adenosine triphosphate
ATP
A high energy compound able to supply energy for metabolic activity.

Adenosine P P
Adenosine diphosphate
ADP
A low energy compound with no available energy to fuel metabolic activity.

P*i*
Inorganic phosphate

Mitochondrion

Cellular respiration
Respiration generates ATP by two methods:

Substrate-level phosphorylation, where an enzyme transfers a phosphate group from a substrate to ADP.

Oxidative phosphorylation, where glucose is oxidised in a series of redox reactions that provide the energy for the formation of ATP.

Cellular Metabolism

1. Describe how ATP acts as a supplier of energy to power metabolic reactions: _____

_____

2. Name the immediate source of energy used to reform ATP from ADP molecules: _____

3. Name the process of re-energizing ADP into ATP molecules: _____

4. Name the ultimate source of energy for plants: _____

5. Name the ultimate source of energy for animals: _____

6. Explain in what way the ADP/ATP system can be likened to a rechargeable battery: _____

_____

7. In the following table, use brief statements to contrast photosynthesis and respiration in terms of the following:

Feature	Photosynthesis	Cellular respiration
Starting materials		
Waste products		
Role of hydrogen carriers: NAD, NADP		
Role of ATP		
Overall biological role		

Related activities: Energy in Cells, Cellular Respiration, Photosynthesis

RA 2

Cellular Respiration

Cellular respiration is the process by which organisms break down energy rich molecules (e.g. glucose) to release the energy in a useable form (ATP). All living cells respire in order to exist, although the substrates they use may vary. **Aerobic respiration** requires oxygen. Forms of cellular respiration that do not require oxygen are said to be **anaerobic**. Some plants and animals can generate ATP anaerobically for short periods of time. Other organisms use only anaerobic respiration and live in oxygen-free environments. For these organisms, there is some other final electron acceptor other than oxygen (e.g. nitrate or Fe^{2+}).

An Overview of Cellular Respiration

Respiration involves three metabolic stages, summarised below. The first two stages are the catabolic pathways that decompose glucose and other organic fuels. In the third stage, the electron transport chain accepts electrons from the first two stages and passes these from one electron acceptor to another. The energy released at each stepwise transfer is used to make ATP. The final electron acceptor in this process is molecular oxygen.

1 **Glycolysis**. This occurs in the cytoplasm and involves the breakdown of glucose into two molecules of pyruvate.

2 **The Krebs cycle**. This occurs in the mitochondrial matrix, and decomposes a derivative of pyruvate to carbon dioxide.

3 **Electron transport and oxidative phosphorylation**. This occurs in the inner membranes of the mitochondrion and accounts for almost 90% of the ATP generated by respiration.

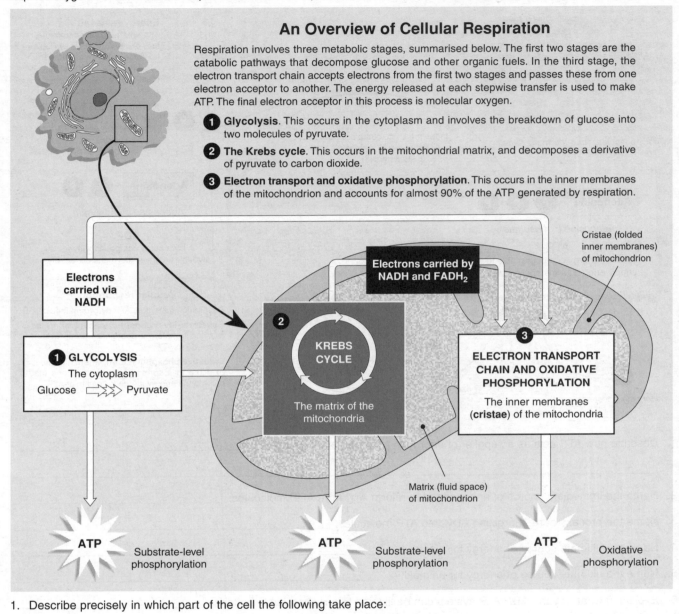

1. Describe precisely in which part of the cell the following take place:

   (a) Glycolysis: _____

   (b) Krebs cycle reactions: _____

   (c) Electron transport chain: _____

2. Provide a clear explanation of what is involved in each of the following processes:

   (a) Substrate-level phosphorylation: _____

   _____

   _____

   (b) Oxidative phosphorylation: _____

   _____

   _____

Related activities: The Role of ATP in Cells, The Biochemistry of Respiration

The Biochemistry of Respiration

Cellular respiration is a catabolic, energy yielding pathway. The breakdown of glucose and other organic fuels (such as fats and proteins) to simpler molecules is **exergonic** and releases energy for the synthesis of ATP. As summarised in the previous activity, respiration involves glycolysis, the Krebs cycle, and electron transport. The diagram below provides a more detailed overview of the events in each of these stages. Glycolysis and the Krebs cycle supply electrons (via NADH) to the electron transport chain, which drives **oxidative phosphorylation**. Glycolysis nets two ATP, produced by **substrate-level phosphorylation**.

The conversion of pyruvate (the end product of glycolysis) to **acetyl CoA** links glycolysis to the Krebs cycle. One "turn" of the cycle releases carbon dioxide, forms one ATP by substrate level phosphorylation, and passes electrons to three NAD^+ and one FAD. Most of the ATP generated in cellular respiration is produced by oxidative phosphorylation when NADH and $FADH_2$ donate electrons to the series of electron carriers in the electron transport chain. At the end of the chain, electrons are passed to molecular oxygen, reducing it to water. Electron transport is coupled to ATP synthesis by **chemiosmosis** (see next page).

Glycolysis (cytoplasm)

The first part of respiration that involves the breakdown of glucose in the cytoplasm. Glucose (a 6-carbon sugar) is broken into two molecules of pyruvate (also called pyruvic acid), a 3-carbon acid. A total of 2 ATP and $2NADH + 2H^+$ are generated from this stage. No oxygen is required (the process is anaerobic).

*   2 molecules of pyruvate are produced per glucose molecule. From this stage, the processing of only one pyruvate is shown.

Transition reaction (matrix)

Pyruvate enters the mitochondrion and carbon dioxide is removed. **Coenzyme A** (CoA) picks up the remaining 2-carbon fragment of the pyruvate to form acetyl coenzyme A.

NAD and FAD are hydrogen acceptors, transporting hydrogens to the electron transport chain (below).

Krebs cycle (matrix)

The acetyl group passes into a cyclic reaction and combines with a 4-carbon molecule to form a 6-carbon molecule. The CoA is released for reuse. Successive steps in the cycle remove carbon as carbon dioxide.

Other molecules (above)

When glucose is in short supply, other organic molecules can provide alternative respiratory substrates.

Electron transport chain (cristae)

Hydrogen pairs are transferred to the electron transport chain, a series of hydrogen and electron carriers, located on the membranes of the **cristae**. The hydrogens or electrons are passed from one carrier to the next, losing energy as they go. The energy released in this stepwise process is used to produce ATP. Oxygen is the final electron acceptor and is reduced to water.

*Note FAD enters the electron transport chain at a lower energy level than NAD, and only 2ATP are generated per FAD.H2.

Total ATP yield per glucose

Glycolysis: 2 ATP, *Krebs cycle*: 2 ATP, *Electron transport*: 34 ATP

Cellular Metabolism

Related activities: Cellular Respiration
Web links: Glycolysis and the Krebs Cycle

A 3

Chemiosmosis

Chemiosmosis is the process whereby the synthesis of ATP is coupled to electron transport and the movement of protons (H^+ ions). **Electron transport carriers** are arranged over the inner membrane of the mitochondrion and oxidise NADH + H^+ and $FADH_2$. Energy from this process forces protons to move, against their concentration gradient, from the mitochondrial matrix into the space between the two membranes. Eventually the protons flow back into the matrix via ATP synthetase molecules in the membrane. As the protons flow down their concentration gradient, energy is released and ATP is synthesised. Chemiosmotic theory also explains the generation of ATP in the light dependent phase of photosynthesis.

Mitochondrion

Inter-membrane space

The energy from the electrons is used to transport hydrogen ions across the membrane.

ATP synthetase

Matrix

Reduced NAD (NADH) provides a source of electrons:

$$NADH + H^+ \longrightarrow NAD^+ + 2e^-$$

$$2H^+ + \tfrac{1}{2}O_2 \longrightarrow H_2O$$

The flow of protons down their concentration gradient in the ATP synthetase enzyme gives energy for:

$$ADP + P_i \longrightarrow ATP$$

1. On the diagram of cellular respiration (previous page), state the number of carbon atoms in each of the molecules (a) – (f):

2. Determine how many ATP molecules **per molecule of glucose** are generated during the following stages of respiration:

   (a) Glycolysis: _____ (b) Krebs cycle: _____ (c) Electron transport chain: _____ (d) Total: _____

3. Explain what happens to the carbon atoms lost during respiration: _____

   _____

4. Describe the role of the following in aerobic cellular respiration:

   (a) Hydrogen atoms: _____

   _____

   _____

   (b) Oxygen: _____

   _____

   _____

5. (a) Identify the process by which ATP is synthesised in respiration: _____

   (b) Briefly summarise this process: _____

   _____

   _____

   _____

   _____

   _____

Anaerobic Pathways

All organisms can metabolise glucose anaerobically (without oxygen) using glycolysis in the cytoplasm, but the energy yield from this process is low and few organisms can obtain sufficient energy for their needs this way. In the absence of oxygen, glycolysis soon stops unless there is an alternative acceptor for the electrons produced from the glycolytic pathway. In yeasts and the root cells of higher plants this acceptor is ethanal, and the pathway is called alcoholic fermentation. In the skeletal muscle of mammals, the acceptor is pyruvate itself and the end product

is lactic acid. In both cases, the duration of the fermentation is limited by the toxic effects of the organic compound produced. Although fermentation is often used synonymously with anaerobic respiration, they are not the same. Respiration always involves hydrogen ions passing down a chain of carriers to a terminal acceptor, and this does not occur in fermentation. In anaerobic respiration, the terminal H^+ acceptor is a molecule other than oxygen, e.g. Fe^{2+} or nitrate.

Alcoholic Fermentation

In alcoholic fermentation, the H^+ acceptor is ethanal which is reduced to ethanol with the release of CO_2. Yeasts respire aerobically when oxygen is available but can use alcoholic fermentation when it is not. At levels above 12-15%, the ethanol produced by alcoholic fermentation is toxic to the yeast cells and this limits their ability to use this pathway indefinitely. The root cells of plants also use fermentation as a pathway when oxygen is unavailable but the ethanol must be converted back to respiratory intermediates and respired aerobically.

Glucose
$C_6H_{12}O_6$

2 ADP

2 ATP Net

NAD.H_2

2 x pyruvate
$CH_3COCOOH$

Alcoholic fermentation
Yeast, higher plant cells

Ethanol
CH_3CH_2OH
waste product

NAD$^+$ NAD.H_2

CO_2 + Ethanal
CH_3CHO
gaseous waste product

Lactic Acid Fermentation

In the absence of oxygen, the skeletal muscle cells of mammals are able to continue using glycolysis for ATP production by reducing pyruvate to lactic acid (the H^+ acceptor is pyruvate itself). This process is called lactic acid fermentation. Lactic acid is toxic and this pathway cannot continue indefinitely. The lactic acid must be removed from the muscle and transported to the liver, where it is converted back to respiratory intermediates and respired aerobically.

Glucose
$C_6H_{12}O_6$

2 ADP

2 ATP Net

NAD.H_2

2 x pyruvate
$CH_3COCOOH$

Lactic Acid Fermentation
Animal tissues

Pyruvate ⟶ **Lactic acid**
$CH_3CHOHCOOH$
waste product

NAD.H_2 NAD$^+$

Some organisms respire only in the absence of oxygen and are known as **obligate anaerobes**. Many of these organisms are bacterial pathogens and cause diseases such as tetanus (*above*), gangrene, and botulism.

Vertebrate skeletal muscle is **facultatively anaerobic** because it has the ability to generate ATP for a short time in the absence of oxygen. The energy from this pathway comes from glycolysis and the yield is low.

The products of alcoholic fermentation have been utilised by humans for centuries. The alcohol and carbon dioxide produced from this process form the basis of the brewing and baking industries.

---

1. Describe the key difference between aerobic respiration and fermentation: _____

   _____

2. (a) Refer to page 25 and determine the efficiency of fermentation compared to aerobic respiration: _____ %

   (b) In simple terms, explain why the efficiency of anaerobic pathways is so low: _____

   _____

3. Explain why fermentation cannot go on indefinitely: _____

   _____

   _____

Related activities: The Biochemistry of Respiration

RDA 2

Measuring Respiration

In small animals or germinating seeds, the rate of cellular respiration can be measured using a simple respirometer: a sealed unit where the carbon dioxide produced by the respiring tissues is absorbed by soda lime and the volume of oxygen consumed is detected by fluid displacement in a manometer. Germinating seeds are also often used to calculate the **respiratory quotient** (RQ): the ratio of the amount of carbon dioxide produced during cellular respiration to the amount of oxygen consumed. RQ provides a useful indication of the respiratory substrate being used.

Respiratory substrates and RQ

The respiratory quotient (RQ) can be expressed simply as:

$$RQ = \frac{CO_2 \text{ produced}}{O_2 \text{ consumed}}$$

When pure carbohydrate is oxidised in cellular respiration, the RQ is 1.0; more oxygen is required to oxidise fatty acids (RQ = 0.7). The RQ for protein is about 0.9. Organisms usually respire a mix of substrates, giving RQ values of between 0.8 and 0.9 (see table 1, below).

Table 1: RQ values for the respiration of various substrates

RQ	Substrate
> 1.0	Carbohydrate with some anaerobic respiration
1.0	Carbohydrates, e.g. glucose
0.9	Protein
0.7	Fat
0.5	Fat with associated carbohydrate synthesis
0.3	Carbohydrate with associated organic acid synthesis

Using RQ to determine respiratory substrate

Fig. 1: RQ in relation to germination stage in wheat

Respiratory substrate is largely fat during early germination

Mainly carbohydrates are used later in germination

Modified after Clegg and MacKean 1994

Fig. 1, above, shows how experimental RQ values have been used to determine the respiratory substrate utilised by germinating wheat seeds (*Triticum sativum*) over the period of their germination.

Table 2: Rates of O_2 consumption and CO_2 production in crickets

Time after last fed/ h	Temperature/ °C	Rate of O_2 consumption/ $mlg^{-1}h^{-1}$	Rate of CO_2 production/ $mlg^{-1}h^{-1}$
1	20	2.82	2.82
48	20	2.82	1.97
1	30	5.12	5.12
48	30	5.12	3.57

Table 2 shows the rates of oxygen consumption and carbon dioxide production of crickets kept under different experimental conditions.

1. Table 2 above shows the results of an experiment to measure the rates of oxygen consumption and carbon dioxide production of crickets 1 hour and 48 hours after feeding at different temperatures:

   (a) Calculate the RQ of a cricket kept at 20°C, 48 hours after feeding (show working): _____

   (b) Compare this RQ to the RQ value obtained for the cricket 1 hour after being fed (20°C). Explain the difference:

   _____

   _____

2. The RQs of two species of seeds were calculated at two day intervals after germination. Results are tabulated to the right:

   (a) Plot the change in RQ of the two species during early germination:

   (b) Explain the values in terms of the possible substrates being respired:

Days after germination	RQ	
	Seedling A	Seedling B
2	0.65	0.70
4	0.35	0.91
6	0.48	0.98
8	0.68	1.00
10	0.70	1.00

_____

_____

_____

_____

_____

_____

Metabolic Pathways

Metabolism is all the chemical activities of life. The myriad enzyme-controlled **metabolic pathways** that are described as metabolism form a tremendously complex network that is necessary in order to 'maintain' the organism. Errors in the step-wise regulation of enzyme-controlled pathways can result in metabolic disorders that in some cases can be easily identified. An example of a well studied metabolic pathway, the metabolism of phenylalanine, is described below.

A Metabolic Pathway

Gene A

Expression of Gene A (by protein synthesis) produces enzyme A

Enzyme A

Enzyme A transforms the *precursor chemical* into the *intermediate chemical* by altering its chemical structure

Precursor chemical

Intermediate chemical

Gene B

Expression of Gene B (by protein synthesis) produces enzyme B

Enzyme B

Enzyme B transforms the *intermediate chemical* into the *end product*

End product

Case Study: The Metabolism of Phenylalanine

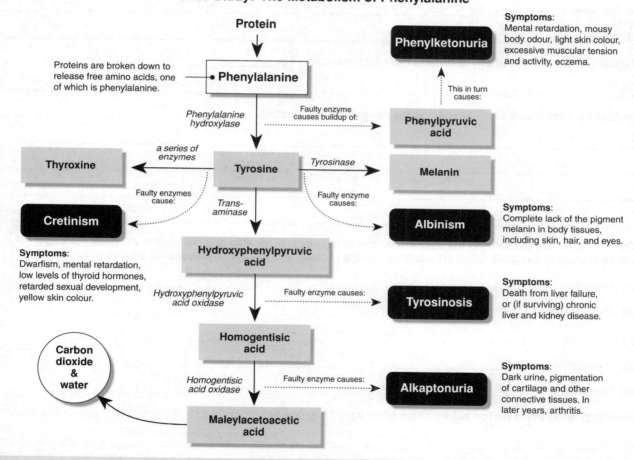

Protein

Proteins are broken down to release free amino acids, one of which is phenylalanine.

• Phenylalanine

Phenylalanine hydroxylase

Faulty enzyme causes buildup of:

Phenylpyruvic acid

This in turn causes:

Phenylketonuria

Symptoms:
Mental retardation, mousy body odour, light skin colour, excessive muscular tension and activity, eczema.

a series of enzymes

Thyroxine

Tyrosine

Tyrosinase

Melanin

Faulty enzymes cause:

Cretinism

Trans-aminase

Faulty enzyme causes:

Albinism

Symptoms:
Complete lack of the pigment melanin in body tissues, including skin, hair, and eyes.

Symptoms:
Dwarfism, mental retardation, low levels of thyroid hormones, retarded sexual development, yellow skin colour.

Hydroxyphenylpyruvic acid

Hydroxyphenylpyruvic acid oxidase

Faulty enzyme causes:

Tyrosinosis

Symptoms:
Death from liver failure, or (if surviving) chronic liver and kidney disease.

Homogentisic acid

Carbon dioxide & water

Homogentisic acid oxidase

Faulty enzyme causes:

Alkaptonuria

Symptoms:
Dark urine, pigmentation of cartilage and other connective tissues. In later years, arthritis.

Maleylacetoacetic acid

A well studied metabolic pathway is the metabolic breakdown of the essential amino acid **phenylalanine**. The first step is carried out by an enzyme produced in the liver, called phenylalanine hydroxylase. This enzyme converts phenylalanine to the amino acid **tyrosine**. Tyrosine, in turn, through a series of intermediate steps, is converted into **melanin**, the skin pigment, and other substances. If phenylalanine hydroxylase is absent, phenylalanine is in part converted into phenylpyruvic acid, which accumulates, together with phenylalanine, in the blood stream. Phenylpyruvic acid and phenylalanine are toxic to the central nervous system and produce some of the symptoms of the genetic disease **phenylketonuria**. Other genetic metabolic defects in the tyrosine pathway are also known. As indicated above, absence of enzymes operating between tyrosine and melanin, is a cause of **albinism**. **Tyrosinosis** is a rare defect that causes hydroxyphenylpyruvic acid to accumulate in the urine. **Alkaptonuria** makes urine turn black on exposure to air, causes pigmentation to appear in the cartilage, and produces symptoms of arthritis. A different block in another pathway from tyrosine produces thyroid deficiency leading to goiterous **cretinism** (due to lack of thyroxine).

1. Using the metabolism of phenyalanine as an example, discuss the role of enzymes in **metabolic pathways**:

_____

_____

_____

_____

_____

_____

2. Identify three **products** of the metabolism of phenylalanine: _____

_____

3. Identify the enzyme failure (faulty enzyme) responsible for each of the following conditions:

   (a) Albinism: _____

   (b) Phenylketonuria: _____

   (c) Tyrosinosis: _____

   (d) Alkaptonuria: _____

4. Explain why people with **phenylketonuria** have light skin colouring: _____

_____

5. Discuss the consequences of disorders in the metabolism of **tyrosine**: _____

_____

_____

_____

_____

6. The five conditions illustrated in the diagram are due to too much or too little of a chemical in the body. For each condition listed below, state which chemical causes the problem and whether it is absent or present in excess:

   (a) Albinism: _____

   (b) Phenylketonuria: _____

   (c) Cretinism: _____

   (d) Tyrosinosis: _____

   (e) Alkaptonuria: _____

7. If you suspected that a person suffered from phenylketonuria, how would you test for the condition if you were a doctor:

_____

_____

_____

8. The diagram at the top of the previous page represents the normal condition for a simple metabolic pathway. A starting chemical, called the **precursor**, is progressively changed into a final chemical called the **end product**.

   Consider the effect on this pathway if **gene A** underwent a mutation and the resulting **enzyme A** did not function:

   (a) Name the chemicals that would be present in **excess**: _____

   (b) Name the chemicals that would be **absent**: _____

Controlling Metabolic Pathways

The **operon** mechanism was proposed by **Jacob and Monod** to account for the regulation of gene activity in response to the needs of the cell. Their work was carried out with the bacterium *Escherichia coli* and the model is not applicable to eukaryotic cells where the genes are not found as operons. An operon consists of a group of closely linked genes that act together and code for the enzymes that control a particular **metabolic pathway**. These may be for the metabolism of an energy source (e.g. lactose) or the synthesis of a molecule such as an amino acid. The structural genes contain the information for the production of the enzymes themselves and they are transcribed as a single **transcription unit**. These structural genes are controlled by a **promoter**, which initiates the formation of the mRNA, and a region of the DNA in front of the structural genes called the **operator**. A gene outside the operon, called the **regulator gene**, produces a **repressor** molecule that can bind to the operator, and block the transcription of the structural genes. It is the repressor that switches the structural genes on or off and controls the metabolic pathway. Two mechanisms operate in the operon model: gene induction and gene repression. **Gene induction** occurs when genes are switched on by an inducer binding to the repressor molecule and deactivating it. In the **Lac operon model** based on *E.coli*, lactose acts as the **inducer**, binding to the repressor and permitting transcription of the structural genes for the utilisation of lactose (an infrequently encountered substrate). **Gene repression** occurs when genes that are normally switched on (e.g. genes for synthesis of an amino acid) are switched off by activation of the repressor.

Control of Gene Expression Through Induction: the *Lac* Operon

Structure of the operon

RNA polymerase

Transcription begins

At least one **structural gene** is present. The structural gene codes for the creation of an enzyme in a metabolic pathway.

Regulator gene · · · Promoter | Operator | Structural gene A · · · **DNA**

The regulator gene, on another part of the DNA, produces the **repressor** molecule by protein synthesis. In the *lac* operon the regulator gene is located next to the promoter.

The **promoter** site is where the RNA polymerase enzyme first attaches itself to the DNA to begin synthesis of the mRNA.

The **operator** is the potential blocking site. It is here that an active repressor molecule will bind, stopping mRNA synthesis from proceeding.

◄——————— **OPERON** ———————►

The operon consists of the structural genes and the promoter and operator sites

Structural genes switched off

RNA polymerase enzyme may not be able to bind to the promoter, or it may be blocked along the DNA.

An active repressor molecule binds to the operator site and suppresses its activity (the gene is "switched off").

Lactose is not a common energy source for *E. coli* and the genes for the metabolism of lactose by the cell are normally switched off. With **lactose absent**, the repressor molecule binds tightly to the operator. This prevents RNA polymerase from transcribing the adjacent structural genes and the enzymes for lactose metabolism are not produced.

Repressor

Transcription is stopped

Regulator gene | Promoter | Operator | Structural gene A · **DNA**

Gene induction

The inducer binds to the repressor altering its shape. It can no longer bind to the DNA, permitting the operator gene to become active (i.e. the gene is "switched on").

When lactose is available, some of it is converted into the **inducer** allolactose. Allolactose binds to the repressor molecule, altering its shape and preventing it from binding to the operator. The structural genes can then be transcribed, and the enzymes for the metabolism of lactose are produced.

Repressor

Inducer

Repressor

Transcription occurs

Regulator gene | Promoter | Operator | Structural gene A · **DNA**

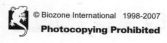
Related activities: Metabolic Pathways
Web links: Gene Control in Eukaryotes

A 3

Cellular Metabolism

Control of Gene Expression Through Repression

Gene repression

RNA polymerase

The repressor requires the presence of an **effector**, which helps the repressor bind to the operator.

The combined effector and repressor molecule bind to the operator and block RNA polymerase, preventing any further transcription.

Repressor

Repressor

Transcription is stopped

| Regulator gene | Promoter | Operator | Structural gene A | | DNA |

Effector in high concentration

The operon is normally "switched on". The genes are "switched off" as a response to the **overabundance** of an **end product** of a metabolic pathway.

In *E. coli*, the enzyme **tryptophan synthetase** synthesises the amino acid tryptophan. The gene for producing this enzyme is normally switched on. When **tryptophan** is present in excess, some of it acts as an **effector** (also called a co-repressor). The effector activates the repressor, and they bind to the operator gene, preventing any further transcription of the structural gene. Once transcription stops, the enzyme tryptophan synthetase is no longer produced. This is an example of **end-product inhibition** (feedback inhibition).

1. Explain the functional role of each of the following in relation to gene regulation in a prokaryote, e.g. *E. coli*:

   (a) Operon: _____

   _____

   (b) Regulator gene: _____

   _____

   (c) Operator: _____

   _____

   (d) Promoter: _____

   _____

   (e) Structural genes: _____

   _____

2. (a) Explain the advantage in having an inducible enzyme system that is regulated by the presence of a substrate:

   _____

   _____

   (b) Suggest when it would **not** be adaptive to have an inducible system for metabolism of a substrate: _____

   _____

   _____

   (c) Giving an example, outline how gene control in a non-inducible system is achieved through **gene repression**:

   _____

   _____

   _____

3. Describe how the two mechanisms of gene control described here are fundamentally different: _____

   _____

   _____

Leaf Structure

The main function of leaves is to collect the radiant energy from the sun and convert it into a form that can be used by the plant. The sugars produced from photosynthesis have two uses: (1) they can be broken down by respiration to release the stored chemical energy to do cellular work or (2) they can provide the plant with building materials for new tissues (growth and repair). The structure of the leaf is adapted to carry out the job of collecting the sun's energy. To an extent, a leaf is also able to control the amount of carbon dioxide entering and the amount of water leaving the plant. Both the external and internal morphology of leaves are related primarily to habitat, especially availability of light and water. Regardless of their varying forms, foliage leaves comprise epidermal, mesophyll, and vascular tissues. The mesophyll (the packing tissue of the leaf) may be variously arranged according to the particular photosynthetic adaptations of the leaf.

Angiosperm Leaves

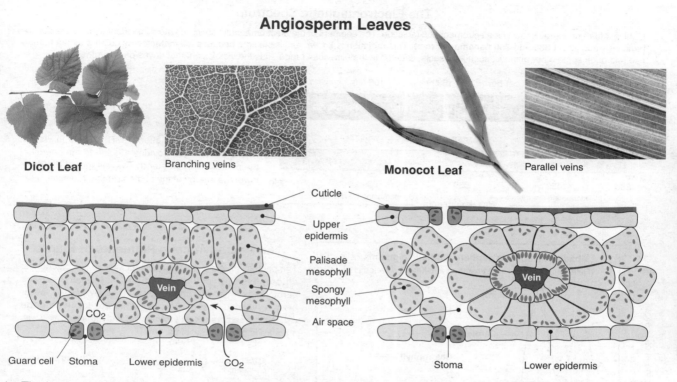

Dicot Leaf Branching veins

Monocot Leaf Parallel veins

Cuticle

Upper epidermis

Palisade mesophyll

Spongy mesophyll

Air space

Vein

CO_2

Guard cell Stoma Lower epidermis CO_2

Stoma Lower epidermis

Cellular Metabolism

1. The internal arrangement of cells and tissues in typical monocot and dicot leaves are illustrated above. Describe two ways in which the internal leaf structure of the dicot differs from that of the monocot:

   (a) _____

   _____

   (b) _____

   _____

2. (a) Explain the purpose of the waxy cuticle that coats the leaf surface: _____

   _____

   (b) Explain why the leaf epidermis is transparent: _____

   _____

   (c) Explain why leaves are usually broad and flat: _____

   _____

3. (a) Identify the region of a dicot leaf where most of the chloroplasts are found: _____

   (b) Name the important process that occurs in the chloroplasts: _____

4. (a) Explain the purpose of the air spaces in the leaf tissue: _____

   _____

   (b) Describe how gases enter and leave the leaf tissue: _____

   _____

Related activities: Photosynthesis, Adaptations for Photosynthesis

RA 1

Pigments and Light Absorption

As light meets matter, it may be reflected, transmitted, or absorbed. Substances that absorb visible light are called **pigments**, and different pigments absorb light of different wavelengths. The ability of a pigment to absorb particular wavelengths of light can be measured with a spectrophotometer. The light absorption vs the wavelength is called the **absorption spectrum** of that pigment. The absorption spectrum of different photosynthetic pigments provides clues to their role in photosynthesis, since light can only perform work if it is absorbed. An **action spectrum** profiles the effectiveness of different wavelength light in fuelling photosynthesis. It is obtained by plotting wavelength against some measure of photosynthetic rate (e.g. CO_2 production). Some features of photosynthetic pigments and their light absorbing properties are outlined below.

The Electromagnetic Spectrum

Light is a form of energy known as electromagnetic radiation. The segment of the electromagnetic spectrum most important to life is the narrow band between about 380 and 750 nanometres (nm). This radiation is known as visible light because it is detected as colours by the human eye (although some other animals, such as insects, can see in the ultraviolet range). It is the visible light that drives photosynthesis.

Electromagnetic radiation (EMR) travels in waves, where wavelength provides a guide to the energy of the photons; the greater the wavelength of EMR, the lower the energy of the photons in that radiation.

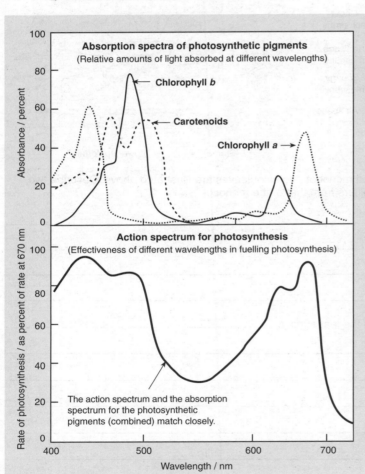

The Photosynthetic Pigments of Plants

The photosynthetic pigments of plants fall into two categories: **chlorophylls** (which absorb red and blue-violet light) and **carotenoids** (which absorb strongly in the blue-violet and appear orange, yellow, or red). The pigments are located on the chloroplast membranes (the thylakoids) and are associated with membrane transport systems.

The pigments of chloroplasts in higher plants (above) absorb blue and red light, and the leaves therefore appear green (which is reflected). Each photosynthetic pigment has its own characteristic **absorption spectrum** (left, top graph). Although only chlorophyll *a* can participate directly in the light reactions of photosynthesis, the **accessory pigments** (chlorophyll *b* and carotenoids) can absorb wavelengths of light that chlorophyll *a* cannot. The accessory pigments pass the energy (photons) to chlorophyll *a*, thus broadening the spectrum that can effectively drive photosynthesis.

Left: Graphs comparing absorption spectra of photosynthetic pigments compared with the action spectrum for photosynthesis.

---

1. Explain what is meant by the absorption spectrum of a pigment: _____

_____

2. Explain why the action spectrum for photosynthesis does not exactly match the absorption spectrum of chlorophyll *a*:

_____

_____

Related activities: Photosynthesis

Photosynthesis

Photosynthesis is of fundamental importance to living things because it transforms sunlight energy into chemical energy stored in molecules. This becomes part of the energy available in food chains. The molecules that trap the energy in their chemical bonds are also used as building blocks to create other molecules. Finally, photosynthesis releases free oxygen gas, essential for the survival of advanced life forms. Below is a diagram summarising the process of photosynthesis.

Summary of Photosynthesis in a C₃ Plant

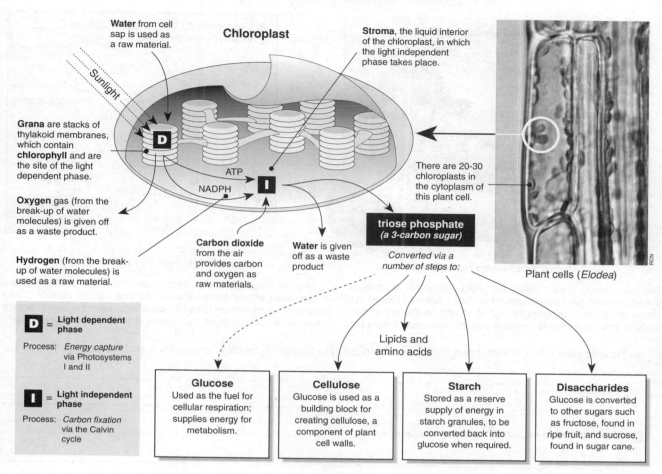

Plant cells (*Elodea*)

D = **Light dependent phase**
Process: *Energy capture via Photosystems I and II*

I = **Light independent phase**
Process: *Carbon fixation via the Calvin cycle*

Glucose
Used as the fuel for cellular respiration; supplies energy for metabolism.

Cellulose
Glucose is used as a building block for creating cellulose, a component of plant cell walls.

Starch
Stored as a reserve supply of energy in starch granules, to be converted back into glucose when required.

Disaccharides
Glucose is converted to other sugars such as fructose, found in ripe fruit, and sucrose, found in sugar cane.

1. Describe the three things of fundamental biological importance provided by photosynthesis:

(a) _____

(b) _____

(c) _____

2. Write the overall chemical equation for photosynthesis using:

(a) Words: _____

(b) Chemical symbols: _____

3. Discuss the potential uses for the end products of photosynthesis: _____

_____

_____

4. Distinguish between the two different regions of a chloroplast and describe the biochemical processes that occur in each:

_____

_____

_____

Related activities: The Biochemistry of Photosynthesis

A 2

Photosynthetic Rate

The rate at which plants can make food (the photosynthetic rate) is dependent on environmental factors, particularly the amount of **light** available, the level of **carbon dioxide** (CO_2) and the **temperature**. The effect of these factors can be tested experimentally by altering one of the factors while holding others constant (a controlled experiment). In reality, a plant is subjected to variations in all three factors at the same time. The interaction of the different factors can also be examined in the same way, as long as only one factor at a time is altered. The results can be expressed graphically.

Factors Affecting Photosynthetic Rate

The two graphs above illustrate the effect of different variables on the rate of photosynthesis in cucumber plants. Graph A (above, left) shows the effect of different intensities of light. In this experiment, the level of carbon dioxide available and the temperature were kept constant. Graph B (above, right) shows the effect of different light intensities at two temperatures and two carbon dioxide (CO_2) concentrations. In each of these experiments either the carbon dioxide level or the temperature was raised at each light intensity in turn.

1. (a) Describe the effect of increasing light intensity on the rate of photosynthesis (temperature and CO_2 constant):

_____

_____

(b) Give a possible explanation for the shape of the curve: _____

_____

2. (a) Describe the effect of increasing the temperature on the rate of photosynthesis: _____

_____

(b) Suggest a reason for this response: _____

_____

3. Explain why the rate of photosynthesis declines when the CO_2 level is reduced: _____

_____

_____

4. (a) In the graph above right, explain how the effects of CO_2 level were separated from the effects of temperature:

_____

_____

(b) State which of the two factors, CO_2 level or temperature, has the greatest effect on photosynthetic rate:

_____

(c) Explain how you can tell this from the graph: _____

_____

The Biochemistry of Photosynthesis

Like cellular respiration, photosynthesis is a redox process, but the electron flow evident in respiration is reversed. In photosynthesis, water is split and electrons are transferred together with hydrogen ions from water to CO_2, reducing it to sugar. The electrons increase in potential energy as they move from water to sugar. The energy to do this is provided by light. Photosynthesis comprises two phases. In the **light dependent phase**, light energy is converted to chemical energy (ATP and reducing power). In the **light independent phase** (or **Calvin cycle**), the chemical energy is used for the synthesis of carbohydrate. The light dependent phase illustrated below shows **non-cyclic phosphorylation**. In **cyclic phosphorylation**, the electrons lost from photosystem II are replaced by those from photosystem I. ATP is generated, but not NADPH.

Light Dependent Phase
(Energy capture)

- This diagram shows **non-cyclic phosphorylation**.

- Photosystem complexes comprise hundreds of pigment molecules, including *chlorophyll a* and *b*.

- **Photosystem II** absorbs light energy to elevate electrons to a moderate energy level.

- **Photosystem I** absorbs light energy to elevate electrons to an even higher level. Its electrons are replaced by electrons from photosystem II.

Chloroplast

Water
Stroma
Grana are stacks of thylakoid membranes (enlarged below)

D
I
ATP
Oxygen
NADPH
Hydrogen
Carbon dioxide
Water
triose phosphate

D	Light dependent
I	Light independent

Cellular Metabolism

Detail of Thylakoid Membrane

When chlorophyll molecules absorb light, an electron is excited to a higher level. This electron 'hole' must be filled.

Electron transport chain: Each electron is passed from one electron carrier to another, losing energy as it goes. This energy is used to pump hydrogen ions across the thylakoid membrane.

Light energy

Light energy

Light energy

$NADP^+ + 2H^+$

$NADPH + H^+$

Thylakoid membrane

$2e^-$
$2e^-$
$2e^-$
$2e^-$

Photosystem II
Photosystem I

$NADP^+$ reductase

Thylakoid space: hydrogen reservoir, low pH

$\frac{1}{2}O_2$
$2H^+$
H_2O

H^+

Flow of H^+ back across the membrane is coupled to ATP synthesis (by **chemiosmosis**).

Photolysis of water: In non-cyclic phosphorylation, the electrons lost to the electron transport chain are replaced by splitting a water molecule (photolysis), releasing oxygen gas and hydrogen ions.

ADP + P_i

ATP

H^+

ATP synthetase converts ADP and inorganic phosphate (P_i) into ATP

NADP is a hydrogen carrier, picking up H^+ from the thylakoid and transporting them to the Calvin cycle.

$NADPH + H^+$

Light Independent Phase
(Carbon fixation)

The light independent reaction, called the **Calvin cycle**, has also been labelled the 'dark phase' of photosynthesis. This is not a good label as it is not necessary that the phase occur in darkness; it simply does not require light to proceed. In the Calvin cycle, hydrogen (H^+) is added to CO_2 and a 5C intermediate to make carbohydrate. The H^+ and ATP are supplied by the light dependent phase above.

CO_2

(a)
RuBP: Ribulose bisphosphate

Ribulose bisphosphate carboxylase (RuBisCO)

(b)
G3P: Glycerate 3-phosphate

ADP + P_i

ATP

NADPH + H^+

Calvin cycle

ATP

ADP + P_i

NADP

(d)
Ribulose phosphate

(c)
Triose phosphate

Hexose sugars

Related activities: Photosynthesis
Web links: Harvesting Light

RA 3

1. Describe the role of the carrier molecule **NADP** in photosynthesis: _____
   _____
   _____

2. Explain the role of chlorophyll molecules in the process of photosynthesis: _____
   _____
   _____
   _____

3. On the previous diagram, write the number of carbon atoms that each molecule has at each stage of the Calvin cycle:

4. Summarise the events in each of the two phases in photosynthesis and identify where each phase occurs:

   (a) **Light dependent phase (D)**: _____
   _____
   _____
   _____

   (b) **Calvin cycle**: _____
   _____
   _____

5. The final product of photosynthesis is triose phosphate. Describe precisely where the carbon, hydrogen and oxygen molecules originate from to make this molecule:
   _____
   _____
   _____

6. Explain how ATP is produced as a result of light striking chlorophyll molecules during the light dependent phase:
   _____
   _____
   _____

7. (a) The diagram of the light dependent phase (top of previous page) describes **non-cyclic phosphorylation**. Explain what you understand by this term:
   _____
   _____

   (b) Suggest why this process is also known as non-cyclic **photo**phosphorylation: _____
   _____

   (c) Explain how photophosphorylation differs from the oxidative phosphorylation occurring in cellular respiration:
   _____
   _____

8. Explain how **cyclic photophosphorylation** differs from non-cyclic photophosphorylation: _____
   _____
   _____
   _____

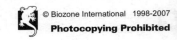

Adaptations for Photosynthesis

In order to photosynthesise, plants must obtain a regular supply of carbon dioxide (CO_2) gas; the raw material for the production of carbohydrate. In green plants, the systems for gas exchange and photosynthesis are linked; without a regular supply of CO_2, photosynthesis ceases. The leaf, as the primary photosynthetic organ, is adapted to maximise light capture and facilitate the entry of CO_2, while minimising water loss. There are various ways in which plant leaves are adapted to do this. The ultimate structure of the leaf reflects the environment of the leaf (sun or shade, terrestrial or aquatic), its resistance to water loss, and the importance of the leaf relative to other parts of the plant that may be photosynthetic, such as the stem.

Sun plant

A **sun leaf**, when exposed to high light intensities, can absorb much of the light available to the cells.

Intense light

Thick leaves

Palisade mesophyll layer often 2 or 3 cells thick

Chloroplasts are mostly restricted to palisade mesophyll cells (few in spongy mesophyll).

Sun leaves

Rate of formation and usage of sugars

Compensation point

Production of sugars by photosynthesis

Net gain in sugars

Loss of sugars used in respiration

Light Intensity

Plants adapted for full sunlight have higher levels of respiration and much higher **compensation points**. **Sun plants** include many weed species found on open ground. They expend much more energy on the construction and maintenance of thicker leaves than do shade plants. The benefit of this investment is that they can absorb the higher light intensities available and grow more quickly.

Shade plant

A **shade leaf** can absorb the light available at lower light intensities. If exposed to high light, most would pass through.

Low light intensity

Thin leaves

Palisade mesophyll layer only 1 cell thick

Chloroplasts occur throughout the mesophyll (as many in the spongy as in the palisade mesophyll).

Shade leaves

Rate of formation and usage of sugars

Compensation point

Production of sugars by photosynthesis

Net gain in sugars

Loss of sugars used in respiration

Light Intensity

Shade plants typically grow in forested areas, partly shaded by the canopy of larger trees. They have lower rates of respiration than sun plants, mainly because they build thinner leaves. The fewer number of cells need less energy for their production and maintenance. As a result, shade plants reach their compensation point at a low light intensity; much sooner than sun plants do.

Cellular Metabolism

1. (a) From the diagrams above, determine what is meant by the **compensation point** in terms of sugar production:

_____

_____

(b) State which type of plant (sun or shade adapted) has the highest level of respiration: _____

(c) Explain how the plant compensates for the higher level of respiration: _____

_____

2. Discuss the adaptations of leaves in **sun** and **shade plants**: _____

_____

_____

_____

_____

Related activities: Leaf Structure

RDA 2

Adaptations for Photosynthesis and Gas Exchange in Plants

Surface view of stomata on a monocot leaf (grass). The parallel arrangement of stomata is a typical feature of monocot leaves. Grass leaves show properties of **xerophytes**, with several water conserving features (see right).

Cross section through a grass leaf showing the stomata housed in grooves. When the leaf begins to dehydrate, it may fold up, closing the grooves and thus preventing or reducing water loss through the stomata.

Oleander (above) is a xerophyte that displays many water conserving features. The stomata are found at the bottom of pits on the underside of the leaf. The pits restrict water loss to a greater extent than they reduce CO_2 uptake.

Some plants (e.g. buttercup above) have photosynthetic stems, and CO_2 enters freely into the stem tissue through stomata in the epidermis. The air spaces in the cortex are more typical of leaf mesophyll than stem cortex.

Hydrophytes, such as *Potamogeton*, above, have stems with massive air spaces. The presence of air in the stem means that they remain floating in the zone of light availability and photosynthesis is not compromised.

This transverse view of the twin leaves of a two-needle pine shows the sunken stomata and substomatal spaces. This adaptation for arid conditions reduces water loss by creating a region of high humidity around the stoma.

3. Describe two adaptations in plants for reducing water loss while maintaining entry of gas into the leaf:

   (a) _____

   _____

   (b) _____

   _____

4. Describe two adaptations of photosynthetic stems that are not present in non-photosynthetic stems, and explain the reasons for these:

   (a) _____

   _____

   (b) _____

   _____

5. The example of a photosynthetic stem above is from a buttercup, a plant in which the leaves are still the primary organs of photosynthesis.

   (a) Identify an example of a plant where the stem is the **only** photosynthetic organ: _____

   (b) Describe the structure of the leaves in your example and suggest a reason for their particular structure:

   _____

   _____

6. Describe one role of the air spaces in the stems of *Potamogeton* related to maintaining photosynthesis:

   _____

Plant Nutritional Requirements

Plants normally obtain minerals from the soil. The availability of mineral ions to plant roots depends on soil texture, since this affects the permeability of the soil to air and water. Mineral ions may be available to the plant in the soil water, adsorbed on to clay particles, or via release from humus and soil weathering. **Macronutrients** (e.g. nitrogen, sulfur, phosphorus) are required in large amounts for building basic constituents such as proteins. **Trace elements** (e.g. manganese, copper, and zinc) are required in small amounts. Many are components of, or activators for, enzymes. After being absorbed, mineral ions diffuse into the endodermis and may diffuse, or be actively transported, to the xylem for transport to other regions of the plant.

Plant Macronutrients

CARBON (as CO_2):

A component of all organic compounds.

NITROGEN (as NO_3^- or NH_4^+):

POTASSIUM (as K^+):

MAGNESIUM (as Mg^{2+}):

CALCIUM (as Ca^{2+}):

SULFUR (as SO_4^{2-}):

Mycorrhizal Associations

Fungal hyphae penetrate the host tissues, forming a network of hyphae (**H**) through the root.

Mycorrhizae are associations between a fungus and a higher plant root. More than 80% of plants have these associations, which increase the effective root surface area for nutrient absorption and act as extended root hair substitutes. Nutrient exchange occurs through a network of hyphae around and within the root. The host plant supplies the fungus with sugars. In return, the host plant is supplied with phosphorus, which is often in very low concentrations in the soil.

PHOSPHORUS (as $H_2PO_4^-$ or HPO_4^{2-}):

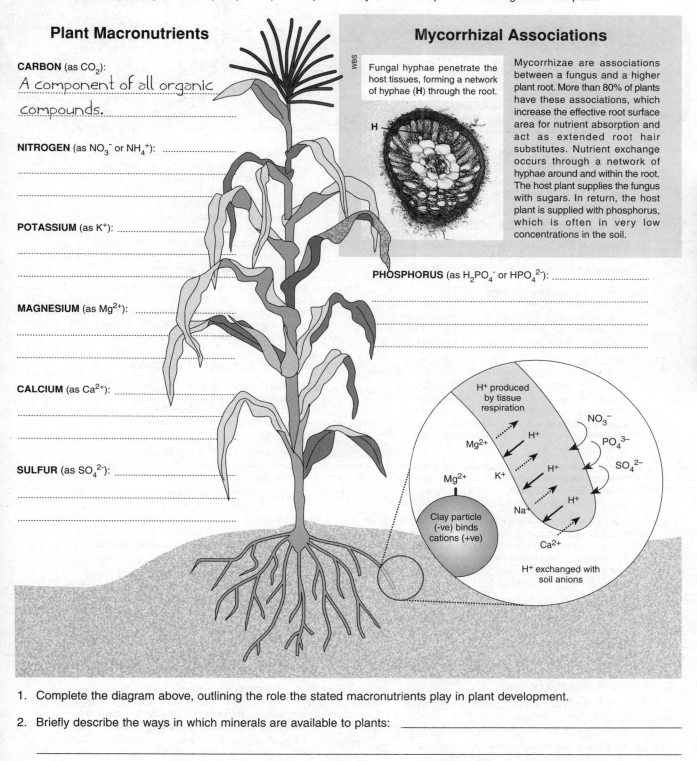

H+ produced by tissue respiration

Mg^{2+}

H^+

NO_3^-

PO_4^{3-}

SO_4^{2-}

Mg^{2+}

K^+

H^+

H^+

Clay particle (-ve) binds cations (+ve)

Na^+

H^+

Ca^{2+}

H+ exchanged with soil anions

Cellular Metabolism

1. Complete the diagram above, outlining the role the stated macronutrients play in plant development.

2. Briefly describe the ways in which minerals are available to plants: _____

_____

3. (a) Describe the role of mycorrhizal associations in plant nutrition: _____

_____

   (b) Suggest why plants without mycorrhizal associations might show poor growth: _____

_____

Related activities: Fertilisers and Land Use

RA 2

Populations and Interactions

AQA-A	AQA-B	SNAB	Edexcel	OCR
Complete:	Complete:	Complete:	Complete:	Complete:
1-5, 7-10, 14, 20-21	1, 7-10, 13, 15-19, 23, 25-26, 29(c)-(e), 32-34	1, 7-14	1, 7-12, 15-17, 19, 21, 23, 25-26, 28-29(a)-(d), 30, 33-34	1, 7-10, 13, 15-17, 21-24, 26, 28-29(a), (e), 30, 33-34

Learning Objectives

□ 1. Compile your own glossary from the **KEY WORDS** displayed in **bold type** in the learning objectives below.

The Diversity and Stability of Ecosystems

□ 2. Recognise that the collection of quantitative population data provides the means by which to study ecosystems in a meaningful way (see the topic *Practical Ecology*).

Ecosystem diversity *(pages 74-77)*

□ 3. Explain what is meant by the **stability** of an ecosystem and identify its components. Explain the relationship between ecosystem stability and diversity. Recognise the role of **keystone species** in ecosystem function.

□ 4. Explain the basis for the following relationships:
- The usually low stability and low diversity of extreme environments where abiotic factors predominate in determining population distribution and abundance.
- The usually high stability and high diversity of more moderate environments where biotic factors predominate in determining population distribution and abundance.

□ 5. Demonstrate an ability to calculate and interpret an **index of diversity**. Appreciate the applications and disadvantages of diversity indices in natural ecosystems and in those modified by human activity.

□ 6. With reference to a specific example, describe the main features of a common type of distributional variation within a community, such as **zonation**. Explain how this pattern arises and how it increases the amount of community diversity. Identify the factors determining species distribution within these communities.

Ecological succession *(pages 71-73)*

□ 7. Recognise **ecological succession** as a community pattern in time that is the result of the interaction of species with their environment.

□ 8. Describe primary succession from **pioneer species** to a **climax community**. Identify the features of species typical of each **seral stage**.

□ 9. Explain how each seral stage alters the physical environment such that conditions are made more favourable for the establishment of the next seral stage.

□ 10. Distinguish between **primary** and **secondary succession**, outlining the features of each type. Include reference to the time scale over which these successions take place.

□ 11. Explain how the **climax** vegetation varies according to local, climate, latitude, and altitude.

□ 12. Explain what is meant by a **deflected succession** and describe how such a succession arises. Explain the nature of the **plagioclimax** that develops as a result of a deflected succession.

□ 13. Describe an example of ecological succession:
(a) **Primary succession**: For example, the development of a climax community after volcanic activity or following glacial retreat.
(b) **Secondary succession**: For example, the change occurring in abandoned agricultural land to established climax woodland.
(c) **Deflected succession**: As when a particular community composition is maintained through human intervention (e.g. mowing or grazing).

□ 14. Describe how community diversity changes during the course of a succession (or **sere**). Comment on the stability of the pioneer and climax communities and relate this to the relative importance of abiotic and biotic factors at each stage (also see #4).

The Ecological Impact of Farming

Agricultural ecosystems *(pages 65-68)*

□ 15. Understand that humans manipulate the natural environment to create **agricultural ecosystems** for the production of food and other resources.

□ 16. Explain what is meant by **intensive farming**. Appreciate the implications of intensive farming practices with respect to their impact on the environment and their contribution to world food production.

□ 17. Using examples, discuss potential conflicts of interest between conservation and the need for increased production of resources such as food. Include reference to the use and effects of nitrogen containing fertilisers, and to alternatives to their use (also see #19).

□ 18. Discuss the impact of **monoculture** and the removal of **hedgerows** on the environment. Describe how intensive agricultural ecosystems differ from natural ecosystems with respect to **stability** and **diversity**.

□ 19. Using appropriate contexts, explain how farming practices can enhance **biodiversity** and help to ensure **sustainability**. Include reference to any of: **organic farming** and **intercropping**, **crop rotation** and the use of **organic fertilisers**, and biological pest control.

Land management issues *(pages 69-70)*

□ 20. Discuss the causes and effects of **deforestation**, emphasising the impact on the **biodiversity** and stability of forest ecosystems, and on carbon and nitrogen cycling. Identify regions (globally or locally) where deforestation is a major problem.

□ 21. Appreciate the conflict between the need to **conserve** forests and the rights of humans to obtain a living from the land. Describe methods by which forests may be conserved to ensure the **sustainable provision** of forestry resources. Comment on the feasibility of these.

□ 22. Describe the management of woodland habitats to maintain or increase biodiversity, as illustrated by practices such as **coppicing**.

Features of Populations (pages 44-46, 53-54)

☐ 23. Giving examples, distinguish between **population** and **community**. Explain the term population density and distinguish it from population size.

☐ 24. Understand that populations are dynamic and that they exhibit attributes not shown by the individuals themselves. Recognise the following attributes of populations and understand their importance to the study of populations: **density**, **distribution**, birth rate (**natality**), mean (average) age, death rate (**mortality**), survivorship (age specific survival), migration rate, age structure. Recognise that these attributes are population specific and represent population patterns.

☐ 25. Discuss the distribution patterns (uniform, clumped, random) of organisms within their range. Suggest which factors govern each type of distribution.

Population Growth and Size (pages 46-52)

☐ 26. Explain the role of **births**, **deaths**, and **migration** in regulating population growth. Express the relationship in an equation.

☐ 27. Explain the role of a **life tables** and **survivorship curves** in providing information on patterns of population survival and mortality. Describe the features of Type I, II, and III **survivorship curves**. Distinguish between *r* and K selection, idetifying the characteristics of *r* and K selected species.

☐ 28. Describe how the trends in population change can be shown in a **population growth curve** of population numbers (Y axis) against time (X axis).

☐ 29. Understand the factors that affect final population size, explaining clearly how they operate and providing examples where necessary. Include reference to:
   (a) **Carrying capacity** of the environment.
   (b) **Environmental resistance**.
   (c) **Density dependent factors**, e.g. intraspecific competition, interspecific competition, predation.
   (d) **Density independent factors**, e.g. catastrophic climatic events.
   (e) **Limiting factors**, e.g. soil nutrients.

☐ 30. Distinguish between **exponential** and **sigmoidal growth curves**. Create labelled diagrams of these curves, indicating the different phases of growth and the factors regulating population growth at each stage.

☐ 31. Recognise patterns of population growth in colonising, stable, declining, and oscillating populations and provide some examples.

Species Interactions (pages 55-64)

☐ 32. Explain the nature of the **interspecific interactions** occurring in communities: **competition**, **mutualism**, **commensalism**, **amensalism**, **allelopathy**, and **exploitation (parasitism, predation, herbivory)**.

☐ 33. Describing at least one example, explain the possible effects of predator-prey interactions on the **population sizes** of both predator and prey.

☐ 34. Describe, and give examples of, **interspecific** and **intraspecific competition**. Explain the effects of inter- or intraspecific competition on the distribution, population size, or niche size of two species.

See the 'Textbook Reference Grid' on pages 8-9 for textbook page references relating to material in this topic.

Supplementary Texts

See pages 5-7 for additional details of these texts:

■ Adds, J., *et al.*, 2004. **Genetics, Evolution and Biodiversity**, (NelsonThornes), chpt. 3-5 (in part).

■ Allen *et al.*, 2001. **Applied Ecology** (Collins), chpt. 1, 4, and 5 (reading as required).

■ Reiss, M. & J. Chapman, 2000. **Environmental Biology** (CUP), chpt. 2 (reading as required).

See page 7 for details of publishers of periodicals:

STUDENT'S REFERENCE

Ecosystem diversity and succession

■ **Save those Woods!** Biol. Sci. Rev., 12 (2) Nov. 1999, pp. 31-35. *Conservation of woodlands involves understanding succession. This article covers primary and secondary succession and discusses the influence of deflecting factors.*

Presentation MEDIA to support this topic:
ECOLOGY:
• Populations
• Communities
• Human Impact

■ **Plant Succession** Biol. Sci. Rev., 14 (2) November 2001, pp. 2-6. *Thorough coverage of primary and secondary succession, including the causes of different types of succession.*

■ **Tropical Rainforest Regeneration** Biol. Sci. Rev., 17(2) Nov. 2004, pp. 34-37. *Tropical rainforests: causes of and reasons for their destruction, and the role of the many complex biotic interactions in forest regeneration.*

■ **Down on the Farm: The Decline in Farmland Birds** Biol. Sci. Rev., 16(4) April 2004, pp. 17-20. *Factors in the decline of bird populations in the UK.*

■ **Biodiversity and Ecosystems** Biol. Sci. Rev., 11(4) March 1999, pp. 18-23. *Ecosystem diversity and its relationship to ecosystem stability.*

■ **Insecticides and the Conservation of Hedgerows** Biol. Sci. Rev., 16(4) April 2004, pp. 28-31. *Well managed hedgerows can reduce the need for insecticides in adjacent crops by encouraging natural pest control agents.*

■ **Razing Amazonia** New Scientist, 15 Oct 2005, pp. 34-39. *The current state of the Amazon rainforest, outlining the causes of deforestation.*

Species interactions

■ **The Other Side of Eden** Biol. Sci. Rev., 15(3) Feb. 2003, pp. 2-7. *An account of the Eden Project; its role in modelling ecosystem dynamics, including the interactions between species, is discussed.*

■ **The Ecology of Newt Populations** Biol. Sci. Rev., 11(5) May 1999, pp. 5-8. *The ecology of newt populations in Britain (includes factors involved in the regulation of population size).*

■ **Inside Story** New Scientist, 29 April 2000, pp. 36-39. *Ecological interactions between fungi and plants and animals: what are the benefits?*

■ **The Future of Red Squirrels in Britain** Biol. Sci. Rev., 16(2) Nov. 2003, pp. 8-11. *A further account of the impact of the grey squirrel on Britain's native red squirrel populations.*

■ **Reds vs Greys: Squirrel Competition** Biol. Sci. Rev., 10(4) March 1998, pp. 30-31. *The nature of the competition between red and grey squirrels in the UK: an example of competitive exclusion?*

■ **Batesian Mimicry in Your Own Backyard** Biol. Sci. Rev., 17(3) Feb. 2005, pp. 25-27. *Batesian mimicry is seen in even the most common species.*

■ **Logarithms and Life** Biol. Sci. Rev., 13(4) March 2001, pp. 13-15. *The basics of logarithmic growth and its application to real populations.*

■ **Predator-Prey Relationships** Biol. Sci. Rev., 10(5) May 1998, pp. 31-35. *Predator-prey relationships, and the defence strategies of prey.*

■ **Fish Predation** Biol. Sci. Rev., 14(1) Sept. 2001, pp. 10-14. *Some fish species in freshwater systems in the UK are important top predators and influence the dynamics of an entire ecosystem.*

■ **Ecology and Nature Conservation** Biol. Sci. Rev., 18(1) Sept. 2005, pp. 11-15. *Further research is needed in order to best restore or manage the natural ecosystems of lowland and upland Britain.*

TEACHER'S REFERENCE

■ **Where Have All the Flowers Gone** Biologist, 48(4) August 2001. *Changes in Britain's woodlands as a consequence of past deforestation.*

■ **Life Support** New Scientist, 15 August 1998, pp. 30-34. *What is the true cost of losing species diversity in ecosystems? The importance of biodiversity in the stability of ecosystems and their ability to recover from human impacts.*

See pages 10-11 for details of how to access **Bio Links** from our web site: **www.biozone.co.uk**. From Bio Links, access sites under the topics:

ECOLOGY: • Introduction to biogeography and ecology ... *and others* > **Ecosystems:** • Freshwater ecosystems • Intertidal zones ... *and others* > **Populations and Communities:** • Communities • Competition • Gap dynamics • Interactions: Biotic factors • Intraspecific relations: Cooperation and competition • Population ecology • Population growth and balance ... *and others*

Features of Populations

Populations have a number of attributes that may be of interest. Usually, biologists wish to determine **population size** (the total number of organisms in the population). It is also useful to know the **population density** (the number of organisms per unit area). The density of a population is often a reflection of the **carrying capacity** of the environment, i.e. how many organisms a particular environment can support. Populations also have structure; particular ratios of different ages and sexes. These data enable us to determine whether the population is declining or increasing in size. We can also look at the **distribution** of organisms within their environment and so determine what particular aspects of the habitat are favoured over others. One way to retrieve information from populations is to **sample** them. Sampling involves collecting data about features of the population from samples of that population (since populations are usually too large to examine in total). Sampling can be done directly through a number of sampling methods or indirectly (e.g. monitoring calls, looking for droppings or other signs). Some of the population attributes that we can measure or calculate are illustrated on the diagram below.

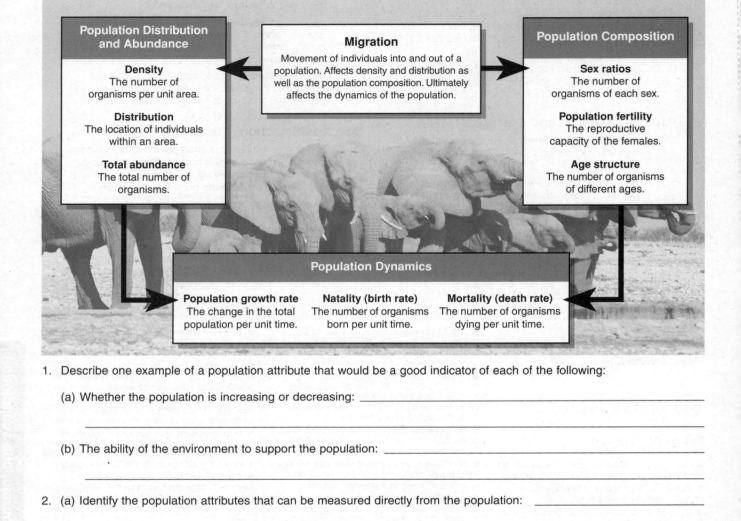

Population Distribution and Abundance

Density
The number of organisms per unit area.

Distribution
The location of individuals within an area.

Total abundance
The total number of organisms.

Migration
Movement of individuals into and out of a population. Affects density and distribution as well as the population composition. Ultimately affects the dynamics of the population.

Population Composition

Sex ratios
The number of organisms of each sex.

Population fertility
The reproductive capacity of the females.

Age structure
The number of organisms of different ages.

Population Dynamics

Population growth rate
The change in the total population per unit time.

Natality (birth rate)
The number of organisms born per unit time.

Mortality (death rate)
The number of organisms dying per unit time.

1. Describe one example of a population attribute that would be a good indicator of each of the following:

   (a) Whether the population is increasing or decreasing: _____

   _____

   (b) The ability of the environment to support the population: _____

   _____

2. (a) Identify the population attributes that can be measured directly from the population: _____

   _____

   (b) Identify the population attributes that must be calculated from the data collected: _____

   _____

3. Describe the value of population sampling for each of the following situations:

   (a) Conservation of a population of an endangered species: _____

   _____

   _____

   (b) Management of a fisheries resource: _____

   _____

   _____

   _____

Related activities: Density and Distribution, Population Age Structure

Density and Distribution

Distribution and density are two interrelated properties of populations. Population density is the number of individuals per unit area (for land organisms) or volume (for aquatic organisms). Careful observation and precise mapping can determine the distribution patterns for a species. The three basic distribution patterns are: random, clumped and uniform. In the diagram below, the circles represent individuals of the same species. It can also represent populations of different species.

Low Density

In low density populations, individuals are spaced well apart. There are only a few individuals per unit area or volume (e.g. highly territorial, solitary mammal species).

High Density

In high density populations, individuals are crowded together. There are many individuals per unit area or volume (e.g. colonial organisms, such as many corals).

Tigers are solitary animals, found at low densities.

Termites form well organised, high density colonies.

Random Distribution

Random distributions occur when the spacing between individuals is irregular. The presence of one individual does not directly affect the location of any other individual. Random distributions are uncommon in animals but are often seen in plants.

Clumped Distribution

Clumped distributions occur when individuals are grouped in patches (sometimes around a resource). The presence of one individual increases the probability of finding another close by. Such distributions occur in herding and highly social species.

Uniform Distribution

Regular distribution patterns occur when individuals are evenly spaced within the area. The presence of one individual decreases the probability of finding another individual very close by. The penguins illustrated above are also at a high density.

1. Explain why some organisms may exhibit a clumped distribution pattern because of:

   (a) Resources in the environment: _____

   _____

   (b) A group social behaviour: _____

   _____

2. Describe a social behaviour found in some animals that may encourage a uniform distribution: _____

   _____

3. Describe the type of environment that would encourage uniform distribution: _____

   _____

4. Describe an example of each of the following types of distribution pattern:

   (a) Clumped: _____

   (b) Random (more or less): _____

   (c) Uniform (more or less): _____

Related activities: Features of Populations

A 1

Population Regulation

Very few species show continued exponential growth. Population size is regulated by factors that limit population growth. The diagram below illustrates how population size can be regulated by environmental factors. **Density independent factors** may affect all individuals in a population equally. Some, however, may be better able to adjust to them. **Density dependent factors** have a greater affect when the population density is higher. They become less important when the population density is low.

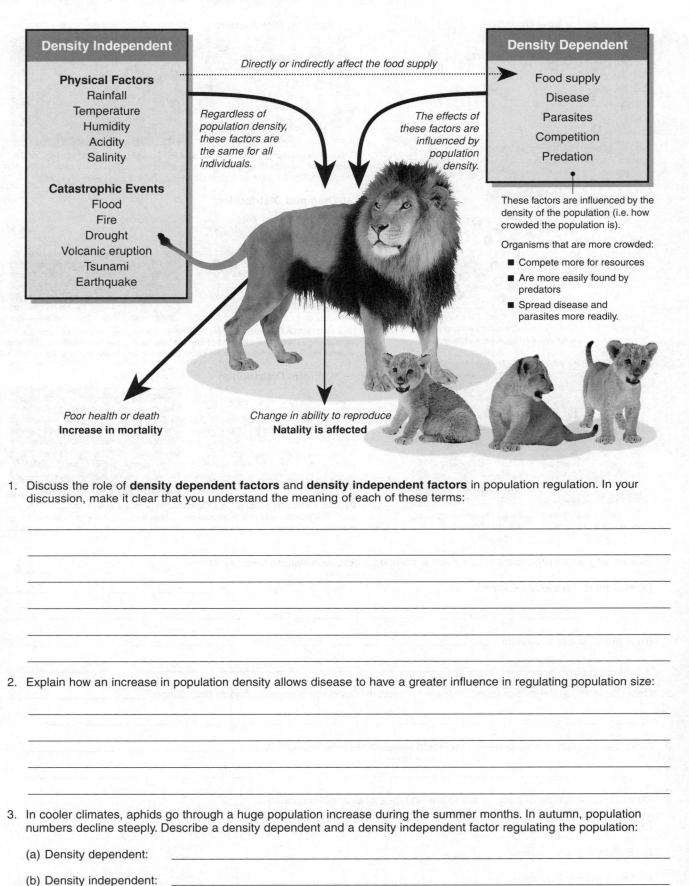

Density Independent

Physical Factors
Rainfall
Temperature
Humidity
Acidity
Salinity

Catastrophic Events
Flood
Fire
Drought
Volcanic eruption
Tsunami
Earthquake

Directly or indirectly affect the food supply

Regardless of population density, these factors are the same for all individuals.

The effects of these factors are influenced by population density.

Density Dependent

Food supply
Disease
Parasites
Competition
Predation

These factors are influenced by the density of the population (i.e. how crowded the population is).

Organisms that are more crowded:

■ Compete more for resources
■ Are more easily found by predators
■ Spread disease and parasites more readily.

Poor health or death
Increase in mortality

Change in ability to reproduce
Natality is affected

1. Discuss the role of **density dependent factors** and **density independent factors** in population regulation. In your discussion, make it clear that you understand the meaning of each of these terms:

_____

_____

_____

_____

_____

_____

_____

2. Explain how an increase in population density allows disease to have a greater influence in regulating population size:

_____

_____

_____

_____

3. In cooler climates, aphids go through a huge population increase during the summer months. In autumn, population numbers decline steeply. Describe a density dependent and a density independent factor regulating the population:

(a) Density dependent: _____

(b) Density independent: _____

Related activities: Density and Distribution

Population Growth

Organisms do not generally live alone. A **population** is a group of organisms of the same species living together in one geographical area. This area may be difficult to define as populations may comprise widely dispersed individuals that come together only infrequently (e.g. for mating). The number of individuals comprising a population may also fluctuate considerably over time. These changes make populations dynamic: populations gain individuals through births or immigration, and lose individuals through deaths and emigration. For a population in **equilibrium**, these factors balance out and there is no net change in the population abundance. When losses exceed gains, the population declines.

Births, *deaths*, *immigrations* (movements into the population) and *emigrations* (movements out of the population) are events that determine the numbers of individuals in a population. Population growth depends on the number of individuals added to the population from births and immigration, minus the number lost through deaths and emigration. This is expressed as:

> **Population growth =**
>
> **Births – Deaths + Immigration – Emigration**
> **(B)   (D)   (I)   (E)**

The difference between immigration and emigration gives *net migration*. Ecologists usually measure the **rate** of these events. These rates are influenced by environmental factors and by the characteristics of the organisms themselves. Rates in population studies are commonly expressed in one of two ways:

- Numbers per unit time, e.g. 20 150 live births per year.

- Per capita rate (number per head of population), e.g. 122 live births per 1000 individuals per year (12.2%).

Calculating change in population numbers

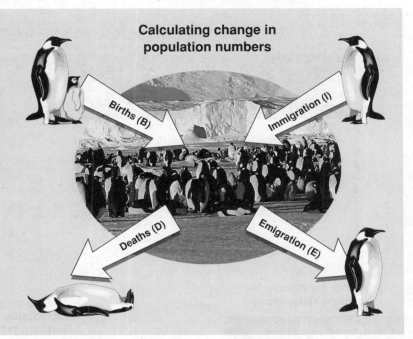

1. Define the following terms used to describe changes in population numbers:

   (a) Death rate (mortality): _____

   (b) Birth rate (natality): _____

   (c) Immigration: _____

   (d) Emigration: _____

   (e) Net migration rate: _____

2. Using the terms, B, D, I, and E (above), construct equations to express the following (the first is completed for you):

   (a) A population in equilibrium: _____ **B + I = D + E** _____

   (b) A declining population: _____

   (c) An increasing population: _____

3. The rate of population change can be expressed as the interaction of all these factors:

   > Rate of population change = birth rate – death rate + net migration rate (positive or negative)

   Using the formula above, determine the annual rate of population change for Mexico and the United States in 1972:

	USA	Mexico			
Birth rate	1.73%	4.3%	Rate of population change for USA	=	_____
Death rate	0.93%	1.0%			
Net migration rate	+0.20%	0.0%	Rate of population change for Mexico	=	_____

4. A population started with a total number of 100 individuals. Over the following year, population data were collected. Calculate birth rates, death rates, net migration rate, and rate of population change for the data below (as percentages):

   (a) Births = 14: Birth rate = _____ (b) Net migration = +2: Net migration rate = _____

   (c) Deaths = 20: Death rate = _____ (d) Rate of population change = _____

   (e) State whether the population is increasing or declining: _____

Related activities: Features of Populations
Web links: Modelling Population Growth 1, Modelling Population Growth 2

DA 1

Populations and Interactions

Life Tables and Survivorship

Life tables, such as those shown below, provide a summary of mortality for a population (usually for a group of individuals of the same age or **cohort**). The basic data are just the number of individuals remaining alive at successive sampling times (the **survivorship** or l*x*). Life tables are an important tool when analysing changes in populations over time. They can tell us the ages at which most mortality occurs in a population and can also provide information about life span and population age structure. From basic life table data, biologists derive survivorship curves, based on the l*x* column. Survivorship curves are standardised as the number of survivors per 1000 individuals so that populations of different types can be easily compared.

Life Table and Survivorship Curve for a Population of the Barnacle *Balanus*

Age in years l*x*			Proportion dying between *x* and *x* +1 /d*x*	Mortality /q*x*
0	142	1.000	0.563	0.563
1	62	0.437	0.198	0.452
2	34	0.239	0.098	0.412
3	20	0.141	0.035	0.250
4	15	0.106	0.028	0.267
5	11	0.078	0.036	0.454
6	6	0.042	0.028	0.667
7	2	0.014	0.0	0.000
8	2	0.014	0.014	1.000
9	0	0.0	0.0	—

Life Table for Female Elk, Northern Yellowstone

x	l*x*	d*x*	q*x*
0	1000	323	.323
1	677	13	.019
2	664	2	.003
3	662	2	.003
4	660	4	.006
5	656	4	.006
6	652	9	.014
7	643	3	.005
8	640	3	.005
9	637	9	.014
10	628	7	.001
11	621	12	.019
12	609	13	.021
13	596	41	.069
14	555	34	.061
15	521	20	.038
16	501	59	.118
17	442	75	.170
18	367	93	.253
19	274	82	.299
20	192	57	.297
21 +	135	135	1.000

Survivorship Curve for Female Elk of Northern Yellowstone National Park

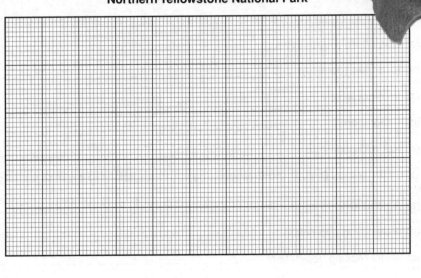

1. (a) In the example of the barnacle *Balanus* above, state when most of the group die: _____

   (b) Identify the type of survivorship curve is represented by these data (see next page): _____

2. (a) Using the grid, plot a survivorship curve for elk hinds (above) based on the life table data provided:

   (b) Describe the survivorship curve for these large mammals: _____

   _____

3. Explain how a biologist might use life table data to manage an endangered population: _____

   _____

   _____

   _____

Survivorship Curves

The survivorship curve depicts age-specific mortality. It is obtained by plotting the number of individuals of a particular cohort against time. Survivorship curves are standardised to start at 1000 and, as the population ages, the number of survivors progressively declines. The shape of a survivorship curve thus shows graphically at which life stages the highest mortality occurs. Survivorship curves in many populations fall into one of three hypothetical patterns (below). Wherever the curve becomes steep, there is an increase in mortality. The convex Type I curve is typical of populations whose individuals tend to live out their physiological life span. Such populations usually produce fewer young and show some degree of parental care. Organisms that suffer high losses of the early life stages (a Type III curve) compensate by producing vast numbers of offspring. These curves are conceptual models only, against which real life curves can be compared. Many species exhibit a mix of two of the three basic types. Some birds have a high chick mortality (Type III) but adult mortality is fairly constant (Type II). Some invertebrates (e.g. crabs) have high mortality only when moulting and show a stepped curve.

Hypothetical Survivorship Curves

Type I
Late loss survivorship curve
Mortality (death rate) is very low in the infant and juvenile years, and throughout most of adult life. Mortality increases rapidly in old age. **Examples**: Humans (in developed countries) and many other large mammals (e.g. big cats, elephants).

Type II
Constant loss survivorship curve
Mortality is relatively constant through all life stages (no one age is more susceptible than another). **Examples**: Some invertebrates such as *Hydra*, some birds, some annual plants, some lizards, and many rodents.

Type III
Early loss survivorship curve
Mortality is very high during early life stages, followed by a very low death rate for the few individuals reaching adulthood. **Examples**: Many fish (not mouth brooders) and most marine invertebrates (e.g. oysters, barnacles).

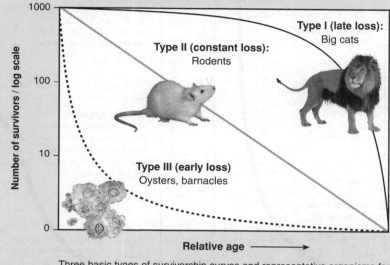

Graph of Survivorship in Relation to Age

Number of survivors / log scale — Relative age →

Type I (late loss): Big cats
Type II (constant loss): Rodents
Type III (early loss): Oysters, barnacles

Three basic types of survivorship curves and representative organisms for each type. The vertical axis may be scaled arithmetically or logarithmically.

Elephants have a close matriarchal society and a long period of parental care. Elephants are long-lived and females usually produce just one calf.

Rodents are well known for their large litters and prolific breeding capacity. Individuals are lost from the population at a more or less constant rate.

Despite vigilant parental care, many birds suffer high juvenile losses (Type III). For those surviving to adulthood, deaths occur at a constant rate.

Populations and Interactions

1. Explain why human populations might not necessarily show a Type I curve: _____

_____

2. Explain how organisms with a Type III survivorship compensate for the high mortality during early life stages:

_____

3. Describe the features of a species with a Type I survivorship that aid in high juvenile survival: _____

_____

4. Discuss the following statement: "There is no standard survivorship curve for a given species; the curve depicts the nature of a population at a particular time and place and under certain environmental conditions.":

_____

_____

_____

Related activities: Life Tables and Survivorship, *r* and K Selection

RA 3

Population Growth Curves

Populations becoming established in a new area for the first time are often termed **colonising populations** (below, left). They may undergo a rapid **exponential** (logarithmic) increase in numbers as there are plenty of resources to allow a high birth rate, while the death rate is often low. Exponential growth produces a J-shaped growth curve that rises steeply as more and more individuals contribute to the population increase. If the resources of the new habitat were endless (inexhaustible) then the population would continue to increase at an **exponential** rate. However, this rarely happens in natural populations. Initially, growth may be exponential (or nearly so), but as the population grows, its increase will slow and it will stabilise at a level that can be supported by the environment (called the carrying capacity or K). This type of growth is called sigmoidal and produces the **logistic growth curve** (below, right). **Established populations** will fluctuate about K, often in a regular way (grey area on the graph below, right). Some species will have populations that vary little from this stable condition, while others may oscillate wildly.

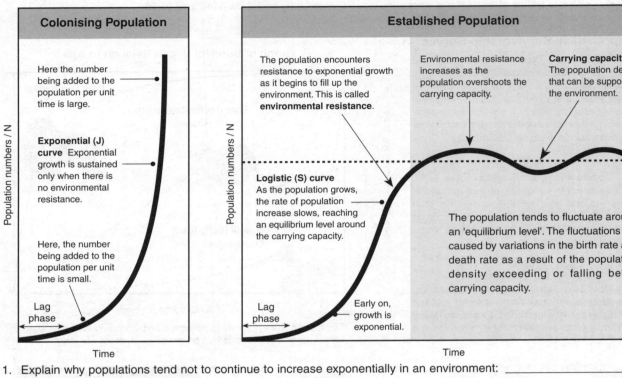

Colonising Population

Here the number being added to the population per unit time is large.

Exponential (J) curve Exponential growth is sustained only when there is no environmental resistance.

Here, the number being added to the population per unit time is small.

Lag phase

Population numbers / N

Time

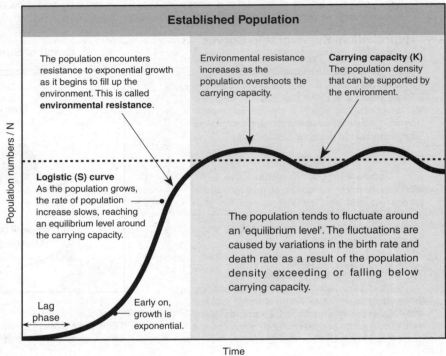

Established Population

The population encounters resistance to exponential growth as it begins to fill up the environment. This is called **environmental resistance**.

Environmental resistance increases as the population overshoots the carrying capacity.

Carrying capacity (K) The population density that can be supported by the environment.

Logistic (S) curve As the population grows, the rate of population increase slows, reaching an equilibrium level around the carrying capacity.

The population tends to fluctuate around an 'equilibrium level'. The fluctuations are caused by variations in the birth rate and death rate as a result of the population density exceeding or falling below carrying capacity.

Lag phase

Early on, growth is exponential.

Population numbers / N

Time

1. Explain why populations tend not to continue to increase exponentially in an environment: _____

_____

2. Explain what is meant by environmental resistance: _____

_____

3. (a) Explain what is meant by carrying capacity: _____

_____

(b) Explain the importance of **carrying capacity** to the growth and maintenance of population numbers: _____

_____

4. Species that expand into a new area, such as rabbits when originally introduced to Britain, typically show a period of rapid population growth followed by a slowing of population growth as density dependent factors become more important and the population settles around a level that can be supported by the carrying capacity of the environment.

(a) Explain why a newly introduced consumer (e.g. rabbit) might initially exhibit a period of exponential population growth:

_____

_____

(b) Describe a likely outcome for a rabbit population after the initial rapid increase had slowed: _____

_____

5. Describe the effect that high rabbit numbers have had on the appearance and carrying capacity of British open lands:

_____

_____

Related activities: *r* and K Selection

Growth in a Bacterial Population

Bacteria normally reproduce by a process called **binary fission**; a simple mitotic cell division that is preceded by cell elongation and involves one cell dividing in two. The time required for a cell to divide is the **generation time** and it varies between organisms and with environmental conditions such as temperature. When a few bacteria are inoculated into a liquid growth medium, and the population is counted at intervals, it is possible to plot a

bacterial growth curve that shows the growth of cells over time. In this activity, you will simulate this for a hypothetical bacterial population with a generation time of 20 minutes. In a bacterial culture with a limited nutrient supply, four growth phases are evident: the early **lag phase**, the **log phase** of exponential growth, the **stationary phase** when growth rate slows, and the **death** phase, when the population goes into logarithmic decline.

One bacterial cell

Binary fission

Two bacterial cells

Time (mins)	Population size
0	1
20	2
40	4
60	8
80	
100	
120	
140	
160	
180	
200	
220	
240	
260	
280	
300	
320	
340	
360	

Populations and Interactions

1. Complete the table (above) by doubling the number of bacteria for every 20 minute interval.

2. Graph the results on the graph grid above. Make sure that you choose suitable scales for each axis. Label the axes and mark out (number) the scale for each axis. Identify the **lag** and **log** phases of growth and mark them on the graph.

3. State how many bacteria were present after: 1 hour: _____ 3 hours: _____ 6 hours: _____

4. Describe the shape of the curve you have plotted: _____

5. Predict what would happen to the shape of the growth curve of this population assuming no further input of nutrients:

   _____

   _____

Related activities: Population Growth Curves

DA 2

r and K Selection

Two parameters govern the logistic growth of populations: the intrinsic rate of natural increase or biotic potential (this is the maximum reproductive potential of an organism, symbolised by an italicised *r*), and the carrying capacity (saturation density) of the environment (represented by the letter **K**). Species can be characterised by the relative importance of *r* and K in their life cycles. Species with a high intrinsic capacity for population increase are called ***r*-selected species**, and include algae, bacteria, rodents, many insects, and most annual plants. These species show life history features associated with rapid growth in disturbed environments. To survive, they must continually invade new areas to compensate for being replaced by more competitive species. In contrast, **K-selected** species, which include most large mammals, birds of prey, and large, long-lived plants, exist near the carrying capacity of their environments and are pushed in competitive environments to use resources more efficiently. These species have fewer offspring and longer lives, and put their energy into nuturing their young to reproductive age. Most organisms have reproductive patterns between these two extremes. Both *r*-selected species (crops) and K-selected species (livestock) are found in agriculture.

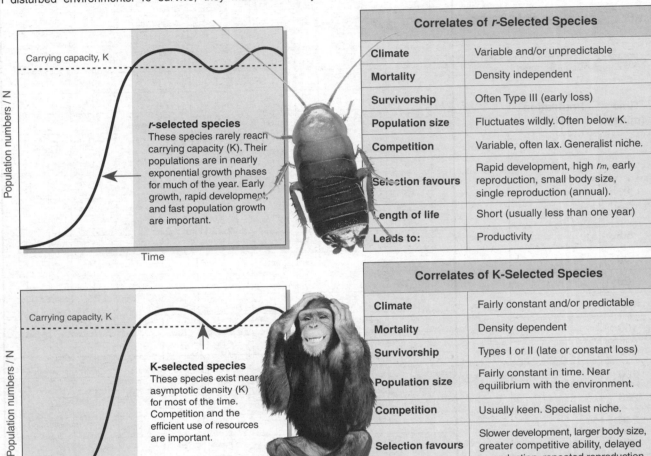

r-selected species
These species rarely reach carrying capacity (K). Their populations are in nearly exponential growth phases for much of the year. Early growth, rapid development, and fast population growth are important.

K-selected species
These species exist near asymptotic density (K) for most of the time. Competition and the efficient use of resources are important.

Correlates of *r*-Selected Species

Climate	Variable and/or unpredictable
Mortality	Density independent
Survivorship	Often Type III (early loss)
Population size	Fluctuates wildly. Often below K.
Competition	Variable, often lax. Generalist niche.
Selection favours	Rapid development, high r_m, early reproduction, small body size, single reproduction (annual).
Length of life	Short (usually less than one year)
Leads to:	Productivity

Correlates of K-Selected Species

Climate	Fairly constant and/or predictable
Mortality	Density dependent
Survivorship	Types I or II (late or constant loss)
Population size	Fairly constant in time. Near equilibrium with the environment.
Competition	Usually keen. Specialist niche.
Selection favours	Slower development, larger body size, greater competitive ability, delayed reproduction, repeated reproduction.
Length of life	Longer (greater than one year)
Leads to:	Efficiency

1. Explain the significance of the *r* and the K notation when referring to *r* and **K selection**: _____

_____

2. Giving an example, explain why *r*-selected species tend to be **opportunists**: _____

_____

_____

3. Explain why K-selected species are also called **competitor species**: _____

_____

_____

4. Suggest why many K-selected species are often vulnerable to extinction: _____

_____

Related activities: Life Tables and Survivorship, Population Growth Curves

Population Age Structure

The **age structure** of a population refers to the relative proportion of individuals in each age group in the population. The age structure of populations can be categorised according to specific age categories (such as years or months), but also by other measures such as life stage (egg, larvae, pupae, instars), of size class (height or diameter in plants). Population growth is strongly influenced by age structure; a population with a high proportion of reproductive and prereproductive aged individuals has a much greater potential for population growth than one that is dominated by older individuals. The ratio of young to adults in a relatively stable population of most mammals and birds is approximately 2:1 (below, left). Growing populations in general are characterised by a large and increasing number of young, whereas a population in decline typically has a decreasing number of young. Population age structures are commonly represented as pyramids, in which the proportions of individuals in each age/size class are plotted with the youngest individuals at the pyramid's base. The number of individuals moving from one age class to the next influences the age structure of the population from year to year. The loss of an age class (e.g. through overharvesting) can profoundly influence a population's viability and can even lead to population collapse.

Age Structures in Animal Populations

These theoretical age pyramids, which are especially applicable to birds and mammals, show how growing populations are characterised by a high ratio of young (white bar) to adult age classes (grey bars). Ageing populations with poor production are typically dominated by older individuals.

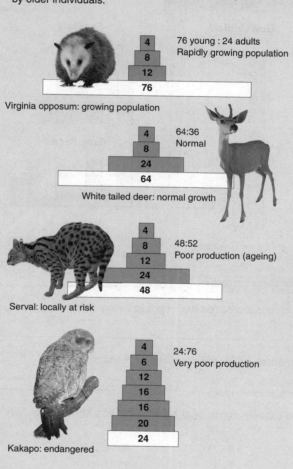

76 young : 24 adults
Rapidly growing population

Virginia opposum: growing population

64:36
Normal

White tailed deer: normal growth

48:52
Poor production (ageing)

Serval: locally at risk

24:76
Very poor production

Kakapo: endangered

Age Structures in Human Populations

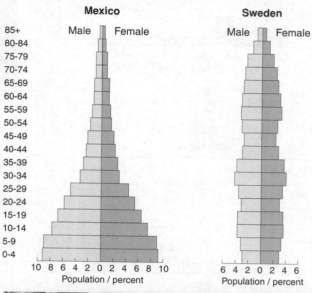

Mexico — Male / Female

Sweden — Male / Female

Age classes: 85+, 80-84, 75-79, 70-74, 65-69, 60-64, 55-59, 50-54, 45-49, 40-44, 35-39, 30-34, 25-29, 20-24, 15-19, 10-14, 5-9, 0-4

Mexico: 10 8 6 4 2 0 2 4 6 8 10 — Population / percent

Sweden: 6 4 2 0 2 4 6 — Population / percent

Extended family: Samoa

Most of the growth in human populations in recent years has occurred in the developing countries in Africa, Asia, and Central and South America. This is reflected in their age structure; a large proportion of the population comprises individuals younger than 15 years (age pyramid above, left). Even if each has fewer children, the population will continue to increase for many years. The stable age structure of Sweden is shown for comparison.

Populations and Interactions

1. For the theoretical age pyramids above left:

(a) State the approximate ratio of young to adults in a rapidly increasing population: _____

(b) Suggest why changes in population age structure alone are not necessarily a reliable predictor of population trends:

_____

_____

2. Explain why the population of Mexico is likely to continue to increase rapidly even if the rate of population growth slows:

_____

_____

Related activities: Features of Populations

RDA 2

Analysis of the age structure of a population can assist in its management because it can indicate where most of the mortality occurs and whether or not reproductive individuals are being replaced. The age structure of both plant and animal populations can be examined; a common method is through an analysis of size which is often related to age in a predictable way.

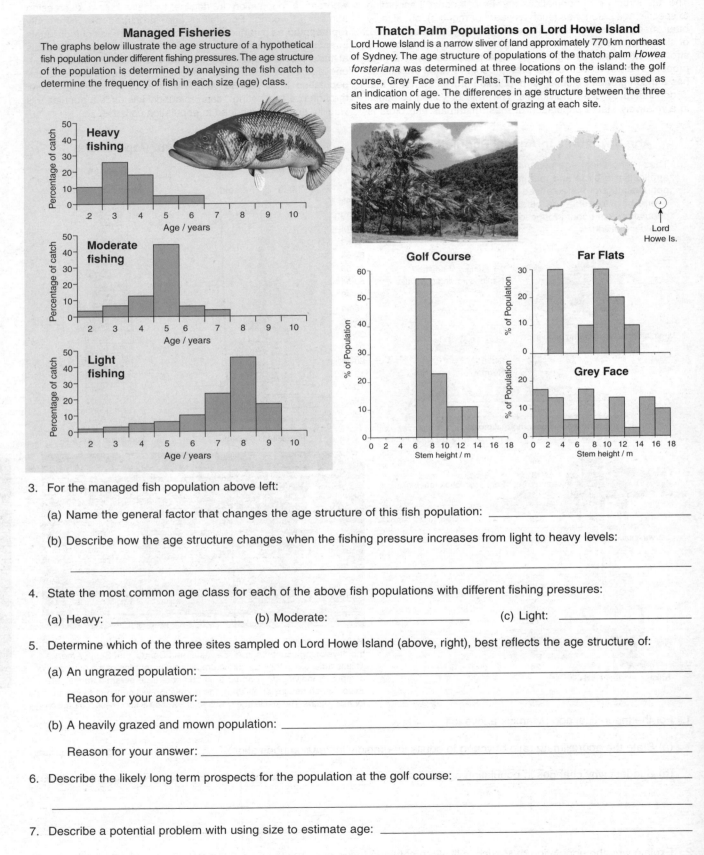

Managed Fisheries

The graphs below illustrate the age structure of a hypothetical fish population under different fishing pressures. The age structure of the population is determined by analysing the fish catch to determine the frequency of fish in each size (age) class.

Thatch Palm Populations on Lord Howe Island

Lord Howe Island is a narrow sliver of land approximately 770 km northeast of Sydney. The age structure of populations of the thatch palm *Howea forsteriana* was determined at three locations on the island: the golf course, Grey Face and Far Flats. The height of the stem was used as an indication of age. The differences in age structure between the three sites are mainly due to the extent of grazing at each site.

3. For the managed fish population above left:

   (a) Name the general factor that changes the age structure of this fish population: _____

   (b) Describe how the age structure changes when the fishing pressure increases from light to heavy levels:

   _____

4. State the most common age class for each of the above fish populations with different fishing pressures:

   (a) Heavy: _____   (b) Moderate: _____   (c) Light: _____

5. Determine which of the three sites sampled on Lord Howe Island (above, right), best reflects the age structure of:

   (a) An ungrazed population: _____

   Reason for your answer: _____

   (b) A heavily grazed and mown population: _____

   Reason for your answer: _____

6. Describe the likely long term prospects for the population at the golf course: _____

   _____

7. Describe a potential problem with using size to estimate age: _____

   _____

8. Explain why a knowledge of age structure could be important in managing a resource: _____

   _____

Species Interactions

No organism exists in isolation. Each takes part in many interactions, both with other organisms and with the non-living components of the environment. Species interactions may involve only occasional or indirect contact (predation or competition) or they may involve close association or **symbiosis**. Symbiosis is a term that encompasses a variety of interactions involving close species contact. There are three types of symbiosis: **parasitism** (a form of exploitation), **mutualism**, and **commensalism**. Species interactions affect population densities and are important in determining community structure and composition. Some interactions, such as allelopathy, may even determine species presence or absence in an area.

Examples of Species Interactions

Parasitism is a common exploitative relationship in plants and animals. A parasite exploits the resources of its host (e.g. for food, shelter, warmth) to its own benefit. The host is harmed, but usually not killed. **Endoparasites**, such as liver flukes (left), tapeworms (centre) and nematodes (right)), are highly specialised to live inside their hosts, attached by hooks or suckers to the host's tissues.

Ectoparasites, such as ticks (above), mites, and fleas, live attached to the outside of the host, where they suck body fluids, cause irritation, and may act as vectors for disease causing microorganisms.

Grouper (the host)

Remora (the commensal)

Mutualism involves an intimate association between two species that offers advantage to both. **Lichens** (above) are the result of a mutualism between a fungus and an alga (or cyanobacterium).

Termites have a mutualistic relationship with the cellulose digesting bacteria in their guts. A similar mutualistic relationship exists between ruminants and their gut microflora of bacteria and ciliates.

In **commensal** relationships, such as between this large grouper and a remora, two species form an association where one organism, the commensal, benefits and the other is neither harmed or helped.

Sea anemone

Shrimp

Stork

Hyaena

Vulture

Many species of decapod crustaceans, such as this anemone shrimp, are commensal with sea anemones. The shrimp gains by being protected from predators by the anemone's tentacles.

Interactions involving **competition** for the same food resources are dominated by the largest, most aggressive species. Here, hyaenas compete for a carcass with vultures and maribou storks.

Predation is an easily identified relationship, as one species kills and eats another (above). Herbivory is similar type of exploitation, except that the plant is usually not killed by the herbivore.

Populations and Interactions

1. Discuss each of the following interspecific relationships, including reference to the species involved, their role in the interaction, and the specific characteristics of the relationship:

(a) **Mutualism** between ruminant herbivores and their gut microflora: _____

_____

_____

_____

_____

Related activities: Interspecific Competition
Web links: Nearctica Ecology: Mutualism

RA 2

(b) **Commensalism** between a shark and a remora: _____

_____

_____

_____

(c) **Parasitism** between a tapeworm and its human host: _____

_____

_____

_____

(d) **Parasitism** between a cat flea and its host: _____

_____

_____

_____

2. Summarise your knowledge of species interactions by completing the following, entering a (+), (−), or (0) for species B, and writing a brief description of each term. Codes: (+): species benefits, (−): species is harmed, (0): species is unaffected.

Interaction	Species A	Species B	Description of relationship
(a) Mutualism	+		
(b) Commensalism	+		
(c) Parasitism	−		
(d) Amensalism	0		
(e) Predation	−		
(f) Competition	−		
(g) Herbivory	+		
(h) Antibiosis	+ / 0		

3. For each of the interactions between two species described below, choose the correct term to describe the interaction and assign a +, − or 0 for each species involved in the space supplied. Use the completed table above to help you:

Description	Term	Species A	Species B
(a) A tiny cleaner fish picking decaying food from the teeth of a much larger fish (e.g. grouper).	Mutualism	Cleaner fish +	Grouper +
(b) Ringworm fungus growing on the skin of a young child.		Ringworm	Child
(c) Human effluent containing poisonous substances killing fish in a river downstream of discharge.		Humans	Fish
(d) Humans planting cabbages to eat only to find that the cabbages are being eaten by slugs.		Humans	Slugs
(e) A shrimp that gets food scraps and protection from sea anemones, which appear to be unaffected.		Shrimp	Anemone
(f) Birds follow a herd of antelopes to feed off disturbed insects, antelopes alerted to danger by the birds.		Birds	Antelope

Predator-Prey Strategies

A predator eating its prey is one of the most conspicuous species interactions. In most cases, the predator and prey are different species, though cannibalism is quite common in some species. Predators have acute senses with which to locate and identify prey. Many also have structures such as teeth, claws, and poison to catch and subdue their prey. Animals can avoid being eaten by using passive defences, such as hiding, or active ones, such as rapid escape or aggressive defence.

Predator Avoidance Strategies Among Animals

Batesian mimicry

Harmless prey gain immunity from attack by mimicking harmful animals (called **Batesian mimicry**). The beetle on the right has the same black and yellow colour scheme as the common wasps, which can give a painful sting.

Müllerian mimicry

Many unpalatable species tend to resemble each other; a phenomenon known as **Müllerian mimicry**. The monarch and queen butterflies are both toxic. By looking alike, these mimics present a common image for predators to avoid.

Poisonous

Highly toxic animals, such as arrow poison frogs and lionfish, may advertise this fact with bright colours. The evolution of toxicity may involve kin selection, as the parent may have to die to educate a future predator of its young.

Chemical defence

Some animals can produce offensive smelling chemicals. American skunks squirt a nauseous fluid at attackers.

Hiding

Animals, such as this clown fish, take refuge in the safety provided by animals with more effective defence.

Detection and escape

Camouflage

Some animals, like this grasshopper, are masters of detecting approaching danger and making a rapid escape.

Cryptic shape and coloration allows some animals, such as this leaf insect, to blend into their background.

Startle display

Effective startle displays must come as a surprise. This stick insect will rear up on its hind legs and fan its wings. The frill necked lizard, native to Australia, will rear up suddenly and erect its frill before retreating off at high speed.

Visual deception

Deceptive markings such as large, fake eyes can apparently deceive predators momentarily, allowing the prey to escape. They may also induce the predator to strike at a nonvital end of the prey. Note the fish's real eye is disguised.

Group defence

Fish often swim in large schools, which move together as one mass in a way that confuses predators.

When birds, such as these flamingoes, form large flocks, the risk of predation to each individual is reduced.

Armoured defence

Tough outer cases that hamper predators are common in both vertebrate and invertebrate prey. Almost all molluscs have protective shells built around their sensitive parts. Some animals can coil up to protect vulnerable parts.

Offensive weapons

Offensive weapons are essential if prey are to actively fend off an attack by a predator. Many grazing mammals have sharp horns and can repel attacks by predators. Some animal bodies have spikes that offer passive defence.

Populations and Interactions

Related activities: Predator-Prey Interactions

RA 3

Prey Capturing Strategies

Concealment	Filter feeding	Lures	Traps

Praying mantis

Manta ray

Angler fish

Web spider

Some animals camouflage themselves in their surroundings, striking when the prey comes within reach.	Many marine animals (e.g. barnacles, baleen whales, sponges, manta rays) filter the water to extract tiny plankton.	This angler fish, glow worms and a type of spider all use lures to attract prey within striking range.	Spiders have developed a unique method of trapping their prey. Strong, sticky silk threads trap flying insects.

Tool use	Stealth	Speed	Group attack

Chimpanzee

Rattlesnake · **Infrared pit**

Cheetah

Pelicans

Some animals are gifted tool users. Chimpanzees use carefully prepared twigs to extract termites from mounds.	The night hunting ability of some poisonous snakes is greatly helped by the presence of infrared senses.	Some animals, such as cheetahs and some predatory birds, can simply outrun or outfly their prey.	Cooperative group behaviour may make prey capture much easier. Pelicans herd fish into 'killing zones'.

1. Explain why poisonous (unpalatable) animals are often brightly coloured so that they are easily seen:

   _____

2. Describe the purpose of large, fake eyes on some butterflies and fish: _____

   _____

   _____

3. Describe a behaviour typical of a (named) prey species that makes them difficult to detect by a predator:

   _____

4. Describe a behaviour of prey that is actively defensive: _____

5. Describe two behaviours of (named) predators that facilitate prey capture:

   (a) _____

   (b) _____

6. Discuss **Batesian and Müllerian mimicry**, explaining the differences between them and any benefits to the species involved directly or indirectly in the relationship:

   _____

   _____

   _____

   _____

7. Describe two possible ways in which toxicity could evolve in prey species, given the prey has to be eaten in order for the predator to learn from the experience:

   (a) _____

   _____

   (b) _____

   _____

Predator-Prey Interactions

Some mammals, particularly in highly seasonal environments, exhibit regular cycles in their population numbers. Snowshoe hares in Canada exhibit such a cycle of population fluctuation that has a periodicity of 9–11 years. Populations of lynx in the area show a similar periodicity. Contrary to early suggestions that the lynx controlled the size of the hare population, it is now known that the fluctuations in the hare population are governed by other factors, most probably the availability of palatable grasses. The fluctuations in the lynx numbers however, do appear to be the result of fluctuations in the numbers of hares (their principal food item). This is true of most **vertebrate** predator-prey systems: predators do not usually control prey populations, which tend to be regulated by other factors such as food availability and climatic factors. Most predators have more than one prey species, although one species may be preferred. Characteristically, when one prey species becomes scarce, a predator will "switch" to another available prey item. Where one prey species is the principal food item and there is limited opportunity for prey switching, fluctuations in the prey population may closely govern predator cycles.

Oscillations in snowshoe hare and Canadian lynx populations

Canadian lynx and snowshoe hare

Regular trapping records of Canadian lynx (left) over a 90 year period revealed a cycle of population increase and decrease that was repeated every 10 years or so. The oscillations in lynx numbers closely matched those of the snowshoe hare (right), their principal prey item. There is little opportunity for prey switching in this system and the lynx are very dependent on the hares for food. Consequently, the oscillations in the two populations have a similar periodicity, with the lynx numbers lagging slightly behind those of the hare.

1. (a) From the graph above, determine the lag time between the population peaks of the hares and the lynx:

    _____

    (b) Explain why there is this time lag between the increase in the hare population and the response of the lynx:

    _____

2. Suggest why the lynx populations appear to be so dependent on the fluctuations on the hare: _____

    _____

    _____

3. (a) In terms of birth and death rates, explain how the availability of palatable food might regulate the numbers of hares:

    _____

    _____

    (b) Explain how a decline in available palatable food might affect their ability to withstand predation pressure:

    _____

    _____

Populations and Interactions

Related activities: Predator-Prey Strategies, Population Growth

DA 2

Competition and Niche Size

Niche size is affected by competition. The effect on niche size will vary depending on whether the competition is weak, moderate or intense, and whether it is between members of the same species (**intraspecific**) or between different species (**interspecific**). The theoretical effects of competition on niche size are outlined in the diagram below.

Realised niche of species

Possible tolerance range

Narrow niche

Possible tolerance range

Broad niche

Possible tolerance range

Moderate interspecific competition

The tolerance range represents the potential (**fundamental**) niche a species could exploit. The actual or **realised** niche of a species is narrower than this because of competition. Niches of closely related species may overlap at the extremes, resulting in competition for resources in the zones of overlap.

Intense interspecific competition

When the competition from one or more closely related species becomes intense, there is selection for a more limited niche. This severe competition prevents a species from exploiting potential resources in the more extreme parts of its tolerance range. As a result, niche breadth decreases (the niche becomes narrower).

Intense intraspecific competition

Competition is most severe between individuals of the same species, because their resource requirements are usually identical. When intraspecific competition is intense, individuals are forced to exploit resources in the extremes of their tolerance range. This leads to expansion of the realised niche to less preferred areas.

Overlap in resource use between competing species

From the concept of the niche arose the idea that two species with the same niche requirements could not coexist, because they would compete for the same resources, and one would exclude the other. This is known as Gause's "**competitive exclusion principle**". If two species compete for some of the same resources (e.g. food items of a particular size), their resource use curves will overlap. Within the zone of overlap competition between the two species will be intense.

1. Distinguish between interspecific competition and intraspecific competition, and contrast the effect of these two types of competition on niche breadth:

   _____

   _____

   _____

   _____

2. Study the diagram above illustrating niche overlap between competing species and then answer the following questions:

   (a) Explain what you would expect to happen if the overlap in resource use of species A and B was to become very large (i.e. they utilised almost the same sized food item):

   _____

   (b) Describe how the degree of resource overlap might change seasonally: _____

   _____

   (c) If the zone of overlap between the resource use of the two species increased a little more from what is shown on the diagram, explain what is likely to happen to the breadth of the **realised** niche of each species:

   _____

3. Niche breadth can become broader in the presence of intense competition between members of the same species (diagram, top right). Describe one other reason why niche breadth could be very wide:

   _____

   _____

Interspecific Competition

In naturally occurring populations, direct competition between different species (**interspecific competition**) is usually less intense than intraspecific competition because coexisting species have evolved slight differences in their realised niches, even though their fundamental niches may overlap (a phenomenon termed **niche differentiation**). However, when two species with very similar niche requirements are brought into direct competition through the introduction of a foreign species, one usually benefits at the expense of the other. The inability of two species with the same described niche to coexist is referred to as the **competitive exclusion principle**. In Britain, introduction of the larger, more aggressive, grey squirrel in 1876 has

contributed to a contraction in range of the native red squirrel (below), and on the Scottish coast, this phenomenon has been well documented in barnacle species (see next page). The introduction of ecologically aggressive species is often implicated in the displacement or decline of native species, although there may be more than one contributing factor. Displacement of native species by introduced ones is more likely if the introduced competitor is also adaptable and hardy. It can be difficult to provide evidence of decline in a species as a direct result of competition, but it is often inferred if the range of the native species contracts and that of the introduced competitor shows a corresponding increase.

1940 **1984**

Red squirrel

Grey squirrel

Region of range overlap

Today, healthy populations of red squirrels exist only in the forests of Ireland, Scotland, and in the north-east and north-west of England.

Red squirrel

The **European red squirrel**, *Sciurus vulgaris*, was the only squirrel species in Britain until the introduction of the **American grey squirrel**, *Sciurus carolinesis*, in 1876. In 44 years since the 1940 distribution survey (above left), the more adaptable grey squirrel has displaced populations of the native red squirrels over much of the British Isles, particularly in the south (above right). Whereas the red squirrels once occupied both coniferous and broad leafed woodland, they are now almost solely restricted to coniferous forest and are completely absent from much of their former range.

Grey squirrel

Populations and Interactions

1. Outline the evidence to support the view that the red-grey squirrel distributions in Britain are an example of the competitive exclusion principle:

_____

_____

_____

2. Some biologists believe that competition with grey squirrels is only one of the factors contributing to the decline in the red squirrels in Britain. Explain the evidence from the 1984 distribution map that might support this view:

_____

_____

_____

_____

Related activities: Competition and Niche Size

RA 2

Competitive Exclusion in Barnacles

High tide mark

Chthamalus
Fundamental niche

A

Inset enlarged, right

Settling *Balanus* larvae die from desiccation at low tide

Chthamalus adults

Low tide mark

Balanus
Fundamental = realised niche

Settling *Chthamalus* larvae are crowded out by *Balanus*

Balanus adults

On the Scottish coast, two species of barnacles, *Balanus balanoides* and *Chthalamus stellatus*, coexist in the same general environment. The barnacles naturally show a stratified distribution, with *Balanus* concentrated on the lower region of the shore, and *Chthalamus* on the upper shore. When *Balanus* were experimentally removed from the lower strata, *Chthalamus* spread into that area. However, when *Chthalamus* were removed from the upper strata, *Balanus* failed to establish any further up the shore than usual.

3. The ability of red and grey squirrels to coexist appears to depend on the diversity of habitat type and availability of food sources (reds appear to be more successful in regions of coniferous forest). Suggest why careful habitat management is thought to offer the best hope for the long term survival of red squirrel populations in Britain:

_____

_____

_____

_____

_____

_____

4. Suggest other conservation methods that might aid the survival of viable red squirrel populations:

_____

_____

_____

5. (a) In the example of the barnacles (above), describe what is represented by the zone labelled with the arrow **A**:

_____

(b) Outline the evidence for the barnacle distribution being the result of competitive exclusion: _____

_____

_____

6. Describe two aspects of the biology of a named introduced species that have helped its success as an invading competitor:

Species: _____

(a) _____

(b) _____

Intraspecific Competition

Some of the most intense competition occurs between individuals of the same species (**intraspecific competition**). Most populations have the capacity to grow rapidly, but their numbers cannot increase indefinitely because environmental resources are finite. Every ecosystem has a **carrying capacity** (K), defined as the number of individuals in a population that the environment can support. Intraspecific competition for resources increases with increasing population size and, at carrying capacity, it reduces the per capita growth rate to zero. When the demand for a particular resource (e.g. food, water, nesting sites, nutrients, or light) exceeds supply, that resource becomes a **limiting factor**. Populations respond to resource limitation by reducing their population growth rate (e.g. through lower birth rates or higher mortality). The response of individuals to limited resources varies depending on the organism. In many invertebrates and some vertebrates such as frogs, individuals reduce their growth rate and mature at a smaller size. In some vertebrates, territoriality spaces individuals apart so that only those with adequate resources can breed. When resources are very limited, the number of available territories will decline.

Intraspecific Competition

Scramble competition in caterpillars

Contest competition in wolves

Display of a male anole

Direct competition for available food between members of the same species is called **scramble competition**. In some situations where scramble competition is intense, none of the competitors gets enough food to survive.

In some cases, competition is limited by hierarchies existing within a social group. Dominant individuals receive adequate food, but individuals low in the hierarchy must **contest** the remaining resources and may miss out.

Intraspecific competition may be for mates or breeding sites, as well as for food. In anole lizards (above), males have a bright red throat pouch and use much of their energy displaying to compete with other males for available mates.

Competition Between Tadpoles of *Rana tigrina*

Both figures: Smith & Smith: Ecology and field biology, 2001

Food shortage reduces both individual growth rate and survival, and population growth. In some organisms, where there is a metamorphosis or a series of moults before adulthood (e.g. frogs, crustacean zooplankton, and butterflies), individuals may die before they mature.

The graph (left) shows how the growth rate of tadpoles (*Rana tigrina*) declines as the density increases from 5 to 160 individuals (in the same sized space).

- At high densities, tadpoles grow more slowly, taking longer to reach the minimum size for metamorphosis (0.75 g), and decreasing their chances of successfully metamorphosing from tadpoles into frogs.
- Tadpoles held at lower densities grow faster, to a larger size, metamorphosing at an average size of 0.889 g.
- In some species, such as frogs and butterflies, the adults and juveniles reduce the intensity of intraspecific competition by exploiting different food resources.

Populations and Interactions

1. Using an example, predict the likely effects of **intraspecific competition** on each of the following:

(a) Individual growth rate: _____

_____

_____

(b) Population growth rate: _____

_____

_____

(c) Final population size: _____

_____

_____

Related activities: Population Growth Curves

RA 2

Golden eagle breeding territories in Northern Scotland, 1967

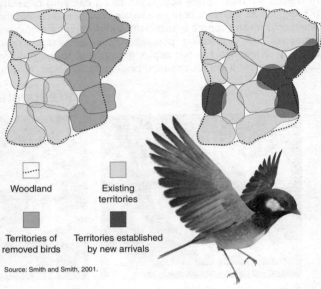

- ● Single site
- ◐ Group of sites belonging to one pair
- ○ Marginal site, not regularly occupied
- ● Breeding, year of survey 1967
- ▨ Low ground unsuitable for breeding eagles.

Territoriality in birds and other animals is usually a result of intraspecific competition. It frequently produces a pattern of uniform distribution over an area of suitable habitat, although this depends somewhat on the distribution of resources. The diagram above shows the territories of golden eagles (*Aquila chrysaetos*) in Scotland. Note the relatively uniform distribution of the breeding sites.

Territoriality in Great Tits (*Parus major*)

Six breeding pairs of great tits were removed from an oak woodland (below). Within three days, four new pairs had moved into the unoccupied areas (below, right) and some residents had expanded their territories. The new birds moved in from territories in hedgerows, considered to be suboptimal habitat. This type of territorial behaviour limits the density of breeding animals in areas of optimal habitat.

- ⬚ Woodland
- ▢ Existing territories
- ▨ Territories of removed birds
- ◼ Territories established by new arrivals

Source: Smith and Smith, 2001.

2. In the tank experiment with *Rana* (see previous page), the tadpoles were contained in a fixed volume with a set amount of food:

   (a) Describe how *Rana* tadpoles respond to resource limitation: _____

   _____

   (b) Categorise the effect on the tadpoles as density-dependent / density-independent (delete one).

   (c) Comment on how much the results of this experiment are likely to represent what happens in a natural population:

   _____

   _____

3. Identify two ways in which animals can reduce the intensity of intraspecific competition:

   (a) _____

   (b) _____

4. (a) Suggest why carrying capacity of an ecosystem might decline: _____

   _____

   (b) Predict how a decline in carrying capacity might affect final population size: _____

   _____

   _____

5. Using appropriate examples, discuss the role of territoriality in reducing intraspecific competition:

   _____

   _____

   _____

   _____

   _____

   _____

   _____

The Impact of Farming

The English countryside has been shaped by many hundreds of years of agriculture. The landscape has changed as farming practices evolved through critical stages. Farming has always had an impact on Britain's rich biodiversity (generally in a negative manner). Modern farming practices, such as increasing mechanisation and the move away from mixed farming operations, have greatly accelerated this decline. In recent years active steps to conserve the countryside, such as **hedgerow legislation**, policies to increase woodland cover, and schemes to promote environmentally sensitive farming practices are slowly meeting their objectives. Since 1990, expenditure on agri-environmental measures has increased, the area of land in organic farming has increased, and the overall volume of inorganic fertilisers and pesticides has decreased. The challenge facing farmers, and those concerned about the countryside, is to achieve a balance between the goals of production and conservation.

Intensive Farming

Intensive farming techniques flourished after World War II. Using **high-yielding hybrid cultivars** and large inputs of **inorganic fertilisers**, **chemical pesticides**, and **farm machinery**, crop yields increased to 3 or 4 times those produced using the more extensive (low-input) methods of 5 decades ago. Large areas planted in monocultures (single crops) are typical. Irrigation and fertiliser programmes are often extensive to allow for the planting of several crops per season. Given adequate irrigation and continued fertiliser inputs, yields from intensive farming are high. Over time, these yields decline as soils are eroded or cannot recover from repeated cropping.

Intensive agriculture relies on the heavy use of irrigation, inorganic fertilisers (produced using fossil fuels), pesticides, and farm machinery. Such farms may specialise in a single crop for many years.

Impact on the environment

- Pesticide use is escalating yet pesticide effectiveness is decreasing. This causes a reduction in species diversity, particularly among the invertebrates.

- Mammals and birds may be affected by **bioaccumulation** of pesticides in the food chain and loss of food sources as invertebrate species diminish.

- Fertiliser use is increasing, resulting in a continued decline in soil and water quality.

- More fertiliser leaches from the soil and enters groundwater as a pollutant, relative to organic farming practices.

- Large fields lacking hedgerows create an impoverished habitat and cause the isolation of remaining wooded areas.

- A monoculture regime leads to reduced biodiversity.

Sustainable Agricultural Practices

Organic farming is a sustainable form of agriculture based on the avoidance of chemicals and applied *inorganic* fertilisers. It relies on mixed (crop and livestock) farming and crop management, combined with the use of environmentally friendly pest controls (e.g., biological controls and flaming), and livestock and green manures. Organic farming uses **crop rotation** and **intercropping**, in which two or more crops are grown at the same time on the same plot, often maturing at different times. If well cultivated, these plots can provide food, fuel, and natural pest control and fertilisers on a sustainable basis. Yields are typically lower than on intensive farms, but the produce can fetch high prices, and pest control and fertiliser costs are reduced.

Some traditional farms in the UK use low-input agricultural practices similar to those used in modern organic farming. However, many small farming units find it difficult to remain economically viable.

Impact on the environment

- Pesticides do not persist in the environment nor accumulate in the food chain.

- Produce is pesticide free and produced in a sustainable way.

- Alternative pest control measures, such as using natural predators and pheromone traps, reduce the dependence on pesticides.

- The retention of hedgerows increases habitat diversity and produces corridors for animal movement between forested areas.

- Crop rotation (alternation of various crops, including legumes) prevents pests and disease species building up to high levels.

- Conservation tillage (ploughing crop residues into the topsoil) as part of the crop rotation cycle improves soil structure.

1. Discuss the conflict of interest between the need for high agricultural production and the need for habitat conservation:

_____

_____

_____

_____

Populations and Interactions

Related activities: Fertilisers and Land Use

RA 2

The Hedgerow Issue

A particularly significant factor of landscape change in recent years has been the amalgamation of fields and the removal of traditional hedgerows. Many traditional, mixed farms (right), which required hedgerows to contain livestock, have been converted to arable farms, and fields have become larger to accommodate modern machinery. In Britain, this conversion has resulted in the loss of thousands of kilometres of hedgerows each year.

Hedgerows are ecologically important because they increase the diversity of wildlife by:

- Providing food and habitats for birds and other animals.
- Acting as corridors, along which animals can move.
- Providing habitat for predators of pest species.

2. From an environmental perspective, outline two advantages of using hedgerows as a form of farm fencing:

(a) _____

(b) _____

3. From the perspective of the farmer, outline two disadvantages of using hedgerows:

(a) _____

(b) _____

4. Discuss the differences between intensive farming and organic farming, identifying the specific features of each:

_____

_____

_____

_____

_____

_____

_____

Diversity of Wild Farmland Bird Populations in the UK, 1970 - 1999.

The hawfinch (*Coccothraustes coccothraustes)* is a common, but often hidden, member of the woodland communities in Britain.

Diversity Index (1970 = 100)

110
100
90
80
70
60
50

1970   1975   1980   1985   1990   1995   1999

Year

Modified from: The Countryside Agency -State of the Countryside 2001 Report.

Populations of wild farmland bird species, because of their wide distribution and position near the top of the food chain, provide good indicators of the state of other wildlife species and of environmental health in general. Over the last 25 years, there has been a marked net decline in the diversity of farmland bird populations. However, since 1986, diversity has ceased to decline further and, in recent years, has actually showed an increase.

5. Suggest two possible reasons for this decline:

(a) _____

(b) _____

6. List three initiatives local and national government have implemented in an attempt to reverse this decline:

_____

Fertilisers and Land Use

A variety of minerals are required by plants. **Macronutrients** are required in large quantities, whilst **trace elements** are needed in only very small amounts. Harvesting interrupts the normal recycling of nutrients and contributes to nutrient losses. These nutrients can be replaced by the addition of fertilisers: materials that supply nutrients to plants. Some plants form symbiotic associations that aid in their nutrition. Legumes in particular (e.g. peas, beans, clover), are well known for their association with nitrogen-fixing bacteria that share their rich source of fixed nitrogen with their host plants. Legumes have been extensively used to maintain pasture productivity for grazing livestock, as well as in crop rotation to rejuvenate nitrogen-depleted soils. Historically, crop rotation was an important method to manage soil exhaustion on farms in the United Kingdom. Such systems have become relatively obsolete as demand for production during the 1980s and early 1990s increased. The availability of low cost artificial fertilisers and pesticides (**agrochemicals**) have made continuous single-crop farms economically viable.

The supply of nutrients to plants (in soil) depends on **soil fertility**. This refers to the condition of the soil relative to the amount and availability to plants of elements required for growth. Soil fertility depends both on the supply of chemical plant nutrients (e.g. nitrogen, phosphorus, and potassium), and on the supply of moisture and oxygen required to provide the appropriate environment for mineral uptake. Nutrients lost through harvesting, erosion, and denitrification can be replenished through the addition of organic and inorganic fertilisers, the decomposition of organic matter, and nitrogen fixation.

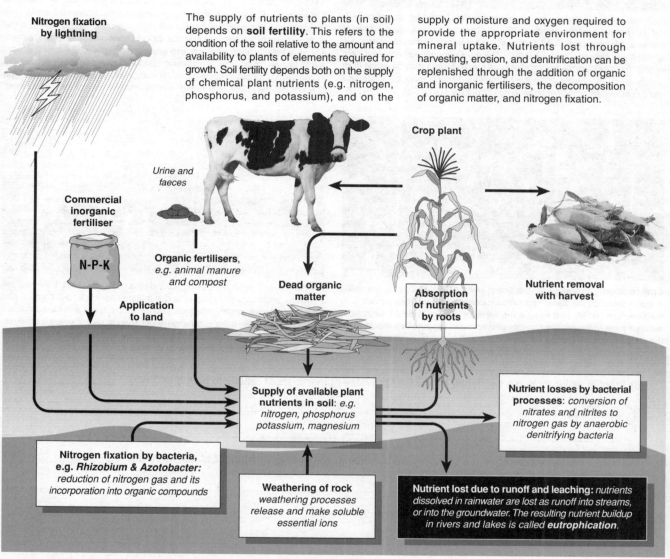

1. (a) Identify three sources of minerals for plants, apart from artificially applied fertilisers:

   _____

   (b) Suggest why these sources alone provide inadequate soil nutrients for plant growth in most cropping situations:

   _____

   _____

2. (a) Distinguish between an organic fertiliser and an inorganic fertiliser: _____

   _____

   _____

   _____

   _____

Populations and Interactions

Related activities: The Impact of Farming, Plant Nutritional Requirements

RA 2

Crop harvesting (above) removes nutrients from the land. If left unreplenished, the soil becomes nutrient deficient. The addition of organic fertilisers (**animal manure**, **green manure**, and **compost**) and inorganic fertilisers (chemically synthesised) restores the soil's fertility. The production of inorganic fertilisers consumes considerable amounts of fossil fuel (e.g. urea may be made from natural gas). In contrast, the use of organic fertilisers, reduces waste and has the added benefit of improving soil structure.

Tilling (above) prepares the land for crop sowing. Sometimes, the crop residue from the previous crop is left on the ground and mixed into the surface layers of the soil. This practice, called **conservation tillage**, improves soil structure through aeration and returns nutrients to the soil. It is best suited to crop rotations where the crop residue involved changes seasonally. In continuous cropping systems, conservation tillage leaves the same type of residue in the soil all year round, and this may harbour pests and disease.

Crop rotation is an agricultural practice in which different crops are cultivated in succession on the same ground in successive years. Its purpose is to maintain soil fertility and reduce the adverse effects of pests. Legumes, such as peas, beans, clover, vetch, and soybean (above) are important in the rotation as they restore nitrogen to the soil. In the UK, other crops that may be included in a typical rotation are wheat, barley, and root crops.

Animal manure, seen here being spread by a muck spreader, improves soil structure, adds organic nitrogen, and stimulates beneficial soil bacteria and fungi. Sustainable farming practices, such as crop rotation and the application of animal manures, have substantially declined in recent years as they do not accommodate the shift to large-scale farming practices and the move away from mixed crop and livestock farming.

(b) Identify which of these two types of fertiliser provides nutrients most readily to plants and why: _____

_____

_____

3. (a) Predict the likely consequence of continued harvesting without replenishment of soil nutrients: _____

_____

(b) Describe ways in which this consequence could be avoided: _____

_____

4. Discuss the benefits and drawbacks of **crop rotation** and **conservation tillage** as agricultural management tools:

_____

_____

_____

_____

5. Explain how overuse of inorganic and organic fertilisers may contribute to excessive **leaching** and **eutrophication**:

_____

_____

_____

_____

_____

Tropical Deforestation

Tropical rainforests prevail in places where the climate is very moist throughout the year (200 to 450 cm of rainfall per year). Almost half of the world's rainforests are in just three countries: **Indonesia** in Southeast Asia, **Brazil** in South America, and **Zaire** in Africa. Much of the world's biodiversity resides in rainforests. Destruction of the forests will contribute towards global warming through a large reduction in photosynthesis. In the Amazon, 75% of deforestation has occurred within 50 km of Brazil's roads. Many potential drugs could still be discovered in rainforest plants, and loss of species through deforestation may mean they will never be found. Rainforests can provide economically sustainable crops (rubber, coffee, nuts, fruits, and oils) for local people.

During the 1980s, the increased use of printers, photocopiers and fax machines nearly doubled office paper consumption in the USA.

Siberian forests have become a new major source of timber for USA and South Korean timber multinationals

Japan is the largest importer of tropical hardwood, receiving 50% of the world's supply

Mexico 6%
Central America 5%
Columbia 14%
Ecuador 9%
Peru 64%
Bolivia 33%
Venezuela 76%
Guyana Suriname Fr. Guiana 80%
Brazil 42%
Ivory Coast 0%
Nigeria 0%
Camaroon 56%
Gabon 79%
Congo 76%
Zaire 61%
Myanmar (Burma) 17%
India 6%
Thailand 0%
Malaysia 20%
Indonesia 38%
Vietnam 0%
Philippines 4%
Papua New Guinea 68%

Sources: National Geographic Atlas 1996; The State of the Environment Atlas 1995

Tropical forest   Tropical deforestation   Country 50%   Percentage of rainforest remaining

The felling of rainforest trees is taking place at an alarming rate as world demand for tropical hardwoods increases and land is cleared for the establishment of agriculture. The resulting farms and plantations often have shortlived productivity.

Huge forest fires have devastated large amounts of tropical rainforest in Indonesia and Brazil in 1997/98. The fires in Indonesia were started by people attempting to clear the forest areas for farming in a year of particularly low rainfall.

The building of new road networks into regions with tropical rainforests causes considerable environmental damage. In areas with very high rainfall there is an increased risk of erosion and loss of topsoil.

Populations and Interactions

1. Describe three reasons why tropical rainforests should be conserved:

   (a) _____

   (b) _____

   (c) _____

2. Identify the three main human activities that cause tropical deforestation and discuss their detrimental effects:

   _____

   _____

   _____

A 2

Managing Woodlands

The United Kingdom was once largely covered with woodland, but over many centuries it was cleared to meet the needs of a growing population. Currently only one tenth of the land area of the UK is covered by forest or other wooded areas. Although the percentage cover is still low, it has doubled since World War I as a result of major **afforestation programmes**. Currently, more than 80% of all forests are available for wood supply, with the remainder protected for conservation reasons. In the UK, commercial forests and woods are mainly composed of conifers and other softwood trees. With careful management it is possible to make use of the forest resource in a sustainable manner without destroying the particular features of the ecosystem.

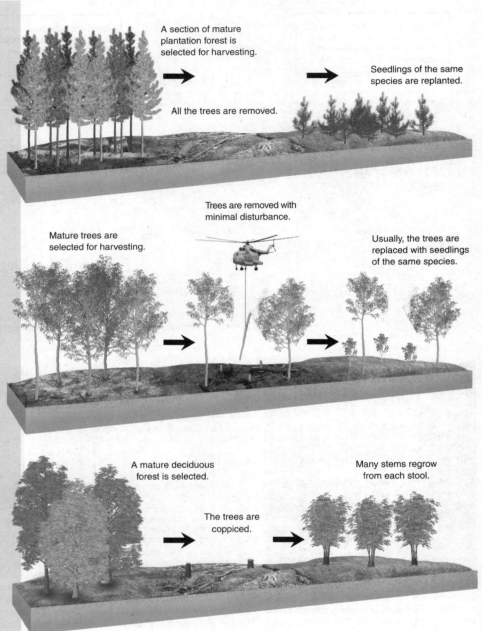

Clear Cutting

A section of a mature forest is selected (based on tree height, girth, or species), and all the trees are removed. During this process the understorey is destroyed. A new forest of economically desirable trees may be planted. In plantation forests, the trees are generally of a single species and may even be clones. Clear cutting is a very productive and economical method of managing a forest, however it is also the most damaging to the natural environment. In plantation forests this may not be of concern, but clear cutting of old growth forests causes enormous ecological damage.

A section of mature plantation forest is selected for harvesting.

All the trees are removed.

Seedlings of the same species are replanted.

Selective Logging

A mature forest is examined, and trees are selected for removal based on height, girth, or species. These trees are felled individually and directed to fall in such a way as to minimise the damage to the surrounding younger trees. The forest is managed in such a way as to ensure continual regeneration of young seedlings and provide a balance of tree ages that mirrors the natural age structure.

Mature trees are selected for harvesting.

Trees are removed with minimal disturbance.

Usually, the trees are replaced with seedlings of the same species.

Coppicing

Coppicing is the ancient practice of harvesting wood for weaving, thatching, firewood, or for making charcoal. A selection of deciduous trees is **coppiced** (cut close to the ground), leaving stumps known as **stools**. Most deciduous trees have the ability to regrow from their bases when they are cut down. However their growth form changes from a single stem to producing many stems. It is these stems that provide the wood for harvesting in the future.

A mature deciduous forest is selected.

The trees are coppiced.

Many stems regrow from each stool.

1. Using the terms *high*, *medium*, *low*, and *no change*, complete the following table.

Method of logging	Volume of logs harvested	Habitat destruction	Risk of soil erosion	Damage to surrounding trees	Biodiversity of regenerating forest
Coppicing					
Selective logging					
Clear cutting					

2. Suggest which logging method best balances the aims of productivity and conservation: _____

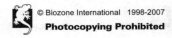

Ecological Succession

Ecological succession is the process by which communities in a particular area change over time. Succession takes place as a result of complex interactions of biotic and abiotic factors. Early communities modify the physical environment. This change results in a change in the biotic community which further alters the physical environment and so on. Each successive community makes the environment more favourable for the establishment of new species. A succession (called a **sere**) proceeds in stages, called seral stages, until the formation of the final climax community, which is stable until further disturbance. Early successional communities are characterised by having a low species diversity, a simple structure, and broad niches. In contrast, community structure in climax communities is complex, with a large number of species interactions. Niches are usually narrow and species diversity is high.

Composition of the community changes with time

| Past community | → | Present community | → | Future community |

Some species in the **past community** were outcompeted, and/or did not tolerate altered abiotic conditions

The **present community** modifies such abiotic factors as:
- Light intensity
- Wind speed
- Air temperature
- Soil composition
- Light quality
- Wind direction
- Soil water
- Humidity

Changing conditions in the **present community** will allow new species to become established. These will make up the **future community**

Primary Succession

Primary succession refers to colonisation of regions where there is no preexisting community. Examples include the emergence of new volcanic islands, new coral atolls, or islands where the previous community has been extinguished by a volcanic eruption (e.g. the Indonesian island of Krakatau). A classic set colonisation sequence (as depicted below) is relatively uncommon. In reality, the rate at which plants colonise bare ground and the sequence of plant communities that develop are influenced by the local conditions and the dispersal mechanisms of plants in the surrounding region.

Climax community

| Bare rock or volcanic ash | → | Lichens | → | Mosses and liverworts | → | Ferns and grasses | → | Shrubs, includng nitrogen fixers | → | Mature trees |

Secondary Succession: Catastrophe Cycle
(800 years for full richness to develop again)

Events that can cause localised or widespread destruction of forest include cyclone damage, forest fires, and hillside slips. The degree of destruction may vary from the loss of canopy trees in storms, to severe species depletion in the case of fire or landslides. The period of time for the canopy trees to reach their former status is considerable. The photograph on the far left shows a forest fire, while the other shows a large number of tree ferns on a landslide in a mid-successional stage of regeneration.

Climax community

| Cyclone, fire or landslide | → | Grasses and bracken fern | → | Abundant saplings of canopy tree species | → | Dense stand of canopy trees suppresses scrub underneath | → | Canopy trees increase in height, with natural thinning | → | Mature trees |

Secondary Succession: Gap Regeneration
(up to 500 years)

Large canopy trees have a profound effect on the make-up of the forest community immediately below. The reduced sunlight impedes the growth of most saplings. When a large tree falls, it opens a crucial hole in the canopy that lets in sunlight. There begins a race between the saplings to grow fast enough to fill the gap. The photograph on the left shows a large canopy tree in temperate rainforest that has recently fallen, leaving a gap in the canopy through which light can penetrate.

Canopy tree removed

Gap created by fall of large tree is colonised by tree ferns and palms

↓

Climax community

| Growth of subcanopy trees suppresses tree ferns. Seedlings of canopy trees grow beneath the subcanopy. | → | Rapid growth of saplings to occupy the gap | → | Mature trees develop to form climax community |

Populations and Interactions

Related activities: Tropical Deforestation
Web links: Mount St Helens

RA 2

Secondary Succession in Cleared Land
(150+ years for mature woodland to develop again)

A secondary succession takes place after a land clearance (e.g. from fire or landslide). Such events do not involve loss of the soil and so tend to be more rapid than primary succession, although the time scale depends on the species involved and on climatic and edaphic (soil) factors. Humans may deflect the natural course of succession (e.g. by mowing) and the climax community that results will differ from the natural community. A climax community arising from a **deflected succession** is called a **plagioclimax**.

Pioneer community

Mature woodland

Climax community

| Primarily bare earth | → | Open pioneer community (annual grasses) | → | Grasses and low growing perennials | → | Scrub: shrubs and small trees | → | Young broad-leaved woodland | → | Mature woodland mainly oak |

Time to develop (years)   1–2   3–5   16–30   31-150   150 +

1. Distinguish between **primary** succession and **secondary** succession: _____

_____

_____

_____

2. Suggest why primary successions rarely follow the classic sequence depicted on the previous page: _____

_____

_____

_____

3. (a) Identify some early colonisers during the establishment phase of a community on bare rock: _____

_____

(b) Describe two important roles of the species that are early colonisers of bare slopes: _____

_____

_____

_____

4. Describe a possible catastrophic event causing succession in a rainforest ecosystem: _____

5. (a) Describe the effect of selective logging on the composition of a forest community: _____

_____

_____

(b) Suggest why selective logging could be considered preferable (for forest conservation) to clear felling of trees:

_____

_____

6. (a) Explain what is meant by a **deflected succession**: _____

(b) Discuss the role that deflected successions might have in maintaining managed habitats: _____

_____

_____

_____

Wetland Succession

Wetland areas present a special case of ecological succession. Wetlands are constantly changing as plant invasion of open water leads to siltation and infilling. This process is accelerated by **eutrophication**. In well drained areas, pasture or **heath** may develop as a result of succession from freshwater to dry land. When the soil conditions remain non-acid and poorly drained, a swamp will eventually develop into a seasonally dry **fen**. In special circumstances (see below) an acid **peat bog** may develop. The domes of peat that develop produce a hummocky landscape with a unique biota. Wetland peat ecosystems may take more than 5000 years to form but are easily destroyed by excavation and lowering of the water table.

Emergent macrophyte species are larger plants that colonise the water margin

Deep water submergent plant species

Encroaching margins

Peat dome

Wetland Succession

An open body of water, with time, becomes silted up and is invaded by aquatic plants. Emergent macrophyte species colonise the accumulating sediments, driving floating plants towards the remaining deeper water.

↓

The increasing density of rooted emergent, submerged, and floating macrophytes encourages further sedimentation by slowing water flows and adding organic matter to the accumulating silt.

↓

The resulting **swamp** is characterised by dense growths of emergent macrophytes and permanent (although not necessarily deep) standing water. As sediment continues to accumulate the swamp surface may, in some cases, dry off in summer.

↓

In colder climates, low evaporation rates, high rainfall, and invasion by *Sphagnum* moss leads to development of a peat **bog**; a low pH, nutrient poor environment where acid-tolerant plants such as sundew replace the less acid-tolerant swamp species.

| Bedrock | Open water | Lake muds | Lake clay | Swamp peat | Bog peat |

Populations and Interactions

1. Identify the two most important abiotic factors in determining whether or not a wetland will develop into bog:

   (a) _____   (b) _____

2. Describe where these conditions encouraging bog development are found: _____

   _____

3. In many areas, bogs are threatened by drainage of land for pasture. With respect to the abiotic environment, suggest why land drainage threatens the viability of bog ecosystems:

   _____

   _____

   _____

4. Describe how *Sphagnum* alters the environment to favour the establishment of other bog species: _____

   _____

   _____

Related activities: Ecological Succession

A 2

Shoreline Zonation

Zonation refers to the division of an ecosystem into distinct zones that experience similar abiotic conditions. In a more global sense, differences in latitude and altitude create distinctive zones of vegetation type, or **biomes**. Zonation is particularly clear on a rocky seashore, where assemblages of different species form a banding pattern approximately parallel to the waterline. This effect is marked in temperate regions where the prevailing weather comes from the same general direction. Exposed shores show the clearest zonation. On sheltered rocky shores there is considerable species overlap and it is only on the upper shore that distinct zones are evident. Rocky shores exist where wave action prevents the deposition of much sediment. The rock forms a stable platform for the secure attachment of organisms such as large seaweeds and barnacles. Sandy shores are less stable than rocky shores and the organisms found there are adapted to the more mobile substrate.

Rocky shore at Sleahead, Ireland.

Seashore Zonation Patterns

The zonation of species distribution according to an environmental gradient is well shown on rocky shorelines. In Britain, exposed rocky shores occur along much of the western coasts. Variations in low and high tide affect zonation, and in areas with little tidal variation, zonation is restricted. High on the shore, some organisms may be submerged only at spring high tide. Low on the shore, others may be exposed only at spring low tide. There is a gradation in extent of exposure and the physical conditions associated with this. Zonation patterns generally reflect the vertical movement of seawater. Sheer rocks can show marked zonation as a result of tidal changes with little or no horizontal shift in species distribution. The profiles below left, show zonation patterns on an exposed rocky shore (left profile) with an exposed sandy shore for comparison (right profile). **SLT** = spring low tide mark, **MLT** = mean low tide mark, **MHT** = mean high tide mark, **SHT** = spring high tide mark.

Key to species

1. Lichen: sea ivory
2. Small periwinkle *Littorina neritoides*
3. Lichen *Verrucaria maura*
4. Rough periwinkle *Littorina saxatilis*
5. Common limpet *Patella vulgaris*
6. Laver *Porphyra*
7. Spiral wrack *Fucus spiralis*
8. Australian barnacle
9. Common mussel *Mytilus edulis*
10. Common whelk *Buccinum undatum*
11. Grey topshell *Gibbula cineraria*
12. Carrageen (Irish moss) *Chondrus crispus*
13. Thongweed *Himanthalia elongata*
14. Toothed wrack *Fucus serratus*
15. Dabberlocks *Alaria esculenta*
16. Common sandhopper
17. Sandhopper *Bathyporeia pelagica*
18. Common cockle *Cerastoderma edule*
19. Lugworm *Arenicola marina*
20. Sting winkle *Ocinebra erinacea*
21. Common necklace shell *Natica alderi*
22. Rayed trough shell *Mactra corallina*
23. Sand mason worm *Lanice conchilega*
24. Sea anemone *Halcampa*
25. Pod razor shell *Ensis siliqua*
26. Sea potato *Echinocardium* (a heart urchin)

Note: Where several species are indicated within a single zonal band, they occupy the entire zone, not just the position where their number appears.

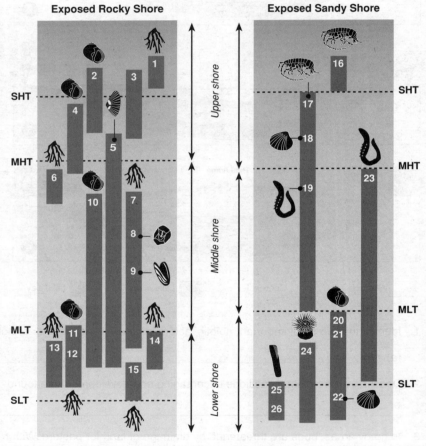

Exposed Rocky Shore   **Exposed Sandy Shore**

Upper shore / Middle shore / Lower shore

SHT, MHT, MLT, SLT

1. (a) Suggest why the time of exposure above water is a major factor controlling species distribution on a rocky shore: _____

(b) Identify two other abiotic factors that might influence species distribution on a rocky shore: _____

(c) Identify two biotic factors that might influence species distribution on a rocky shore: _____

2. Describe the zonation pattern on a rocky shore: _____

Related activities: Interspecific Competition
Web links: Tide Pool Ecology

A 2

Ecosystem Stability

Ecological theory suggests that all species in an ecosystem contribute in some way to ecosystem function. Therefore, species loss past a certain point is likely to have a detrimental effect on the functioning of the ecosystem and on its ability to resist change (its stability). Although many species still await discovery, we do know that the rate of species extinction is increasing. Scientists estimate that human destruction of natural habitat is implicated in the extinction of up to 100 000 species every year. This substantial loss of biodiversity has serious implications for the long term stability of many ecosystems.

The Concept of Ecosystem Stability

The stability of an ecosystem refers to its apparently unchanging nature over time. Ecosystem stability has various components, including **inertia** (the ability to resist disturbance) and **resilience** (ability to recover from external disturbances). Ecosystem stability is closely linked to the biodiversity of the system, although it is difficult to predict which factors will stress an ecosystem beyond its range of tolerance. It was once thought that the most stable ecosystems were those with the greatest number of species, since these systems had the greatest number of biotic interactions operating to buffer them against change. This assumption is supported by experimental evidence but there is uncertainty over what level of biodiversity provides an insurance against catastrophe.

Monoculture / Natural grassland

Rainforest / Deforestation

Single species crops (monocultures), such as the soy bean crop (above, left), represent low diversity systems that can be vulnerable to disease, pests, and disturbance. In contrast, natural grasslands (above, right) may appear homogeneous, but contain many species which vary in their predominance seasonally. Although they may be easily disturbed (e.g. by burning) they are very resilient and usually recover quickly.

Tropical rainforests (above, left) represent the highest diversity systems on Earth. Whilst these ecosystems are generally resistant to disturbance, once degraded, (above, right) they have little ability to recover. The biodiversity of ecosystems at low latitudes is generally higher than that at high latitudes, where climates are harsher, niches are broader, and systems may be dependent on a small number of key species.

Community Response to Environmental Change

Environmental change or community response — Time or space

- Environmental variation
- Response of a low diversity community
- Response of a high diversity community

Modified from Biol. Sci. Rev., March 1999 (p. 22)

In models of ecosystem function, higher species diversity increases the stability of ecosystem functions such as productivity and nutrient cycling. In the graph above, note how the low diversity system varies more consistently with the environmental variation, whereas the high diversity system is buffered against major fluctuations. In any one ecosystem, some species may be more influential than others in the stability of the system. Such **keystone (key) species** have a disproportionate effect on ecosystem function due to their pivotal role in some ecosystem function such as nutrient recycling or production of plant biomass.

Elephants can change the entire vegetation structure of areas into which they migrate. Their pattern of grazing on taller plant species promotes a predominance of lower growing grasses with small leaves.

Termites are amongst the few larger soil organisms able to break down plant cellulose. They shift large quantities of soil and plant matter and have a profound effect on the rates of nutrient processing in tropical environments.

The starfish *Pisaster* is found along the coasts of North America where it feeds on mussels. If it is removed, the mussels dominate, crowding out most algae and leading to a decrease in the number of herbivore species.

Populations and Interactions

Calculation and Use of Diversity Indices

One of the best ways to determine the health of an ecosystem is to measure the variety (rather than the absolute number) of organisms living in it. Certain species, called **indicator species**, are typical of ecosystems in a particular state (e.g. polluted or pristine). An objective evaluation of an ecosystem's biodiversity can provide valuable insight into its status, particularly if the species assemblages have changed as a result of disturbance.

Diversity can be quantified using a **diversity index (DI)**. Diversity indices attempt to quantify the degree of diversity and identify indicators for environmental stress or degradation. Most indices of diversity are easy to use and they are widely used in ecological work, particularly for monitoring ecosystem change or pollution. One example, which is a derivation of **Simpson's index**, is described below. Other indices produce values ranging between 0 and almost 1. These are more easily interpreted because of the more limited range of values, but no single index offers the "best" measure of diversity: they are chosen on their suitability to different situations.

Simpson's Index for finite populations

This diversity index (DI) is a commonly used inversion of Simpson's index, suitable for finite populations.

$$ DI = \frac{N(N-1)}{\Sigma n(n-1)} $$

After Smith and Smith as per IOB.

Where:

DI = Diversity index
N = Total number of individuals (of all species) in the sample
n = Number of individuals of each species in the sample

This index ranges between 1 (low diversity) and infinity. The higher the value, the greater the variety of living organisms. It can be difficult to evaluate objectively without reference to some standard ecosystem measure because the values calculated can, in theory, go to infinity.

Example of species diversity in a stream

The example describes the results from a survey of stream invertebrates. The species have been identified, but this is not necessary in order to calculate diversity as long as the different species can be distinguished. Calculation of the DI using Simpson's index for finite populations is:

Species	No. of individuals
A (Common backswimmer)	12
B (Stonefly larva)	7
C (Silver water beetle)	2
D (Caddis fly larva)	6
E (Water spider)	5
Total number of individuals = 32	

$$ DI = \frac{32 \times 31}{(12 \times 11) + (7 \times 6) + (2 \times 1) + (6 \times 5) + (5 \times 4)} = \frac{992}{226} = 4.39 $$

A stream community with a high macroinvertebrate diversity (above) in contrast to a low diversity stream community (below).

Photos: Stephen Moore

1. Explain what you understand by the term **ecosystem stability**: _____

_____

_____

2. Suggest one probable reason why high biodiversity provides greater ecosystem stability: _____

_____

_____

3. Describe a situation where a species diversity index may provide useful information: _____

_____

_____

4. An area of forest floor was sampled and six invertebrate species were recorded, with counts of 7, 10, 11, 2, 4, and 3 individuals. Using Simpson's index for finite populations, calculate DI for this community:

(a) DI= _____ DI = _____

(b) Comment on the diversity of this community: _____

_____

_____

Keystone Species

A **keystone species** is one that has an effect on ecosystem function that is out of proportion to its abundance or biomass (e.g. a top predator or a species that creates habitat for a large number of other species). The role of keystone species is usually brought to attention when they are removed or disappear from an ecosystem and there is a dramatic shift in community structure and function as a result. The phenomenon has been observed in a wide range of ecosystems and for a wide range of organisms. The concept of the keystone species has become very popular in conservation biology and has been used as an argument for the reintroduction of some species (such as grey wolves and European beavers) into parts of their former range.

Keystone Species in Europe and the UK

Grey wolf

European beaver, *Castor fiber*

Scots pine

Grey or **timber wolves** (*Canis lupus*) are a keystone predator and were once widespread through North America, Europe, and Eurasia. Historically, they have been eliminated because of their perceived threat to humans and livestock and now occupy only a small fraction of their former range. As a top predator, the wolf is a keystone species in the ecosystems to which it belongs. When they are absent, populations of their prey (e.g. red deer) increase to the point that they adversely affect other flora and fauna.

The **European beaver** was originally distributed throughout most of Europe and northern Asia but populations have been decimated as a result of both hunting and habitat loss. Where they occur, beavers are critical to ecosystem function and a number of species depend partly or entirely on beaver ponds for survival. Their tree-felling activity is akin to a natural coppicing process and promotes vigorous regrowth, while historically they helped the spread of alder (a water-loving species) in Britain.

Scots pine (*Pinus sylvestris*) is the most widely distributed conifer in the world. In the Scots pine forests in Scotland, this species occupies a unique position, both because of the absence of other native conifers and because it directly or indirectly supports so many other species. Among those dependent on Scots pine for survival are blaeberries, wood ants, pine martens, and a number of bird species, including the capercallie and the UK's only endemic bird, the Scottish crossbill.

1. Explain why **keystone species** are so important to ecosystem function:

   _____

   _____

2. For each of the following species, discuss features of their biology that contribute to their position as keystone species:

   (a) Grey wolf: _____

   _____

   (b) European beaver: _____

   _____

   (c) Scots pine: _____

   _____

3. Using an illustrative example if possible, suggest why the concept of keystone species has been used to argue for the reintroduction of species into regions where they have been removed (e.g. by hunting):

   _____

   _____

   _____

   _____

Populations and Interactions

Related activities: The Impact of Farming, Ecosystem Stability

RA 2

Practical Ecology

AQA-A	AQA-B	SNAB	Edexcel	OCR
Complete:	Complete:	Complete:	Complete:	Complete:
1-3, 5 (b)-(d), 9	1-3, 5, 8-11	1-11	1-3, 5, 7, 9-11	1-5 (b), (c), 9
Extension:	Extension:	as required	Extension:	Extension: 6, 8
4, 6, 8, 10-11	4, 6		4, 6, 8	

Learning Objectives

☐ 1. Compile your own glossary from the **KEY WORDS** displayed in **bold type** in the learning objectives below.

Sampling Populations *(pages 79-80, 83-100)*

☐ 2. Identify the type of information that can be gained from population studies (e.g. **abundance**, **density**, **age structure**, **distribution**). Explain why we **sample** populations and describe the advantages and drawbacks involved.

☐ 3. A field study should enable you to test a hypothesis about a certain aspect of a population. You should provide an outline of your study including reference to the type of data you will collect and the methods you will use, the size of your sampling unit (e.g. quadrat size) and the number of samples you will take, the assumptions of your investigation, and controls.

☐ 4. Explain how and why sample size affects the accuracy of population estimates. Explain how you would decide on a suitable sample size. Discuss the compromise between sampling accuracy and sampling effort.

☐ 5. Describe the following techniques used to study aspects of populations (e.g. **distribution**, **abundance**, **density**). Identify the advantages and limitations of each method with respect to sampling time, cost, and the suitability to the organism and specific habitat type:

(a) **Direct counts**
(b) **Frame** and/or **point quadrats**
(c) **Belt** and/or **line transects**
(d) **Mark and recapture** and the **Lincoln index**
(e) **Netting** and **trapping**

☐ 6. Recognise the value to population studies of **radio-tracking** and **indirect methods** of sampling such as counting nests, and recording calls and droppings.

☐ 7. Describe **qualitative methods** for investigating the distribution of organisms in specific habitats.

☐ 8. Describe the methods used to ensure **random sampling**, and appreciate why this is important.

☐ 9. Recognise the appropriate ways in which different types of data may be recorded, analysed, and presented. Demonstrate an ability to apply statistical tests, as appropriate, to population data. Recognise that the type of data collected and the design of the study will determine how the data can be analysed.

Measuring Abiotic Factors *(pages 81-82)*

☐ 10. Describe methods to measure abiotic factors in a habitat. Include reference to the following (as appropriate): pH, light, temperature, dissolved oxygen, current speed, total dissolved solids, and conductivity.

☐ 11. Appreciate the influence of abiotic factors on the distribution and abundance of organisms in a habitat.

See the 'Textbook Reference Grid' on pages 8-9 for textbook page references relating to material in this topic.

Supplementary Texts

See pages 5-7 for additional details of these texts:

■ Adds, J. *et al.*, 1999. **Tools, Techniques and Assessment in Biology** (NelsonThornes), chpt. 6.

■ Adds, J., *et al.*, 2004. **Genetics, Evolution and Biodiversity**, (NelsonThornes), chpt. 3 as reqd.

■ Allen *et al.*, 2001. **Applied Ecology** (Collins), pp. 6-14 (includes diversity indices).

■ Indge, B., 2004. **Data and Data Handling for A2 Level Biology** (Hodder Arnold H&S), as required.

■ Jones, A., *et al.*, 2007. **Practical Skills in Biology** (Pearson), as required.

■ Reiss, M. & J. Chapman, 2000. **Environmental Biology** (CUP), chpt. 7.

Presentation MEDIA to support this topic:
ECOLOGY:
• Practical Ecology

See page 7 for details of publishers of periodicals:

STUDENT'S REFERENCE

■ **Ecological Projects** Biol. Sci. Rev., 8(5) May 1996, pp. 24-26. *A thorough guide to planning and carrying out a field-based project. Essential reading for A level students.*

■ **Fieldwork - Sampling Animals** Biol. Sci. Rev., 10(4) March 1998, pp. 23-25. *Appropriate methodology for collecting animals in the field. Includes a synopsis of mark and recapture.*

■ **Fieldwork Sampling - Plants** Biol. Sci. Rev., 10(5) May 1998, pp. 6-8. *Excellent article covering the methodology for sampling plant communities (line and belt transects and quadrat techniques).*

■ **Bird Ringing** Biol. Sci. Rev., 14(3) Feb. 2002, pp. 14-19. *Techniques used in investigating populations of highly mobile organisms: mark and recapture methods, ringing techniques, and application of diversity indices.*

■ **British Butterflies in Decline** Biol. Sci. Rev., 14(4) April 2002, pp. 10-13. *Documented changes in the distribution of British butterfly species. This account includes a description of the techniques used to monitor changes in population numbers.*

TEACHER'S REFERENCE

■ **Ecology Fieldwork in 16 to 19 Biology** SSR, 84(307) December 2002, pp. 87. *Fieldwork is recognised as an important element of education in biology, yet the anecdotal evidence suggests that it is declining. This article examines the fieldwork opportunities provided to 16-19 students and suggests how these could be enhanced and firmly established for all students studying biology.*

See pages 10-11 for details of how to access **Bio Links** from our web site: **www.biozone.co.uk**. From Bio Links, access sites under the topics:

GENERAL BIOLOGY ONLINE RESOURCES

• Biology I interactive animations • Instructional multimedia, University of Alberta ... and others > **Glossaries**: Ecology glossary ... and others

ECOLOGY: • Introduction to biogeography and ecology > **Environmental Monitoring:** • Remote sensing and monitoring ... and others > **Populations and Communities:** • Communities • Population ecology • Quantitative population ecology • Sirtracking for wildlife research • Tree regeneration and growth in gaps

Designing Your Field Study

The following provides an example and some ideas for designing a field study. It provides a framework which can be modified for most simple comparative field investigations. For reasons of space, the full methodology is not included.

Pill millipede
Glomeris marginata

Oak woodland

Coniferous woodland

Oak or coniferous woodland

8 m

20 m

1. Sampling sites numbered 1-8 at evenly spaced intervals on a 2 x 2 m grid within an area of 20 m x 8 m.

Sampling equipment: leaf litter light trap

Light from a battery operated lamp drives the invertebrates down through the leaf litter.

Large (diameter 300 mm) funnel containing leaf litter resting on a gauze platform.

Gauze allows invertebrates of a certain size to move down the funnel.

Collecting jar placed in the litter on the forest floor traps the invertebrates that fall through the gauze and prevents their escape.

Notes on collection and analysis

- Mean millipede abundance was calculated from the counts from the eight sites. The difference in abundance at the sites was tested using a Student's *t* test.
- After counting and analysis of the samples, all the collected invertebrates were returned to the sites.

Observation

A student read that a particular species of pill millipede (left) is extremely abundant in forest leaf litter, but a search in the litter of a conifer-dominated woodland near his home revealed only very low numbers of this millipede species.

Hypothesis

This millipede species is adapted to a niche in the leaf litter of oak woodlands and is abundant there. However, it is rare in the litter of coniferous woodland. The **null hypothesis** is that there is no difference between the abundance of this millipede species in oak and coniferous woodland litter.

Sampling Programme

A sampling programme was designed to test the prediction that the millipedes are more abundant in the leaf litter of oak woodlands than in coniferous woodlands.

Equipment and Procedure

Sites: For each of the two woodland types, an area 20 x 8 m was chosen and marked out in 2 x 2 m grids. Eight sampling sites were selected, evenly spaced along the grid as shown.

- The general area for the study chosen was selected on the basis of the large amounts of leaf litter present.
- Eight sites were chosen as the largest number feasible to collect and analyse in the time available.
- The two woodlands were sampled on sequential days.

Capture of millipedes: At each site, a 0.4 x 0.4 m quadrat was placed on the forest floor and the leaf litter within the quadrat was collected. Millipedes and other leaf litter invertebrates were captured using a simple gauze lined funnel containing the leaf litter from within the quadrat. A lamp was positioned over each funnel for two hours and the invertebrates in the litter moved down and were trapped in the collecting jar.

- After two hours each jar was labelled with the site number and returned to the lab for analysis.
- The litter in each funnel was bagged, labelled with the site number and returned to the lab for weighing.
- The number of millipedes at each site was recorded.
- The numbers of other invertebrates (classified into major taxa) were also noted for reference.

Assumptions

- The areas chosen in each woodland were representative of the woodland types in terms of millipede abundance.
- Eight sites were sufficient to adequately sample the millipede populations in each forest.
- A quadrat size of 0.4 x 0.4 m contained enough leaf litter to adequately sample the millipedes at each site.
- The millipedes were not preyed on by any of the other invertebrates captured in the collecting jar.
- All the invertebrates within the quadrat were captured.
- Millipedes moving away from the light are effectively captured by the funnel apparatus and cannot escape.
- Two hours was long enough for the millipedes to move down through the litter and fall into the trap.

Note that these last two assumptions could be tested by examining the bagged leaf litter for millipedes after returning to the lab.

Practical Ecology

Related activities: Quadrat-Based Estimates

A 3

A note about sample size

When designing a field study, the size of your sampling unit (e.g. quadrat size) and the sample size (the number of samples you will take) should be major considerations. There are various ways to determine the best quadrat size. Usually, these involve increasing the quadrat size until you stop finding new species. For simple field studies, the number of samples you take (the sample size or *n* value) will be determined largely by the resources and time that you have available to collect and analyse your data. It is usually best to take as many samples as you can, as this helps to account for any natural variability present and will give you greater confidence in your data. For a summary of these aspects of study design as well as coverage of collecting methods see: *Jones, A. et al. (1998) Practical Skills in Biology* (or the earlier 1994 edition).

1. Explain the importance of recognising any assumptions that you are making in your study:

_____

_____

2. Describe how you would test whether the quadrat size of 0.4 x 0.4 m was adequate to effectively sample the millipedes:

_____

_____

3. Suggest why the litter was bagged, returned to the lab and then weighed properly for the analysis:

_____

_____

4. Suggest why the numbers of other invertebrates were also recorded even though it was only millipede abundance that was being investigated:

_____

_____

_____

_____

YOUR CHECKLIST FOR FIELD STUDY DESIGN

The following provides a checklist for a field study. Check off the points when you are confident that you have satisfied the requirements in each case:

1. **Preliminary:**

☐ (a) Makes a hypothesis based on observation(s).

☐ (b) The hypothesis (and its predictions) are testable using the resources you have available (the study is feasible).

☐ (c) The organism you have chosen is suitable for the study and you have considered the ethics involved.

2. **Assumptions and site selection:**

☐ (a) You are aware of any assumptions that you are making in your study.

☐ (b) You have identified aspects of your field design that could present problems (such as time of year, biological rhythms of your test organism, difficulty in identifying suitable habitats etc.).

☐ (c) The study sites you have selected have the features necessary in order for you to answer the questions you have asked in your hypothesis.

3. **Data collection:**

☐ (a) You are happy with the way in which you are going to take your measurements or samples.

☐ (b) You have considered the size of your sampling unit and the number of samples you are going to take (and tested for these if necessary).

☐ (c) You have given consideration to how you will analyse the data you collect and made sure that your study design allows you to answer the questions you wish to answer.

Monitoring Physical Factors

Most ecological studies require us to measure the physical factors (parameters) in the environment that may influence the abundance and distribution of organisms. In recent years there have been substantial advances in the development of portable, light-weight meters and dataloggers. These enable easy collection and storage of data in the field.

Quantum light meter: Measures light intensity levels. It is not capable of measuring light quality (wavelength).

Dissolved oxygen meter: Measures the amount of oxygen dissolved in water (expressed as mgl^{-1}).

pH meter: Measures the acidity of water or soil, if it is first dissolved in pure water (pH scale 0 to 14).

Total dissolved solids (TDS) meter: Measures content of dissolved solids (as ions) in water in mgl^{-1}.

Current meter: The electronic sensor is positioned at set depths in a stream or river on the calibrated wading rod as current readings are taken.

Multipurpose meter: This is a multi-functional meter, which can measure salinity, conductivity and temperature simply by pushing the MODE button.

Secchi disc: This simple device is used to provide a crude measure of water clarity (the maximum depth at which the disc can just be seen).

Collecting a water sample: A Nansen bottle is used to collect water samples from a lake for lab analysis, testing for nutrients, oxygen and pH.

This photo JDG. Others. Campus photography, University of Waikato

Dataloggers and Environmental Sensors

Dataloggers are electronic instruments that record measurements over time. They are equipped with a microprocessor, data storage facility, and sensor. Different sensors are employed to measure a range of variables in water (photos A and B) or air (photos C and D), as well as make physiological measurements. The datalogger is connected to a computer, and software is used to set the limits of operation (e.g. the sampling interval) and initiate the logger. The logger is then disconnected and used remotely to record and store data. When reconnected to the computer, the data are downloaded, viewed, and plotted. Dataloggers, such as those pictured here from PASCO, are being increasingly used in professional and school research. They make data collection quick and accurate, and they enable prompt data analysis.

Dataloggers are now widely used to monitor conditions in aquatic environments. Different variables such as pH, temperature, conductivity, and dissolved oxygen can be measured by changing the sensor attached to the logger.

Dataloggers fitted with sensors are portable and easy to use in a wide range of terrestrial environments. They are used to measure variables such as air temperature and pressure, relative humidity, light, and carbon dioxide gas.

Datalogger photos courtesy of PASCO

Practical Ecology

Related activities: Physical Factors and Gradients

RA 2

1. The physical factors of an exposed rocky shore and a sheltered estuarine mudflat differ markedly. For each of the factors listed in the table below, briefly describe how they may differ (if at all):

Environmental parameter	Exposed rocky coastline	Estuarine mudflat
Severity of wave action		
Light intensity and quality		
Salinity/ conductivity		
Temperature change (diurnal)		
Substrate/ sediment type		
Oxygen concentration		
Exposure time to air (tide out)		

Legend:
- Red stem moss
- Fern moss
- Snake moss
- Star moss
- Eye brow moss
- Broad leaved star moss
- Tree moss
- Lichens (various species)

Percentage cover (0, 50, 100) for Quadrat 1 through Quadrat 5.

QUADRAT	1	2	3	4	5
Height / m	0.4	0.8	1.2	1.6	2.0
Light / arbitrary units	40	56	68	72	72
Humidity / percent	99	88	80	76	78
Temperature / °C	12.1	12.2	13	14.3	14.2

Lichen

Moss

2. The figure (above) shows the changes in vegetation cover along a 2 m vertical transect up the trunk of an oak tree (*Quercus*). Changes in the physical factors light, humidity, and temperature along the same transect were also recorded. From what you know about the ecology of mosses and lichens, account for the observed vegetation distribution:

_____

_____

_____

Sampling Populations

Information about the populations of rare organisms in isolated populations may, in some instances, be collected by direct measure (direct counts and measurements of all the individuals in the population). However, in most cases, populations are too large to be examined directly and they must be sampled in a way that still provides information about them. Most practical exercises in population ecology involve the collection or census of living organisms, with a view to identifying the species and quantifying their abundance and other population features of interest. Sampling techniques must be appropriate to the

community being studied and the information you wish to obtain. Some of the common strategies used in ecological sampling, and the situations for which they are best suited, are outlined in the table below. It provides an overview of points to consider when choosing a sampling regime. One must always consider the time and equipment available, the organisms involved, and the impact of the sampling method on the environment. For example, if the organisms involved are very mobile, sampling frames are not appropriate. If it is important not to disturb the organisms, observation alone must be used to gain information.

Method	Equipment and procedure	Information provided and considerations for use
Point sampling Random Systematic (grid)	Individual points are chosen on a map (using a grid reference or random numbers applied to a map grid) and the organisms are sampled at those points. Mobile organisms may be sampled using traps, nets etc.	**Useful for**: Determining species abundance and community composition. If samples are large enough, population characteristics (e.g. age structure, reproductive parameters) can be determined. **Considerations**: Time efficient. Suitable for most organisms. Depending on method, environmental disturbance is minimal. Species occurring in low abundance may be missed.
Transect sampling 0.5 m Environmental gradient	Lines are drawn across a map and organisms occurring along the line are sampled. **Line transects**: Tape or rope marks the line. The species occurring on the line are recorded (all along the line or, more usually, at regular intervals). Lines can be chosen randomly (left) or may follow an environmental gradient. **Belt transects**: A measured strip is located across the study area to highlight any transitions. Quadrats are used to sample the plants and animals at regular intervals along the belt. Plants and immobile animals are easily recorded. Mobile or cryptic animals need to be trapped or recorded using appropriate methods.	**Useful for**: Well suited to determining changes in community composition along an environmental gradient. When placed randomly, they provide a quick measure of species occurrence. **Considerations for line transects**: Time efficient. Most suitable for plants and immobile or easily caught animals. Disturbance to the environment can be minimised. Species occurring in low abundance may be missed. **Considerations for belt transects**: Time consuming to do well. Most suitable for plants and immobile or easily caught animals. Good chance of recording most or all species. Efforts should be made to minimise disturbance to the environment.
Quadrat sampling	Sampling units or quadrats are placed randomly or in a grid pattern on the sample area. The occurrence of organisms in these squares is noted. Plants and slow moving animals are easily recorded. Rapidly moving or cryptic animals need to be trapped or recorded using appropriate methods.	**Useful for**: Well suited to determining community composition and features of population abundance: species density, frequency of occurrence, percentage cover, and biomass (if harvested). **Considerations**: Time consuming to do well. Most suitable for plants and immobile or easily caught animals. Quadrat size must be appropriate for the organisms being sampled and the information required. Disturbing if organisms are removed.
Mark and recapture (capture-recapture) First sample: marked Second sample: proportion recaptured	Animals are captured, marked, and then released. After a suitable time period, the population is resampled. The number of marked animals recaptured in a second sample is recorded as a proportion of the total.	**Useful for**: Determining total population density for highly mobile species in a certain area (e.g. butterflies). Movements of individuals in the population can be tracked (especially when used in conjunction with electronic tracking devices). **Considerations**: Time consuming to do well. Not suitable for immobile species. Population should have a finite boundary. Period between samplings must allow for redistribution of marked animals in the population. Marking should present little disturbance and should not affect behaviour.

1. Explain why we **sample** populations: _____

_____

2. Describe a sampling technique that would be appropriate for determining each of the following:

(a) The percentage cover of a plant species in pasture: _____

(b) The density and age structure of a plankton population: _____

(c) Change in community composition from low to high altitude on a mountain: _____

Related activities: Quadrat Sampling, Transect Sampling, Mark and Recapture Sampling

Practical Ecology

A 2

Indirect Sampling

If populations are small and easily recognised they may be monitored directly quite easily. However, direct measurement of elusive, easily disturbed, or widely dispersed populations is not always feasible. In these cases, indirect methods can be used to assess population abundance, provide information on habitat use and range, and enable biologists to link habitat quality to species presence or absence. Indirect sampling methods provide less reliable measures of abundance than direct sampling methods, such as mark and recapture, but are widely used nevertheless. They rely on recording the signs of a species, e.g. scat, calls, tracks, and rubbings or markings on vegetation, and using these to assess population abundance. In Australia, the Environmental Protection Agency (EPA) provides a Frog Census Datasheet (below) on which volunteers record details about frog populations and habitat quality in their area. This programme enables the EPA to gather information across Australia.

Recording a date and accurate map reference is important

Population estimates are based on the number of frog calls recorded by the observer

To sample nocturnal, highly mobile species, e.g. bats, electronic devices, such as the bat detector above, can be used to estimate population density. In this case, the detector is tuned to the particular frequency of the hunting clicks emitted by specific bat species. The number of calls recorded per unit time can be used to estimate numbers per area.

The analysis of animal tracks allows wildlife biologists to identify habitats in which animals live and to conduct population surveys. Interpreting tracks accurately requires considerable skill as tracks may vary in appearance even when from the same individual. Tracks are particularly useful as a way to determine habitat use and preference.

Wombat scat

All animals leave scats (faeces) which are species specific and readily identifiable. Scats can be a valuable tool by which to gather data from elusive, nocturnal, easily disturbed, or highly mobile species. Faecal analyses can provide information on diet, movements, population density, sex ratios, age structure, and even genetic diversity.

1. (a) Describe the kind of information that could be gathered from the Frog Census Datasheet:

_____

_____

(b) Identify the benefits of linking a measure of abundance to habitat assessment: _____

_____

2. Describe one other indirect method of population sampling and outline its advantages and drawbacks:

_____

_____

_____

Quadrat Sampling

Quadrat sampling is a method by which organisms in a certain proportion (sample) of the habitat are counted directly. As with all sampling methods, it is used to estimate population parameters when the organisms present are too numerous to count in total. It can be used to estimate population **abundance** (number), **density**, **frequency of occurrence**, and **distribution**. Quadrats may be used without a transect when studying a relatively uniform habitat. In this case, the quadrat positions are chosen randomly using a random number table.

The general procedure is to count all the individuals (or estimate their percentage cover) in a number of quadrats of known size and to use this information to work out the abundance or percentage cover value for the whole area. The number of quadrats used and their size should be appropriate to the type of organism involved (e.g. grass vs tree).

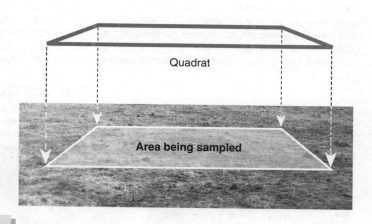

Quadrat

Area being sampled

$$\text{Estimated average density} = \frac{\text{Total number of individuals counted}}{\text{Number of quadrats X area of each quadrat}}$$

Guidelines for Quadrat Use:

1. The **area of each quadrat** must be known exactly and ideally quadrats should be the same shape. The quadrat does not have to be square (it may be rectangular, hexagonal etc.).

2. **Enough quadrat samples** must be taken to provide results that are representative of the total population.

3. The **population of each quadrat** must be known exactly. Species must be distinguishable from each other, even if they have to be identified at a later date. It has to be decided beforehand what the count procedure will be and how organisms over the quadrat boundary will be counted.

4. The size of the quadrat should be appropriate to the organisms and habitat, e.g. a large size quadrat for trees.

5. The quadrats must be **representative of the whole area**. This is usually achieved by **random sampling** (right).

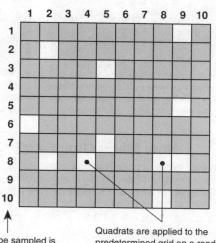

The area to be sampled is divided up into a grid pattern with indexed coordinates

Quadrats are applied to the predetermined grid on a random basis. This can be achieved by using a random number table.

Sampling a centipede population

A researcher by the name of Lloyd (1967) carried out a sampling of centipedes in Wytham Woods, near Oxford in England. A total of 37 hexagon–shaped quadrats were used, with a diameter of 30 cm (see diagram on right). These were arranged in a pattern so that they were all touching each other. Use the data in the diagram to answer the following questions.

1. Determine the average number of centipedes captured per quadrat:

   _____

2. Calculate the estimated average density of centipedes per square metre (remember that each quadrat is 0.08 square metres in area):

   _____

3. Looking at the data for individual quadrats, describe in general terms the distribution of the centipedes in the sample area:

   _____

   _____

4. Describe one factor that might account for the distribution pattern:

   _____

   _____

Each quadrat was a hexagon with a diameter of 30 cm and an area of 0.08 square metres.

The number in each hexagon indicates how many centipedes were caught in that quadrat.

Centipede

Practical Ecology

Related activities: Density and Distribution, Sampling a Leaf Litter Population
Web links: Investigating Marine Life

DA 2

Quadrat-Based Estimates

The simplest description of a plant community in a habitat is a list of the species that are present. This qualitative assessment of the community has the limitation of not providing any information about the **relative abundance** of the species present. Quick estimates can be made using **abundance scales**, such as the

ACFOR scale described below. Estimates of percentage cover provide similar information. These methods require the use of **quadrats**. Quadrats are used extensively in plant ecology. This activity outlines some of the common considerations when using quadrats to sample plant communities.

What Size Quadrat?

Quadrats are usually square, and cover 0.25 m² (0.5 m x 0.5 m) or 1 m², but they can be of any size or shape, even a single point. The quadrats used to sample plant communities are often 0.25 m². This size is ideal for low-growing vegetation, but quadrat size needs to be adjusted to habitat type. The quadrat must be large enough to be representative of the community, but not so large as to take a very long time to use.

A quadrat covering an area of 0.25 m² is suitable for most low growing plant communities, such as this alpine meadow, fields, and grasslands.

Larger quadrats (e.g. 1 m²) are needed for communities with shrubs and trees. Quadrats as large as 4 m x 4 m may be needed in woodlands.

Small quadrats (0.01 m² or 100 mm x 100 mm) are appropriate for lichens and mosses on rock faces and tree trunks.

How Many Quadrats?

As well as deciding on a suitable quadrat size, the other consideration is how many quadrats to take (the sample size). In species-poor or very homogeneous habitats, a small number of quadrats will be sufficient. In species-rich or heterogeneous habitats, more quadrats will be needed to ensure that all species are represented adequately.

Determining the number of quadrats needed

- Plot the cumulative number of species recorded (on the *y* axis) against the number of quadrats already taken (on the *x* axis).
- The point at which the curve levels off indicates the suitable number of quadrats required.

Fewer quadrats are needed in species-poor or very uniform habitats, such as this bluebell woodland.

Describing Vegetation

Density (number of individuals per unit area) is a useful measure of abundance for animal populations, but can be problematic in plant communities where it can be difficult to determine where one plant ends and another begins. For this reason, plant abundance is often assessed using **percentage cover**. Here, the percentage of each quadrat covered by each species is recorded, either as a numerical value or using an abundance scale such as the ACFOR scale.

The ACFOR Abundance Scale

A = Abundant (30% +)

C = Common (20-29%)

F = Frequent (10-19%)

O = Occasional (5-9%)

R = Rare (1-4%)

The AFCOR scale could be used to assess the abundance of species in this wildflower meadow. Abundance scales are subjective, but it is not difficult to determine which abundance category each species falls into.

1. Describe one difference between the methods used to assess species abundance in plant and in animal communities:

_____

2. Identify the main consideration when determining appropriate quadrat size: _____

3. Identify the main consideration when determining number of quadrats: _____

4. Explain two main disadvantages of using the ACFOR abundance scale to record information about a plant community:

(a) _____

(b) _____

Related activities: Quadrat Sampling

Sampling a Leaf Litter Population

The diagram on the next page represents an area of leaf litter from a forest floor with a resident population of organisms. The distribution of four animal species as well as the arrangement of leaf litter is illustrated. Leaf litter comprises leaves and debris that have dropped off trees to form a layer of detritus. This exercise is designed to practice the steps required in planning and carrying out a sampling of a natural population. It is desirable, but not essential, that students work in groups of 2–4.

1. Decide on the sampling method

For the purpose of this exercise, it has been decided that the populations to be investigated are too large to be counted directly and a quadrat sampling method is to be used to estimate the average density of the four animal species as well as that of the leaf litter.

2. Mark out a grid pattern

Use a ruler to mark out 3 cm intervals along each side of the sampling area (area of quadrat = 0.03 x 0.03 m). **Draw lines** between these marks to create a 6 x 6 grid pattern (total area = 0.18 x 0.18 m). This will provide a total of 36 quadrats that can be investigated.

3. Number the axes of the grid

Only a small proportion of the possible quadrat positions are going to be sampled. It is necessary to select the quadrats in a random manner. It is not sufficient to simply guess or choose your own on a 'gut feeling'. The best way to choose the quadrats randomly is to create a numbering system for the grid pattern and then select the quadrats from a random number table. Starting at the *top left hand corner*, **number the columns** and **rows** from 1 to 6 on each axis.

4. Choose quadrats randomly

To select the required number of quadrats randomly, use random numbers from a random number table. The random numbers are used as an index to the grid coordinates. Choose 6 quadrats from the total of 36 using table of random numbers provided for you at the bottom of the next page. Make a note of which column of random numbers you choose. Each member of your group should choose a different set of random numbers (i.e. different column: A–D) so that you can compare the effectiveness of the sampling method.

Column of random numbers chosen: _____

NOTE: Highlight the boundary of each selected quadrat with coloured pen/highlighter.

5. Decide on the counting criteria

Before the counting of the individuals for each species is carried out, the criteria for counting need to be established.

There may be some problems here. You must decide before sampling begins as to what to do about individuals that are only partly inside the quadrat. Possible answers include:

(a) Only counting individuals if they are completely inside the quadrat.

(b) Only counting individuals that have a clearly defined part of their body inside the quadrat (such as the head).

(c) Allowing for 'half individuals' in the data (e.g. 3.5 snails).

(d) Counting an individual that is inside the quadrat by half or more as one complete individual.

Discuss the merits and problems of the suggestions above with other members of the class (or group). You may even have counting criteria of your own. Think about other factors that could cause problems with your counting.

6. Carry out the sampling

Carefully examine each selected quadrat and **count the number of individuals** of each species present. Record your data in the spaces provided on the next page.

7. Calculate the population density

Use the combined data TOTALS for the sampled quadrats to estimate the average density for each species by using the formula:

$$\text{Density} = \frac{\text{Total number in all quadrats sampled}}{\text{Number of quadrats sampled} \times \text{area of a quadrat}}$$

Remember that a total of 6 quadrats are sampled and each has an area of 0.0009 m^2. The density should be expressed as the number of individuals *per square metre* (no. m^{-2}).

Woodlouse: [ ]   False scorpion: [ ]

Centipede: [ ]   Leaf: [ ]

Springtail: [ ]

8. (a) In this example the animals are not moving. Describe the problems associated with sampling moving organisms. Explain how you would cope with sampling these same animals if they were really alive and very active:

_____

_____

_____

(b) Carry out a direct count of all four animal species and the leaf litter for the whole sample area (all 36 quadrats). Apply the data from your direct count to the equation given in (7) above to calculate the actual population density (remember that the number of quadrats in this case = 36):

Woodlouse: [ ]   Centipede: [ ]   False scorpion: [ ]   Springtail: [ ]   Leaf: [ ]

Compare your estimated population density to the actual population density for each species:

_____

_____

Practical Ecology

Related activities: Quadrat Sampling, Mark and Recapture Sampling

PDA 2

Coordinates for each quadrat	Woodlouse	Centipede	False scorpion	Springtail	Leaf
1:					
2:					
3:					
4:					
5:					
6:					
TOTAL					

Table of random numbers

A	B	C	D
2 2	3 1	6 2	2 2
3 2	1 5	6 3	4 3
3 1	5 6	3 6	6 4
4 6	3 6	1 3	4 5
4 3	4 2	4 5	3 5
5 6	1 4	3 1	1 4

The table above has been adapted from a table of random numbers from a statistics book. Use this table to select quadrats randomly from the grid above. Choose one of the columns (A to D) and use the numbers in that column as an index to the grid. The first digit refers to the row number and the second digit refers to the column number. To locate each of the 6 quadrats, find where the row and column intersect, as shown below:

Example: **5 2** refers to the 5th row and the 2nd column

Transect Sampling

A **transect** is a line placed across a community of organisms. Transects are usually carried out to provide information on the **distribution** of species in the community. This is of particular value in situations where environmental factors change over the sampled distance. This change is called an **environmental gradient** (e.g. up a mountain or across a seashore). The usual practice for small transects is to stretch a string between two markers. The string is marked off in measured distance intervals, and the species at each marked point are noted. The sampling points along the transect may also be used for the siting of quadrats, so that changes in density and community composition can be recorded. Belt transects are essentially a form of continuous quadrat sampling. They provide more information on community composition but can be difficult to carry out. Some transects provide information on the vertical, as well as horizontal, distribution of species (e.g. tree canopies in a forest).

Point sampling

Sample point — Sample point — Sample point — Sample point — Sample point — Sample point — Sample point — Sample point — Sample point

Continuous belt transect

Some sampling procedures require the vertical distribution of each species to be recorded

Continuous sampling

Quadrats are placed adjacent to each other in a continuous belt

Interrupted belt transect

4 quadrats across each sample point

Line of transect

1. Belt transect sampling uses quadrats placed along a line at marked intervals. In contrast, point sampling transects record only the species that are touched or covered by the line at the marked points.

   (a) Describe one disadvantage of belt transects: _____

   _____

   (b) Explain why line transects may give an unrealistic sample of the community in question: _____

   _____

   _____

   (c) Explain how belt transects overcome this problem: _____

   _____

   (d) Describe a situation where the use of transects to sample the community would be inappropriate: _____

   _____

2. Explain how you could test whether or not a transect sampling interval was sufficient to accurately sample a community:

   _____

   _____

   _____

Practical Ecology

Related activities: Density and Distribution

DA 2

Kite graphs are an ideal way in which to present distributional data from a belt transect (e.g. abundance or percentage cover along an environmental gradient. Usually, they involve plots for more than one species. This makes them good for highlighting probable differences in habitat preference between species. Kite graphs may also be used to show changes in distribution with time (e.g. with daily or seasonal cycles).

3. The data on the right were collected from a rocky shore field trip. Periwinkles from four common species of the genus *Littorina* were sampled in a continuous belt transect from the low water mark, to a height of 10 m above that level. The number of each of the four species in a 1 m² quadrat was recorded.

Plot a **kite graph** of the data for all four species on the grid below. Be sure to choose a scale that takes account of the maximum number found at any one point and allows you to include all the species on the one plot. Include the scale on the diagram so that the number at each point on the kite can be calculated.

Field data notebook

Numbers of periwinkles (4 common species) showing vertical distribution on a rocky shore

Periwinkle species:

Height above low water /m	L. littorea	L. saxatalis	L. neritoides	L. littoralis
0-1	0	0	0	0
1-2	1	0	0	3
2-3	3	0	0	17
3-4	9	3	0	12
4-5	15	12	0	1
5-6	5	24	0	0
6-7	2	9	2	0
7-8	0	2	11	0
8-9	0	0	47	0
9-10	0	0	59	0

Sampling Animal Populations

Unlike plants, most animals are highly mobile and present special challenges in terms of sampling them **quantitatively** to estimate their distribution and abundance. The equipment available for sampling animals ranges from various types of nets and traps (below), to more complex electronic devices, such as those used for radio-tracking large mobile species.

Beating tray

Tree branch is shaken or beaten with a stick

Insects and other invertebrates fall

Canvas stretched over frame

Longworth small mammal trap

Nestbox containing bedding

Tunnel

Entrance to trap (door closed)

Nestbox is supported at an angle so that urine and rain water can drain out

Plankton net

Cone of bolting silk

Plastic container for collecting plankton sample

Metal hoop

Tow rope

Tie cord

Bridle

Canvas sleeve

Direction of current

Sweep net

Stout canvas can withstand rough treatment

Swoop net through low vegetation

Stiff hoop reinforces rim

Kick sampling

Rocks upstream of the net are disturbed

Small aquatic invertebrates are dislodged and collect in the net

Direction of current

Water sampler

Line pulled to remove stopper

Line supporting the sample bottle

Rubber bung

Glass tubing flush with the bottom surface of rubber bung

When stopper is removed, water flows into sample bottle

Plastic tube to allow air to escape when filling with sample water

Weight attached to air-filled bottle to allow it to sink

Pooter or aspirator

Clear plastic tube

Glass collecting tube that sucks up small animals

Rubber or cork bung

Gauze covering opening of tube

Specimen tube

Glass mouthpiece through which operator sucks

Tullgren funnel

Light from a battery operated lamp drives the invertebrates down through the soil or litter

Large diameter funnel containing soil or leaf litter resting on a gauze platform

Gauze allows invertebrates of a certain size to move down the funnel

Pitfall trap

Flat rock

Ground slopes away from trap to assist drainage

Support made of small stones or sticks

Jam jar sunk into ground

3 cm of water or 50% ethanol may be added as immobiliser

Collecting jar traps the invertebrates that fall through the gauze and prevents their escape

Practical Ecology

Related activities: Density and Distribution

RA 2

Electrofishing a stream (Sweden)

Adelie penguins fitted with transmitters

Electronic bat detector showing frequency dial

Electrofishing: An effective, but expensive method of sampling larger stream animals (e.g. fish). Wearing a portable battery backpack, the operator walks upstream holding the anode probe and a net. The electrical circuit created by the anode and the stream bed stuns the animals, which are netted and placed in a bucket to recover. After analysis (measurement, species, weights) the animals are released.

Radio-tracking: A relatively non-invasive method of examining many features of animal populations, including movement, distribution, and habitat use. A small transmitter with an antenna (arrowed) is attached to the animal. The transmitter emits a pulsed signal which is picked up by a receiver. In difficult terrain, a tracking antenna can be used in conjunction with the receiver to accurately fix an animal's position.

Electronic detection devices: To sample nocturnal, highly mobile species such as bats, electronic devices, such as the bat detector illustrated above, can be used to estimate population density. In this case, the detector is tuned to the particular frequency of the hunting clicks emitted by the bat species of interest. The number of calls recorded per unit time can be used to estimate numbers within a certain area.

1. Describe what each the following types of sampling equipment is used for in a sampling context:

   (a) Kick sampling technique: _Provides a semi-quantitative sample of substrate-dwelling stream invertebrates_

   (b) Beating tray: _____

   (c) Longworth small mammal trap: _____

   (d) Plankton net: _____

   (e) Sweep net: _____

   (f) Water sampler: _____

   (g) Pooter or aspirator: _____

   (h) Tullgren funnel: _____

   (i) Pitfall trap: _____

2. Suggest why pitfall traps are not recommended for estimates of population density: _____

3. (a) Suggest what influence mesh size might have on the sampling efficiency of a plankton net: _____

   (b) Explain how this would affect your choice of mesh size when sampling animals in a pond: _____

Mark and Recapture Sampling

The mark and recapture method of estimating population size is used in the study of animal populations where individuals are highly mobile. It is of no value where animals do not move or move very little. The number of animals caught in each sample must be large enough to be valid. The technique is outlined in the diagram below.

First capture

In the first capture, a random sample of animals from the population is selected. Each selected animal is marked in a distinctive way.

Release back into the natural population

The marked animals from the first capture are released back into the natural population and left for a period of time to mix with the unmarked individuals.

Second capture

Only a proportion of the second capture sample will have animals that were marked in the previous capture.

The Lincoln Index

$$\text{Total population} = \frac{\text{No. of animals in 1st sample (all marked)} \quad X \quad \text{Total no. of animals in 2nd sample}}{\text{Number of marked animals in the second sample (recaptured)}}$$

The mark and recapture technique comprises a number of simple steps:

1. The population is sampled by capturing as many of the individuals as possible and practical.

2. Each animal is marked in a way to distinguish it from unmarked animals (unique mark for each individual not required).

3. Return the animals to their habitat and leave them for a long enough period for complete mixing with the rest of the population to take place.

4. Take another sample of the population (this does not need to be the same sample size as the first sample, but it does have to be large enough to be valid).

5. Determine the numbers of marked to unmarked animals in this second sample. Use the equation above to estimate the size of the overall population.

1. For this exercise you will need several boxes of matches and a pen. Work in a group of 2-3 students to 'sample' the population of matches in the full box by using the mark and recapture method. Each match will represent one animal.

(a) Take out 10 matches from the box and mark them on 4 sides with a pen so that you will be able to recognise them from the other unmarked matches later.
(b) Return the marked matches to the box and shake the box to mix the matches.
(c) Take a sample of 20 matches from the same box and record the number of marked matches and unmarked matches.
(d) Determine the total population size by using the equation above.
(e) Repeat the sampling 4 more times (steps b-d above) and record your results:

	Sample 1	Sample 2	Sample 3	Sample 4	Sample 5
Estimated population					

(f) Count the actual number of matches in the matchbox : _____

(g) Compare the actual number to your estimates and state by how much it differs: _____

Related activities: Sampling Animal Populations

PDA 2

Practical Ecology

2. In 1919 a researcher by the name of Dahl wanted to estimate the number of trout in a Norwegian lake. The trout were subject to fishing so it was important to know how big the population was in order to manage the fish stock. He captured and marked 109 trout in his first sample. A few days later, he caught 177 trout in his second sample, of which 57 were marked. Use the Lincoln index (on the previous page) to estimate the total population size:

Size of first sample: _____

Size of second sample: _____

Number marked in second sample: _____

Estimated total population: _____

3. Discuss some of the problems with the mark and recapture method if the second sampling is:

(a) Left too long a time before being repeated: _____

_____

_____

(b) Too soon after the first sampling: _____

_____

_____

4. Describe two important assumptions being made in this method of sampling, which would cause the method to fail if they were not true:

(a) _____

_____

_____

(b) _____

_____

_____

5. Some types of animal would be unsuitable for this method of population estimation (i.e. would not work).

(a) Name an animal for which this method of sampling would not be effective: _____

(b) Explain your answer above: _____

_____

_____

6. Describe three methods for marking animals for mark and recapture sampling. Take into account the possibility of animals shedding their skin, or being difficult to get close to again:

(a) _____

(b) _____

(c) _____

7. Scientists in the UK and Canada have, at various times since the 1950s, been involved in computerised tagging programmes for Northern cod (a species once abundant in Northern Hemisphere waters but now severely depleted). Describe the type of information that could be obtained through such tagging programmes:

_____

_____

_____

_____

Student's t Test Exercise

Data from two flour beetle populations are provided below. The numbers of beetles in each of ten samples were counted. The experimenter wanted to test if the densities of the two populations were significantly different. The exercise below involves manual computation to determine a t value. Follow the steps to complete the test. You can also use a spreadsheet programme such as *Microsoft Excel* to do the computations (see the next page) or perform the entire analysis for you (see the TRC).

1. (a) Complete the calculations to perform the t test for these two populations. Some calculations are provided for you.

X (counts)		x − x̄ (deviation from the mean)		(x − x̄)² (deviation from mean)²	
Popn A	Popn B	Popn A	Popn B	Popn A	Popn B
465	310	9.3	−10.6	86.5	112.4
475	310	19.3	−10.6	372.5	112.4
415	290				
480	355				
436	350				
435	335				
445	295				
460	315				
471	316				
475	330				

$n_A = 10$ $n_B = 10$

The number of samples in each data set

The sum of each column is called the sum of squares

$\Sigma(x - \bar{x})^2$ $\Sigma(x - \bar{x})^2$

Step 1: Summary statistics

Tabulate the data as shown in the first 2 columns of the table (left). Calculate the mean and give the n value for each data set. Compute the standard deviation if you wish.

Popn A $\bar{x}_A = 455.7$
$n_A = 10$
$s_A = 21.76$

Popn B $\bar{x}_B = 320.6$
$n_B = 10$
$s_B = 21.64$

Step 2: State your null hypothesis

Step 3: Decide if your test is one or two-tailed

(b) The variance for population A: $s^2_A =$

The variance for population B: $s^2_B =$

Calculating the t value

Step 4a: Calculate sums of squares

Complete the computations outlined in the table left. The sum of each of the final two columns (left) is called the sum of squares.

Step 4b: Calculate the variances

Calculate the variance (s^2) for each set of data. This is the sum of squares divided by $n-1$ (number of samples in each data set − 1). In this case the n values are the same, but they need not be.

$$s^2_A = \frac{\Sigma(x - \bar{x})^2_{(A)}}{n_A - 1} \qquad s^2_B = \frac{\Sigma(x - \bar{x})^2_{(B)}}{n_B - 1}$$

(c) The difference between the population means

$(\bar{x}_A - \bar{x}_B) =$

Step 4c: Difference between means

Calculate the *actual* difference between the means

$$(\bar{x}_A - \bar{x}_B)$$

(d) t (calculated) =

Step 4d: Calculate t

Calculate the t value. Ask for assistance if you find interpreting the lower part of the equation difficult

$$t = \frac{(\bar{x}_A - \bar{x}_B)}{\sqrt{\dfrac{s^2_A}{n_A} + \dfrac{s^2_B}{n_B}}}$$

(e) Determine degrees of freedom (d.f.)

d.f. $(n_A + n_B - 2) =$

Step 4e: Determine degrees of freedom

Degrees of freedom (d.f.) are defined by the number of samples (e.g. counts) taken: d.f. $= n_A + n_B - 2$ where n_A and n_B are the number of counts in each of populations A and B.

(f) $P =$

t (critical value) =

Step 5: Consult the t table

Consult the t-tables (opposite page) for the critical t value at the appropriate degrees of freedom and the acceptable probability level (e.g. $P = 0.05$).

(g) Your decision is:

Step 5a: Make your decision

Make your decision whether or not to reject H_0. If your t value is large enough you may be able to reject H_0 at a lower P value (e.g. 0.001), increasing your confidence in the alternative hypothesis.

Practical Ecology

Web links: Student's t Tests

DA 3

2. The previous example (manual calculation for two beetle populations) is outlined below in a spreadsheet (created in *Microsoft Excel*). The spreadsheet has been shown in a special mode with the formulae displayed. Normally, when using a spreadsheet, the calculated values will appear as the calculation is completed (entered) and a formula is visible only when you click into an individual cell. When setting up a spreadsheet, you can arrange your calculating cells wherever you wish. What is important is that you accurately identify the cells being used for each calculation. Also provided below is a summary of the spreadsheet notations used and a table of critical values of *t* at different levels of *P*. Note that, for brevity, only some probability values have been shown. To be significant at the appropriate level of probability, calculated values must be greater than those in the table for the appropriate degrees of freedom.

(a) Using the data in question 1, set up a spreadsheet as indicated below to calculate *t*. Save your spreadsheet. Print it out and staple the print-out into your workbook.

Notation	Meaning
Columns and rows	Columns are denoted A, B, C ... at the top of the spreadsheet, rows are 1, 2, 3, ... on the left. Using this notation a cell can be located e.g. C3
=	An "equals" sign *before* other entries in a cell denotes a formula.
( )	Parentheses are used to group together terms for a single calculation. This is important for larger calculations (see cell C21 above)
C3:C12	Cell locations are separated by a colon. C3:C12 means "every cell between and including C3 and C12"
SUM	Denotes that what follows is added up. =SUM(C3:C12) means "add up the values in cells C3 down to C12"
COUNT	Denotes that the number of values is counted =COUNT(C3:C12) means "count up the number of values in cells C3 down to C12"
SQRT	Denotes "take the square root of what follows"
^2	Denotes an exponent e.g. x^2 means that value x is squared.

Above is a table explaining some of the spreadsheet notations used for the calculation of the *t* value for the exercise on the previous page. It is not meant to be an exhaustive list for all spreadsheet work, but it should help you to become familiar with some of the terms and how they are used. This list applies to *Microsoft Excel*. Different spreadsheets may use different notations. These will be described in the spreadsheet manual.

Table of critical values of *t* at different levels of *P*.

Degrees of freedom	Level of Probability		
	0.05	0.01	0.001
1	12.71	63.66	636.6
2	4.303	9.925	31.60
3	3.182	5.841	12.92
4	2.776	4.604	8.610
5	2.571	4.032	6.869
6	2.447	3.707	5.959
7	2.365	3.499	5.408
8	2.306	3.355	5.041
9	2.262	3.250	4.781
10	2.228	3.169	4.587
11	2.201	3.106	4.437
12	2.179	3.055	4.318
13	2.160	3.012	4.221
14	2.145	2.977	4.140
15	2.131	2.947	4.073
16	2.120	2.921	4.015
17	2.110	2.898	3.965
18	2.101	2.878	3.922
19	2.093	2.861	3.883
20	2.086	2.845	3.850

(b) Save your spreadsheet under a different name and enter the following new data values for population B: **425, 478, 428, 465, 439, 475, 469, 445, 421, 438**. Notice that, as you enter the new values, the calculations are updated over the entire spreadsheet. Re-run the *t*-test using the new *t* value. State your decision for the two populations now:

New *t* value: _____   Decision on null hypothesis (circle one):   reject H$_0$ / do not reject H$_0$

Using Chi-Squared in Ecology

The following exercise illustrates the use of chi-squared (χ^2) in ecological studies of habitat preference. In the first example, it is used for determining if the flat periwinkle *(Littorina littoralis)* shows significant preference for any of the four species of seaweeds with which it is found. Using quadrats, the numbers of periwinkles associated with each seaweed species were recorded. The data from this investigation are provided for you in Table 1. In the second example, the results of an investigation into habitat preference in woodlice (also called pillbugs, sowbugs, or slaters) are presented for analysis (Table 2).

1. (a) State your null hypothesis for this investigation (H0):

    _____

    _____

    _____

   (b) State the alternative hypothesis (HA): _____

    _____

    _____

Table 1: Number of periwinkles associated with different seaweed species

Seaweed species	Number of periwinkles
Spiral wrack	9
Bladder wrack	28
Toothed wrack	19
Knotted wrack	64

2. Use the chi-squared (χ^2) test to determine if the differences observed between the samples are significant or if they can be attributed to chance alone. The table of critical values of χ^2 at different *P* values is available in *Advanced biology AS*, on the Teacher Resource CD-ROM, or in any standard biostatistics book.

   (a) Enter the observed values (no. of periwinkles) and complete the table to calculate the χ^2 value:

   (b) Calculate χ^2 value using the equation:

   $$\chi^2 = \sum \frac{(O - E)^2}{E} \qquad \chi^2 = \underline{\hspace{2cm}}$$

   (c) Calculate the degrees of freedom: _____

   (d) Using a reference, state the *P* value corresponding to your calculated χ^2 value:

    _____

Category	O	E	O – E	$(O - E)^2$	$\dfrac{(O - E)^2}{E}$
Spiral wrack					
Bladder wrack					
Toothed wrack					
Knotted wrack					
Σ					Σ

   (e) State whether or not you reject your null hypothesis:

    reject H0 / do not reject H0   *(circle one)*

3. Students carried out an investigation into habitat preference in woodlice. In particular, they were wanting to know if the woodlice preferred a humid atmosphere to a dry one, as this may play a part in their choice of habitat. They designed a simple investigation to test this idea. The woodlice were randomly placed into a choice chamber for 5 minutes where they could choose between dry and humid conditions (atmosphere). The investigation consisted of five trials with ten woodlice used in each trial. Their results are shown on Table 2 (right):

   (a) State the null and alternative hypotheses (H0 and HA):

    _____

    _____

Table 2: Number of woodlice in each habitat

Trial	Atmosphere	
	Dry	Humid
1	2	8
2	3	7
3	4	6
4	1	9
5	5	5

   Use a separate piece of paper to calculate the chi-squared value and summarise your answers below:

   (b) Calculate the χ^2 value: _____

   (c) Calculate the degrees of freedom and state the *P* value corresponding to your calculated χ^2 value: _____

   (d) State whether or not you reject your null hypothesis (H0):   reject H0 / do not reject H0   *(circle one)*

Practical Ecology

Comparing More Than Two Groups

The Student's *t* test is limited to comparing two groups of data. To compare more than two groups at once you need to use a test that is appropriate to this aim. One such test, appropriate for normally distributed data, is an analysis of variance (**ANOVA**), as described in the next activity. A good place to start with any such an analysis is to plot your data, together with some measure of the reliability of the statistic you calculate as an indicator of the true population parameter, For normally distributed data, this is likely to be the mean and the 95% confidence interval (95% CI). See *Advanced Biology AS* for an introduction to this statistic. In the example described below, students recorded the survival of weevil larvae on five different pasture types and calculated descriptive statistics for the data. After you have worked through the analysis, you should be able to enter your own data and calculate descriptive statistics using *Microsoft Excel*. The plot and full analysis of these data are presented in the next activity.

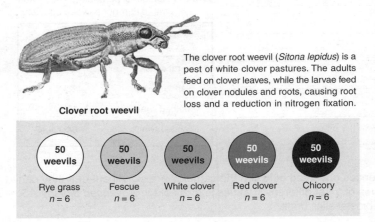

The clover root weevil (*Sitona lepidus*) is a pest of white clover pastures. The adults feed on clover leaves, while the larvae feed on clover nodules and roots, causing root loss and a reduction in nitrogen fixation.

Clover root weevil

50 weevils	50 weevils	50 weevils	50 weevils	50 weevils
Rye grass *n* = 6	Fescue *n* = 6	White clover *n* = 6	Red clover *n* = 6	Chicory *n* = 6

Comparing the Means of More Than Two Experimental Groups

Research has indicated that different pastures have different susceptibility to infestation by a pest insect, the clover root weevil (left). Armed with this knowledge, two students decided to investigate the effect of pasture type on the survival of clover root weevils. The students chose five pasture types, and recorded the number of weevil larvae (from a total of 50) surviving in each pasture type after a period of 14 days. Six experimental pots were set up for each pasture type (*n* = 6). Their results and the first part of their analysis (calculating the descriptive statistics) are presented in this activity.

1 **Calculating Descriptive Statistics**

Enter the data for each category of the independent variable in separate columns.

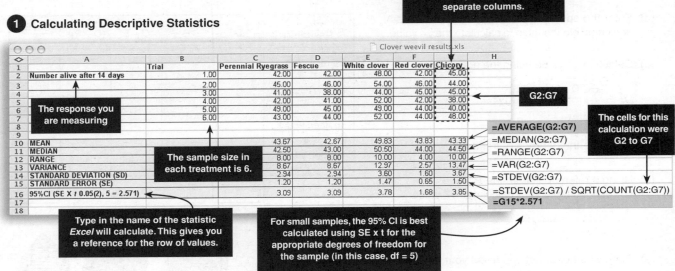

Type in the name of the statistic *Excel* will calculate. This gives you a reference for the row of values.

For small samples, the 95% CI is best calculated using SE x t for the appropriate degrees of freedom for the sample (in this case, df = 5)

	nial Ryegrass	Fescue	White clover	Red clover	Chicory
Mean	43.67	42.67	49.83	43.83	43.33
95% CI	3.09	3.09	3.78	1.68	3.85

Enter the values for means and 95% CI in columns under the categories.

2 To plot a graph in *Excel*, you will need to enter the values for the data you want to plot in a format that *Excel* can use. To do this, enter the values in columns under each category (left). Each column will have two entries: mean and 95% CI. In this case, we want to plot the mean survival of weevils in different pasture types and add the 95% CI as error bars. You are now ready to plot the data set.

1. Identify the number of treatments in this experimental design: _____

2. Identify the independent variable for this experiment: _____

3. Identify the dependent variable for this experiment: _____

4. Identify the type of graph you would use to plot this data set and explain your choice: _____

   _____

5. Explain what the 95% confidence interval tells you about the means for each treatment: _____

   _____

   _____

Analysis of variance (or **ANOVA**) could be considered beyond the scope of the statistical analyses you would do at school. However, using *Excel* (previous activity and below), it is not difficult and is an appropriate test for a number of situations in experimental biology. It is described below for the data on weevil survival described in the previous activity. A plot of the data is also described as an aid to interpreting the statistical analysis. An ANOVA may also be appropriate when the independent variable

is continuous, but in these instances a regression is likely to be a more suitable analysis (see *Advanced Biology AS*). After you have worked through this activity you should be able to:

■ Use *Excel* to plot data appropriately with error bars.
■ Interpret the graphically presented data and reach tentative conclusions about the findings of the experiment.
■ Support your conclusions with an ANOVA, in *Microsoft Excel*, to test the significance of differences in the sample data.

Drawing the Graph

Recall the design of this experiment. The independent variable is categorical, so the correct graph type is a column chart. To plot the graph, select the row of mean values (including column headings) from the small table of results you constructed (screen, right).

1 From the menu bar choose: **Insert > Chart > Column**. This is **STEP 1** in the Chart Wizard. Click **Next**.

2 At **STEP 2** in Chart Wizard: The data range you have selected (the source data for the chart) will appear in the "Data range" window. Click **Next**.

3 At **STEP 3** in Chart Wizard: You have the option to add a title, labels for your X and Y axes, turn off gridlines, and add (or remove) a legend (a legend is useful when you have two or more columns, such as males and females, for each treatment). When you have added all the information you want, click **Next**.

4 At **STEP 4** in Chart Wizard: Specify the chart location. It should appear in the active sheet by default. Click on the chart and move it to reveal the data.

	al Ryegrass	Fescue	White clover	Red clover	Chicory
Mean	43.67	42.67	49.83	43.83	43.33
95% CI	3.09	3.09	3.78	1.68	3.85

Select the row of mean values, including column headings.

A column chart is the appropriate choice for these data.

A preview of the plot appears

Add or remove information for your graph by clicking on each of these tabs ...

... and enter information in the fields as indicated

Specify the chart location. You have the option to put the graph on a new sheet if you wish.

Click finish

Select error bars for both above and below the mean

Click on Custom and enter the row of values for 95% CI

5 **Adding error bars:** A chart will appear. **Right click** (Ctrl-click on Mac) on any part of any column and choose **Format data series** (below). To add error bars, select the **Y error bars** tab, and click on the symbol that shows **Display both**. Click on **Custom**, and use the data selection window to select the row of 95% CI data. Click OK.

Practical Ecology

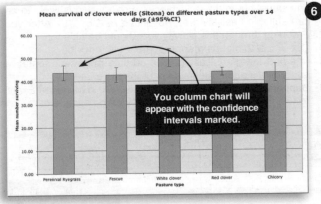

Mean survival of clover weevils (Sitona) on different pasture types over 14 days (±95%CI)

You column chart will appear with the confidence intervals marked.

6 After you have added error bars using the "Format data series" option, your chart will plot with error bars. These error bars represent the 95% confidence intervals for the mean of each treatment. The completed plot gives you a visual representation of your data and an immediate idea as to how confident you can be about the differences between the five treatments. You are now ready to run your analysis of variance.

About ANOVA

ANOVA is a test for normally distributed data, which can determine whether or not there is a real (statistically significant) difference between two or more sample means (an ANOVA for two means is a t test). ANOVA does this by accounting for the amount of variability (in the measured variable) within a group and comparing it to the amount of variability between the groups (treatments). It is an ideal test when you are looking for the effect of a particular treatment on some kind of biological response, e.g. plant growth with different fertilisers. ANOVA is a very useful test for comparing multiple samples because, like the t test, it does not require large sample sizes.

Running the ANOVA

ANOVA is part of the *Excel* "Data Analysis Toolpak", which is part of normal *Excel*, but is not always installed. If there is no Data Analysis option on the Tools menu, you need to run *Excel* from the original installation disk to install the "Analysis Toolpak add-in".

From the Tools menu select Data Analysis and **ANOVA single factor**. This brings up the ANOVA dialogue box (numbered points 1 and 2). Here we are concerned with the effect of one variable on another, so the ANOVA is single factor (also called one way ANOVA).

Click in the **Output Range** box and click on a free cell on the worksheet which will become the left cell of the results table. Click OK.

The output is a large data table (number 3 below) and you will need to adjust the number of significant figures to 2 or 3 (You do this from the Format menu > Cells > Number). The most important cell is the P value, which, as usual is the probability that the null hypothesis (of no difference between the groups) is true.

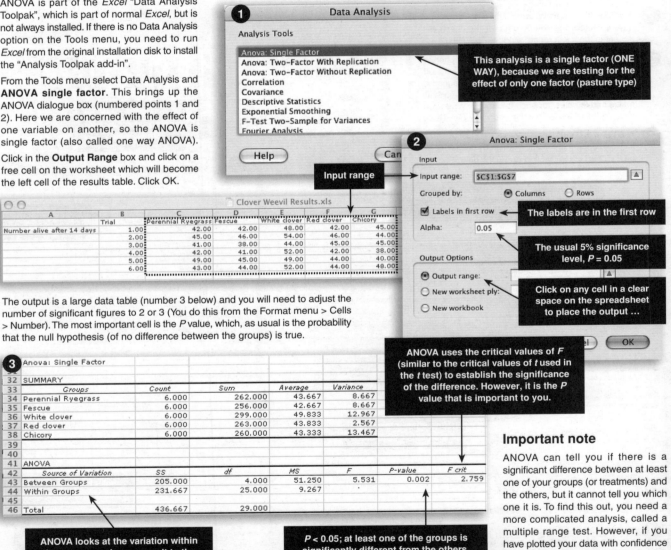

1 Data Analysis

Analysis Tools

Anova: Single Factor
Anova: Two-Factor With Replication
Anova: Two-Factor Without Replication
Correlation
Covariance
Descriptive Statistics
Exponential Smoothing
F-Test Two-Sample for Variances
Fourier Analysis

Help   Can

This analysis is a single factor (ONE WAY), because we are testing for the effect of only one factor (pasture type)

2 Anova: Single Factor

Input

Input range: $CS1:$G$7
Grouped by: ● Columns   ○ Rows
☑ Labels in first row   The labels are in the first row
Alpha: 0.05   The usual 5% significance level, $P = 0.05$

Output Options

● Output range: Click on any cell in a clear space on the spreadsheet to place the output …
○ New worksheet ply:
○ New workbook

Input range

OK

Clover Weevil Results.xls

	Trial	Perennial Ryegrass	Fescue	White clover	Red clover	Chicory
Number alive after 14 days	1.00	42.00	42.00	48.00	42.00	45.00
	2.00	45.00	46.00	54.00	46.00	44.00
	3.00	41.00	38.00	44.00	45.00	45.00
	4.00	42.00	41.00	52.00	42.00	38.00
	5.00	49.00	45.00	49.00	44.00	40.00
	6.00	43.00	44.00	52.00	44.00	48.00

3 Anova: Single Factor

SUMMARY

Groups	Count	Sum	Average	Variance
Perennial Ryegrass	6.000	262.000	43.667	8.667
Fescue	6.000	256.000	42.667	8.667
White clover	6.000	299.000	49.833	12.967
Red clover	6.000	263.000	43.833	2.567
Chicory	6.000	260.000	43.333	13.467

ANOVA

Source of Variation	SS	df	MS	F	P-value	F crit
Between Groups	205.000	4.000	51.250	5.531	0.002	2.759
Within Groups	231.667	25.000	9.267			
Total	436.667	29.000				

ANOVA looks at the variation within the groups and compares it to the variation between groups.

$P < 0.05$; at least one of the groups is significantly different from the others. ANOVA cannot tell you which one.

ANOVA uses the critical values of F (similar to the critical values of t used in the t test) to establish the significance of the difference. However, it is the P value that is important to you.

Important note

ANOVA can tell you if there is a significant difference between at least one of your groups (or treatments) and the others, but it cannot tell you which one it is. To find this out, you need a more complicated analysis, called a multiple range test. However, if you have plotted your data with confidence intervals, it is likely that you will be able to find this out from your graph.

1. The ANOVA cannot tell us which one (or more) of the groups is significantly different from the others. Explain how you could you use your graphical analysis to answer this question:

_____

_____

_____

Sources of Variation

AQA-A	AQA-B	SNAB	Edexcel	OCR
Complete:	Complete:	Complete:	Complete:	Complete:
1-5, 7-13, 18, 24-25 (a)-(b)	1, 5-12 (a)-(c), 13, 18	1, 8-15, 18, 20-24, 27-28	1, 5-13, 18-20, 24-26 Extension: 5, 16, 22, 25	1, 7-13, 24-26, 28
Extension: 6, 14-16	Extension: 2-4, 14-16	Extension: 2-6		

Learning Objectives

☐ 1. Compile your own glossary from the **KEY WORDS** displayed in **bold type** in the learning objectives below.

Phenotypic Variation *(pages 105-106, 151-152)*

☐ 2. Recognise the need for **random sampling** and the importance of **chance** in contributing to differences between samples. Demonstrate an ability to collect and graphically display data pertaining to variation in the phenotype of a natural population.

☐ 3. Understand the concept of **normal distribution** about a mean. Explain how mean and **standard deviation** are used to measure variation in a sample.

☐ 4. Demonstrate an ability to calculate and interpret **standard error** of sample data pertaining to variation in the phenotype of a natural population.

☐ 5. Describe the difference between **continuous** and **discontinuous variation** and provide examples to illustrate each type. Explain the basis of continuous and discontinuous variation by reference to the number of genes controlling the characteristic (also refer to the material on polygeny in the chapter, *Inheritance*).

☐ 6. Collect data pertaining to human traits that show continuous variation, e.g. foot size, hand span, height. From the same set of subjects, collect data pertaining to human traits that show discontinuous variation, e.g. ear lobe shape or mid digit hair. Plot these sets of data separately to illustrate the differences in distribution.

Genes and Environment *(pages 107-108)*

☐ 7. Define: **gene-environment interaction**. Recognise that both genotype and environment contribute to phenotypic variation. Understand that this relationship can be expressed mathematically as: $V_P = V_G + V_E$.

☐ 8. Use examples to describe how the environment may affect the phenotype.

Meiosis, Linkage, and Recombination

Meiosis *(pages 103-104, 109-113)*

☐ 9. Appreciate the general significance of meiosis in generating genetic variation. Identify how meiosis creates new allele combinations in the gametes. Suggest how sexual reproduction confers an evolutionary advantage.

☐ 10. Define: **homologous chromosomes** (=homologues), **genome**, **chromatid**, **centromere**. Describe how chromosome numbers can vary between somatic cells (**diploid** 2N) and gamete cells (**haploid** 1N).

☐ 11. Recognise that meiosis, like mitosis, involves DNA replication during interphase in the parent cell, but that this is followed by *two* cycles of nuclear division.

☐ 12. Using simple diagrams, summarise the principal events associated with meiosis, to include:
   (a) Pairing of **homologous chromosomes** (synapsis) and the formation of **bivalents**.
   (b) **Chiasma** formation and exchange between **chromatids** in the first, (**reduction**) **division**.
   (c) Separation of chromatids in the second division and the production of **haploid cells**.
   (d) The associated behaviour of the **nuclear envelope**, plasma membrane, and **centrioles**.
   (e) Identification of the names of the main stages.

☐ 13. Describe the behaviour of homologous chromosomes (and their associated alleles) during meiosis and fertilisation, with reference to:
   - The **independent assortment** of maternal and paternal chromosomes. Chance governs which pole each chromosome of a bivalent moves to, ensuring random combinations of non-homologous chromosomes in the haploid nuclei.
   - The **recombination** of segments of maternal and paternal homologous chromosomes in **crossing over**.
   - The **random fusion** of gametes during fertilisation.

Linkage and recombination *(pages 103, 112-117)*

☐ 14. With respect to inheritance, explain what is meant by the terms: **linkage** and **linkage group**. Explain the consequences of linkage to the inheritance of alleles.

☐ 15. Recall that **recombination** refers to the exchange of alleles between homologous chromosomes as a result of **crossing over**. Explain the consequences of recombination to the inheritance of alleles.

☐ 16. Explain the effect of linkage and recombination on the phenotypic ratios from dihybrid crosses.

☐ 17. Explain what is meant by the **crossover frequency** (frequency of recombination, **crossover value**, or COV). Describe how crossover values are used as a measure of the relative positions of genes on the same chromosome. Demonstrate the use of crossover values in chromosome (gene) mapping.

Mutation and Variation *(pages 103-104, 118-128, 172 and the TRC: Point Mutation Problems)*

☐ 18. Understand the significance of **mutation** as the ultimate source of all new alleles.

☐ 19. Define **spontaneous mutation**. Recognise that each species has its own frequency of naturally occurring mutations (called the **background mutation rate**).

☐ 20. Define **mutagen** and **induced mutation**. Identify the environmental factors that can cause mutations. Describe the effects of chemical mutagens and radiation on DNA and the rate of mutation.

□ 21. Mutations may occur in different types of cell. The location of a mutation can have a different significance in terms of producing heritable change. Clearly define: **germ line**, **somatic mutation**, **gametic mutation**.

□ 22. Recognise that mutations may have different survival value in different environments. Using examples, explain how mutations may be **harmful**, **beneficial**, or **neutral** (silent) in their effect on the organism. Explain the evolutionary importance of **neutral mutations**.

□ 23. In more detail than above, identify the genetic basis of a beneficial mutation, as illustrated by **antibiotic resistance** in bacteria.

□ 24. Explain what is meant by a **gene mutation** and provide examples. Distinguish between gene mutations involving change in a single nucleotide (often called **point mutations**), and those involving changes to a triplet (e.g. triplet deletion or triplet repeat). Explain why gene mutations have the most evolutionary potential.

□ 25. Understand point mutations as illustrated by:
  (a) **Base deletion**
  (b) **Base substitution**
  (c) **Base insertion**

  Explain how a **frame shift** can arise and explain the consequences of this type of error.

□ 26. Describe the effect of a point mutation on the resulting amino acid sequence and on the phenotype, as illustrated by the **sickle cell mutation** in humans.

□ 27. Recognise other disorders that arise as a result of gene mutations: β-**thalassaemia**, **cystic fibrosis**, and **Huntington disease**. Describe the genetic basis of one or more of these diseases.

□ 28. Outline the implications of the **Human Genome Project** for the detection and diagnosis of genetic disorders. Suggest how our deeper knowledge of genetic inheritance and the basis of genetic disorders will assist in genetic counselling in the future.

 See the 'Textbook Reference Grid' on pages 8-9 for textbook page references relating to material in this topic.

Supplementary Texts

See pages 5-7 for additional details of these texts:

■ Adds, J., *et al.*, 2004. **Genetics, Evolution and Biodiversity**, (NelsonThornes), chpt. 7 as reqd.

■ Clegg, C.J., 1999. **Genetics and Evolution**, (John Murray), pp. 19-22, 34-39.

■ Jones, N., *et al.*, 2001. **The Essentials of Genetics**, chpt. 2, 5, 13-14 and p. 257.

See page 7 for details of publishers of periodicals:

STUDENT'S REFERENCE

■ **What is a Gene?** Biol. Sci. Rev., 15(2) Nov. 2002, pp. 9-11. *A good synopsis of genes and their role in heredity, mutations, and transcriptional control of gene expression.*

■ **Spermatogenesis** Biol. Sci. Rev., 15(4), April 2003, pp. 10-14. *The process and control of sperm production. This account includes a discussion of the cell division processes (both meiotic and mitotic) involved in the production of viable sperm.*

■ **Mechanisms of Meiosis** Biol. Sci. Rev., 15(4), April 2003, pp. 20-24. *A clear and thorough account of the events and mechanisms of meiosis. There is a discussion of what happens when meiosis goes wrong, as well as an examination of synapsis and crossing over.*

■ **Genes, the Genome, and Disease** New Scientist, 17 Feb. 2001, (Inside Science). *Understanding the human genome: producing genome maps, the role of introns in gene regulation, and the future of genomic research.*

■ **A Chromosomal Rearrangement** Biol. Sci. Rev., 18(2), Nov. 2005, pp. 25-27. *How chromosome rearrangements alter genetic balance.*

■ **How do Mutations Lead to Evolution?** New Scientist, 14 June 2003, pp. 32-39, 48-51. *An account of the five most common points of discussion regarding evolution and the mechanisms by which it occurs. This article looks at how mutations to the promoter regions of DNA, which control gene expression, may account for the evolution of morphology and behaviour.*

■ **Secrets of The Gene** National Geographic, 196(4) October 1999, pp. 42-75. *The nature of genes and the location of some human mutations.*

■ **Radiation and Risk** New Scientist, 18 March 2000 (Inside Science). *An excellent account of the biological effects of radiation.*

■ **Antibiotic Resistance** Biol. Sci. Rev., 12(2) November 1999, pp. 28-30. *The genetic basis of antibiotic resistance in bacteria.*

■ **Bioinformatics** Biol. Sci. Rev., 15(3), February 2003, pp. 13-15 and 15(4) April 2003, pp. 2-6. *Two accounts of the burgeoning field of bioinformatics: what it is and what it offers biology and medicine.*

■ **Drosophila: The past, present and future of genetics** Biol. Sci. Rev., 18(2) November 2005, pp. 10-13. *Once again the fruitfly D. melanogster is the organism of choice in studies of genetics.*

TEACHER'S REFERENCE

■ **DNA 50** SSR, 84(308), March 2003, pp. 17-80. *A special issue celebrating 50 years since the discovery of DNA. There are various articles examining the practical and theoretical aspects of teaching molecular genetics and inheritance.*

■ **Radiation Roulette** New Scientist, 11 October 1997, pp. 36-40. *Ionising radiation damages DNA and causes chromosomal defects. Sometimes the damage is not immediately apparent.*

■ **Cystic Fibrosis** Scientific American, December 1995, pp. 36-43. *The basis of the cystic fibrosis mutation: the nature of the mutation and how it brings about the symptoms of the disease.*

■ **The Great Inventors** New Scientist, 21 Feb. 2004, pp. 41-43. *The human genome is riddled with duplications and these provide the raw material for mutations and natural selection. Random mutation can modify one copy of a duplicated gene without interfering with existing biochemical pathways.*

■ **Owning the Stuff of Life** Scientific American, Feb. 2006, pp. 58-65. *Genetic patenting and the issues involved. Nearly 20% of the genes in the human genome have one or more patents on them, and almost 95% of the known cancer genes have been patented but the ethical issues have so far been largely ignored.*

■ **Evolving with the Enemy** New Scientist, 22 Nov. 2003, pp. 44-47. *In the coevolution of pathogens and their human hosts, pathogens have ensured that mutations protecting against infection are preserved in the human genome. Inherited genetic disorders, such as thalassaemias, result when a person inherits a double dose of an otherwise beneficial mutation.*

■ **The Enigma of Huntington's Disease** Scientific American, December 2002, pp. 60-95. *An account of the nature and cause of this inherited disorder. it includes a diagram illustrating the molecular basis of the disease.*

■ **New Movements in Parkinson's** Scientific American, July 2005, pp. 58-65. *Parkinson's disease cannot be slowed, stopped, or prevented. Recent discoveries about how proteins malfunction and the genetic underpinnings of Parkinson's have opened up new avenues for research and optimism for future treatments.*

■ **Life Force** New Scientist, 4 Dec. 2004, pp. 46-49. *Experimental evidence for evolution in bacteria. The findings have implications for the battle against the rise of microbial drug resistance.*

■ **Day of the Mutators** New Scientist, 14 Feb. 1998, pp. 38-42. *The role of mutations in bacterial evolution: bacteria may direct their own evolution through the accumulation of favourable mutations.*

See pages 10-11 for details of how to access **Bio Links** from our web site: **www.biozone.co.uk**. From Bio Links, access sites under the topics:

GENERAL BIOLOGY ONLINE RESOURCES • AP Interactive animation • Biology I interactive animations • Instructional multimedia, University of Alberta ... *and others* > **Online Textbooks and Lecture Notes**: • S-Cool! A level biology revision guide ... *and others* > **Glossaries**: • Genetics glossary ... *and others*
CELL BIOLOGY AND BIOCHEMISTRY: • Cell & molecular biology online • MIT biology hypertextbook • Molecular biology web book > **Cell Division**: • Cell division: Meiosis and sexual reproduction • Comparison of mitosis and meiosis **GENETICS** > **Molecular Genetics**: • Beginner's guide to molecular biology ... *and others* > **Inheritance**: • Online Mendelian inheritance in man • Patterns of inheritance > **Mutations and Genetic Disorders**: • Genetic Mutations • Mutant fruit flies • Mutations • Advances in cystic fibrosis • Mutations causing cystic fibrosis • The lung net: cystic fibrosis • Sickle cell disease • Sickle cell information centre • What are metabolic disorders? • Variation in chromosome number • Visible chromosomal mutations • What is polyploidy? ... *and others*

Presentation MEDIA to support this topic:
GENES & INHERITANCE
• **Mutations**

Sources of Genetic Variation

The genetic variability between individuals is what makes us all different from each other. Brothers and sisters may look similar to each other but there are always significant differences between them (unless they happen to be identical twins). The differences between close relatives is due mostly to a **shuffling** of the existing genetic material into new combinations. In addition to this is the new variation that originates from the **mutation** of existing genes. While most mutations are harmful, there are a significant number that are thought to be 'silent' and do not appear to have any effect on the individual. On rare occasions, a mutation may even prove to be beneficial. Mutations create new **alleles** and form an important part of the evolutionary process.

Mutations

Gene mutations; chromosome mutations

Mutations are the source of all **new** genetic information. Existing genes are modified by base substitutions and deletions, causing the formation of new alleles.

Mutation: Substitute **T** instead of **C**

Original DNA

Mutant DNA

Gene Mutations

Mutations may cause alterations in the genetic instructions coded in the DNA of chromosomes. Most mutations are harmful, some are neutral (no effective change), while a very few may provide some improvement on the earlier version of the gene. Mutations may be accumulated (inherited) over many generations.

Chromosome Mutations

Pieces of chromosome may be rearranged during meiosis. Pieces may be turned upside-down, duplicated, moved from one chromosome to another or lost altogether. Most instances are harmful, but occasionally they may be beneficial.

Sexual Reproduction

Independent assortment; crossing over and recombination; mate selection

Sexual reproduction provides a rearrangement and shuffling of the genetic material into new combinations.

Independent Assortment

Genes are carried on chromosomes, 23 pairs in the case of humans. Each chromosome pair is sorted independently of the other pairs during meiosis. This random shuffling produces a huge variety of gametes from a single individual (parent).

Recombination

Pieces of chromosome are often exchanged with a chromosome's homologue (its paired chromosome with equivalent genes). This increases shuffling of allele combinations.

Mate Selection

Variation is further enhanced by the choice of mate to produce offspring. Different combinations of genes will come together in the offspring, depending on which two parents mate together.

Genotype

Determines the **genetic potential** of an individual

Dominant, recessive, codominant and multiple allele systems, as well as interactions between genes, combine in their effects.

Phenotype

The phenotype expressed in an individual is the result of all the factors listed on this page. The genetic instructions for creating the individual may be modified along the way, or at least modified by environmental influences.

Environmental Factors

Environmental factors may influence the expression of the genotype. These factors may include physical factors such as temperature, light intensity, presence of groundwater, diet or nutrients, wind exposure, and pH. The presence of other organisms may also affect the expression of the genotype.

Related activities: Gene-Environment Interactions, For Harm or Benefit?

RA 2

1. Describe three ways in which sexual reproduction can provide genetic variation in individuals:

    (a) _____

    _____

    (b) _____

    _____

    (c) _____

    _____

2. Explain how the environment of a particular genotype can affect the phenotype:

    _____

    _____

    _____

3. Describe three common ways by which humans can, by choice, alter their phenotype:

    (a) _____

    _____

    (b) _____

    _____

    (c) _____

    _____

4. Explain why siblings (brothers and sisters) have a family similarity but are not identical (unless they happen to be identical twins):

    _____

    _____

    _____

    _____

    _____

5. (a) Explain what is meant by a neutral (silent) mutation: _____

    _____

    _____

    (b) Discuss how neutral mutations can be important in the evolution of populations: _____

    _____

    _____

    _____

    _____

    _____

    _____

    _____

An estimated 25 000 genes determine all human **traits**. While most traits are determined by more than one gene, a number are determined by a single gene system, with dominant/recessive, codominant, or multiple allele inheritance. Single gene traits (below) show **discontinuous variation** in a population, with individuals showing only one of a limited number of phenotypes (usually two or three). Single gene traits may, however, show variable **penetrance**. Penetrance describes the extent to which the properties controlled by a gene will be expressed. Highly penetrant genes will be expressed regardless of the effects of the environment, whereas a gene with low penetrance will only sometimes produce the trait with which it is associated.

Sources of Variation

Trait: Handedness

Dominant	Recessive
Phenotype: Right-handed	**Phenotype:** Left-handed
Allele: R	**Allele:** r

The trait of left or right handedness is genetically determined. Right-handed people have the dominant allele, while left handedness is recessive. People that consider themselves ambidextrous can assume they have the dominant allele for this trait.

Trait: Hand clasp

Dominant	Recessive
Phenotype: Left thumb on top	**Phenotype:** Right thumb on top
Allele: C	**Allele:** c

Like handedness, hand clasping shows dominance/recessiveness. When the hands are clasped together, either the left or the right thumb will naturally come to rest on top. The left thumb on top is the dominant trait (C), while the right thumb on top is recessive (c).

Trait: Dimpled chin

Dominant	Recessive
Phenotype: Chin cleft	**Phenotype:** No chin cleft
Allele: D	**Allele:** d

A cleft or dimple on the chin is inherited. A cleft is dominant (D), while the absence of a cleft is recessive (d), although this gene shows **variable penetrance**, probably as a result of modifier genes.

Trait: Middle digit hair

Dominant	Recessive
Phenotype: Hair on middle segment	**Phenotype:** No hair on mid segment
Allele: M	**Allele:** m

Some people have a dominant allele that causes hair to grow on the middle segment of their fingers. It may not be present on all fingers, and in some cases may be very fine and hard to see.

Trait: Ear lobe shape

Dominant	Recessive
Phenotype: Lobes free	**Phenotype:** Lobes attached
Allele: F	**Allele:** f

In people with only the recessive allele (homozygous recessive), ear lobes are attached to the side of the face. The presence of a dominant allele causes the ear lobe to hang freely.

Trait: Thumb hyperextension

Dominant	Recessive
Phenotype: 'Hitchhiker's thumb'	**Phenotype:** Normal thumb
Allele: H	**Allele:** h

Photos: RA

There is a gene that controls the trait known as 'hitchhiker's thumb' which is technically termed distal hyperextensibility. People with the dominant phenotype are able to curve their thumb backwards without assistance, so that it forms an arc shape.

Related activities: Polygenes

P 1

Your Genotype Profile

Use the descriptions and the symbols on the previous page to determine your own genotype. In situations where you exhibit the dominant form of the trait, it may be helpful to study the features of your family to determine whether you are homozygous dominant or heterozygous. If you do not know whether you are heterozygous for a given trait, assume you are.

Your traits:	Thumb	Ear lobes	Chin cleft	Middle digit hair	Handedness	Hand clasp
Phenotype:						
Genotype:						

1. Enter the details of your own genotype in the table above. The row: 'Phenotype' requires that you write down the version of the trait that is expressed in you (e.g. blue eyes). Each genotype should contain two alleles.

2. Use a piece of paper and cut out 12 squares. Write the symbols for your alleles listed in the table above (each of the two alleles on two separate squares for the six traits) and write your initials on the back.

3. Move about the class, shaking hands with other class members to simulate mating (this interaction does not have to be with a member of the opposite sex).

4. Proceed to determine the possible genotypes and phenotypes for your offspring with this other person by:

   (a) Selecting each of the six characters in turn

   (b) Where a genotype for a person is known to be

homozygous (dominant or recessive) that person will simply place down one of the pieces of paper with their allele for that gene. If they are heterozygous for this trait, toss a coin to determine which gets 'donated' with heads being the dominant allele and tails being the recessive.

   (c) The partner places their allele using the same method as in (b) above to determine their contribution to this trait.

   (d) Write down the resulting genotype in the table below and determine the phenotype for that trait.

   (e) Proceed on to the next trait.

5. Try another mating with a different partner or the same partner and see if you end up with a child of the same phenotype.

Child 1	Thumb	Ear lobes	Chin cleft	Middle digit hair	Handedness	Hand clasp
Phenotype:						
Genotype:						

Child 2	Thumb	Ear lobes	Chin cleft	Middle digit hair	Handedness	Hand clasp
Phenotype:						
Genotype:						

Gene-Environment Interactions

External environmental factors can modify the phenotype coded by genes. This can occur both in the development of the embryo and later in life. Even identical twins, which are essentially clones, have minor differences in their appearance due to factors such as diet. Environmental factors that affect the phenotype of plants and animals include nutrients or diet, temperature, altitude or latitude, and the presence of other organisms.

The Effect of Temperature

The sex of some animals is determined by the temperature at which they were incubated during their embryonic development. Examples include turtles, crocodiles, and the American alligator. In some species, high incubation temperatures produce males and low temperatures produce females. In other species, the opposite is true. The advantages of temperature regulated sex determination may arise through prevention of inbreeding (since all siblings will tend to be of the same sex).

Female   Male

The Effect of Other Organisms

The presence of other individuals of the same species may control the determination of sex in other individuals of the group. Some fish species, including some in the wrasse family (e.g. *Coris sandageri*, above), show such a change in phenotype. The fish live in groups consisting of a single male with attendant females and juveniles. In the presence of a male, all juvenile fish of this species grow into females. When the male dies, the dominant female will undergo physiological changes to become a male for the group. The male has distinctive vertical bands behind the gills. The female is pale in colour and has very faint markings.

Severe stunting (krummholz)

Growth to genetic potential

Cline

The Effect of Altitude

Increasing altitude can stunt the phenotype of plants with the same genotype. In some conifers, e.g. Engelmann spruce (*Picea engelmannii*), plants at low altitude grow to their full genetic potential, but become progressively more stunted as elevation increases, forming krummholz (gnarled bushy growth forms) at the highest, most severe sites. This situation, where there is a continuous, or nearly continuous, gradation in a phenotypic character within a species, associated with a change in an environmental variable, is called a **cline**.

Helmet develops in response to the presence of chemicals released by invertebrate predators. The helmet makes *Daphnia* more difficult to attack and handle.

Spine length increases

Non-helmeted form   **Helmeted form with long tail spine**

Some organisms respond to the presence of other, potentially harmful, organisms by changing their morphology or body shape. Invertebrates such as *Daphnia* will grow a large helmet when a predatory midge larva (*Chaoborus*) is present. Such responses are usually mediated through the action of chemicals produced by the predator (or competitor), and are common in plants as well as animals.

1. Giving appropriate examples, distinguish clearly between **genotype** and **phenotype**: _____

_____

_____

_____

2. Identify some of the physical factors associated with altitude that could affect plant phenotype: _____

_____

3. The hydrangea is a plant that exhibits a change in the colour of its flowers according to the condition of the soil. Identify the physical factor that causes hydrangea flowers to be blue or pink. If you can, find out how this effect is exerted:

_____

_____

_____

4. Colour pointing in some breeds of cats such as the Siamese, involves the activity of a temperature sensitive enzyme that produces the pigment melanin. Explain why the darker patches of fur are found only on the face, paws and tail:

_____

_____

Related activities: Epistasis, The Species Concept

RA 2

At conception (the formation of the zygote), an organism possesses a genetic potential to grow into an adult form with certain characteristics. The exact form it takes is determined largely by the genes in its chromosomes, but it is also strongly influenced by a vast range of environmental factors acting upon it. These factors may subject an organism to stresses and may limit its growth to something less than it is capable of (e.g. plants that are grown at high altitude or in very exposed locations will often have stunted growth). Changes in the phenotype due solely to environmental factors are not inherited. Traumatic events such as the loss of a limb on a tree, or the removal of the tail in a young mammal (e.g. lambs, pups), does not affect the phenotype of the next generation (trees do not grow with limbs missing, and not one lamb has been born without a tail).

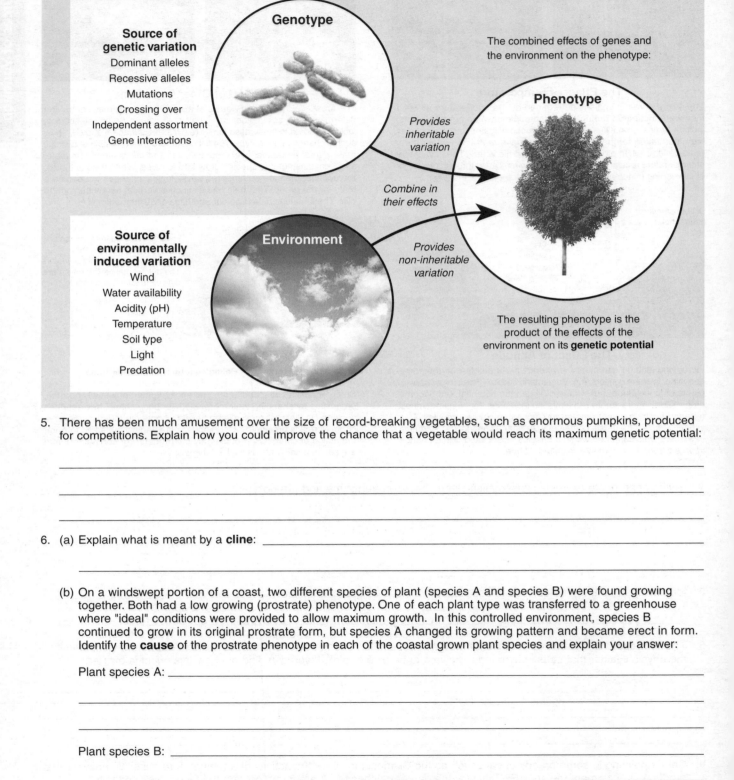

Genotype

Source of genetic variation
Dominant alleles
Recessive alleles
Mutations
Crossing over
Independent assortment
Gene interactions

The combined effects of genes and the environment on the phenotype:

Phenotype

Provides inheritable variation

Combine in their effects

Source of environmentally induced variation
Wind
Water availability
Acidity (pH)
Temperature
Soil type
Light
Predation

Environment

Provides non-inheritable variation

The resulting phenotype is the product of the effects of the environment on its **genetic potential**

5. There has been much amusement over the size of record-breaking vegetables, such as enormous pumpkins, produced for competitions. Explain how you could improve the chance that a vegetable would reach its maximum genetic potential:

_____

_____

_____

6. (a) Explain what is meant by a **cline**: _____

_____

(b) On a windswept portion of a coast, two different species of plant (species A and species B) were found growing together. Both had a low growing (prostrate) phenotype. One of each plant type was transferred to a greenhouse where "ideal" conditions were provided to allow maximum growth. In this controlled environment, species B continued to grow in its original prostrate form, but species A changed its growing pattern and became erect in form. Identify the **cause** of the prostrate phenotype in each of the coastal grown plant species and explain your answer:

Plant species A: _____

_____

_____

Plant species B: _____

_____

_____

(c) Identify which of these species (A or B) would be most likely to exhibit clinal variation: _____

Meiosis

Meiosis is a special type of cell division concerned with producing sex cells (gametes) for the purpose of sexual reproduction. It involves a single chromosomal duplication followed by two successive nuclear divisions, and results in a halving of the diploid chromosome number. Meiosis occurs in the sex organs of plants and animals. If genetic mistakes (**gene** and **chromosome mutations**) occur here, they will be passed on to the offspring (they will be inherited).

Meiosis I
(Reduction division)
The first division separates the homologous chromosomes into two intermediate cells.

Interphase

In a non-dividing cell, the **chromosomes** are not visible as discrete structures because they are uncoiled to make the DNA information available for protein synthesis.

Meiosis is preceded by **DNA replication**, during which each of the chromosomes replicates. For each chromosome, there are now two genetically identical sister **chromatids** (as yet unseparated). It is at this stage that **gene mutations** may occur. These may create new versions of genes (alleles).

Prophase 1

The chromosomes condense. The **homologues**, each consisting of two sister chromatids, pair up in a process called **synapsis** to form **bivalents**. At this stage the arms of the chromatids can become entangled, and segments of chromosome can be exchanged in a process called **crossing over**.

Metaphase 1

The bivalents line up at the 'equator' (the **metaphase plate**) of the cell in a way that is random. This results in **independent assortment** of maternal and paternal chromosomes.

Anaphase 1

Chromosomes in anaphase of meiosis I

Telophase 1

Intermediate cell

Homologues separate

Intermediate Cell

Prophase 2

Meiosis II
('Mitotic' division)
The second division is merely a mitotic one in nature, where the chromatids are pulled apart, but the number of chromosomes remains the same. This allows large numbers of gametes to be produced.

Spindle apparatus forms. Chromosomes migrate towards the metaphase plate.

Metaphase 2

Chromosomes line up on the metaphase plate.

Anaphase 2

Sister chromatids (now individual chromosomes), separate.

Telophase 2

Gametes
(Eggs or sperm)

Related activities: Gene Mutations
Web links: Meiosis Tutorial

RA 1

The meiotic spindle normally distributes chromosomes to daughter cells without error. However, mistakes can occur in which the homologous chromosomes fail to separate properly at anaphase during meiosis I, or sister chromatids fail to separate during meiosis II. In these cases, one gamete receives two of the same type of chromosome and the other gamete receives no copy. This mishap, called **non-disjunction**, results in abnormal numbers of chromosomes passing to the gametes. If either of the aberrant gametes unites with a normal one at fertilisation, the offspring will have an abnormal chromosome number, known as an **aneuploidy**.

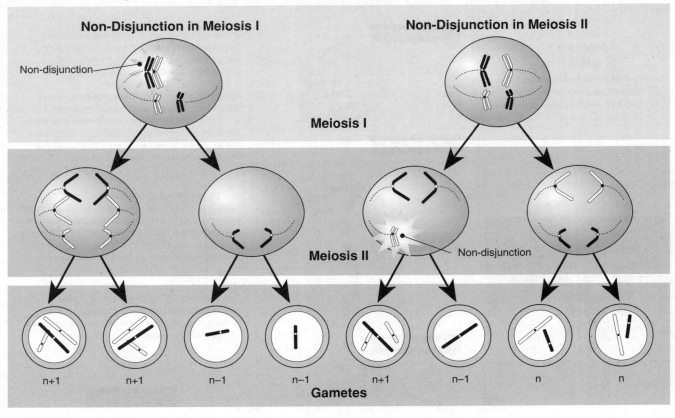

1. Describe the behaviour of the chromosomes in the first division of meiosis: _____

_____

_____

_____

_____

_____

2. Describe the behaviour of the chromosomes in the second division of meiosis: _____

_____

_____

3. Explain how mitosis conserves chromosome number while meiosis reduces the number from diploid to haploid:

_____

_____

4. Both these light micrographs (A and B) show chromosomes in metaphase of meiosis. State in what way they are different:

_____

_____

_____

_____

The Advantages of Sex

Sources of Variation

Asexual and **sexual reproduction** have different outcomes in terms of the inheritance of characteristics. In asexually reproducing organisms, the offspring are **clones** of the parent and clones with favourable mutations compete with other clones. In sexual reproduction, inherited favourable mutations may still compete, but there is the chance for them to combine in a single individual. Sexual reproduction provides many variants. At any one time, the best adapted "allele combinations" will be most successful and produce more offspring. An explanation as to why larger organisms tend to reproduce sexually is offered below.

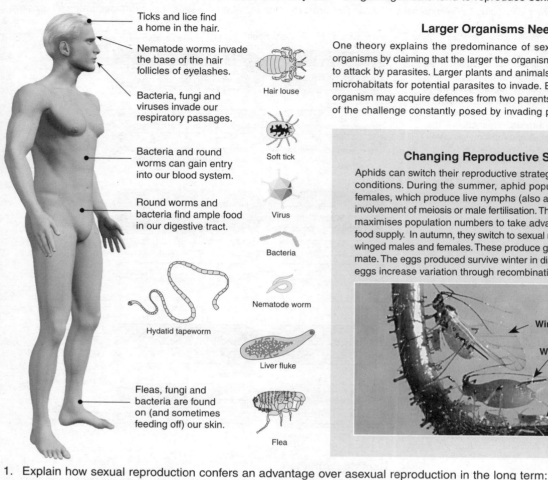

Ticks and lice find a home in the hair.

Nematode worms invade the base of the hair follicles of eyelashes.

Bacteria, fungi and viruses invade our respiratory passages.

Bacteria and round worms can gain entry into our blood system.

Round worms and bacteria find ample food in our digestive tract.

Fleas, fungi and bacteria are found on (and sometimes feeding off) our skin.

Hair louse

Soft tick

Virus

Bacteria

Nematode worm

Hydatid tapeworm

Liver fluke

Flea

Larger Organisms Need Sex

One theory explains the predominance of sexual reproduction in larger organisms by claiming that the larger the organism, the greater its vulnerability to attack by parasites. Larger plants and animals provide a greater range of microhabitats for potential parasites to invade. By reproducing sexually, an organism may acquire defences from two parents and keep one jump ahead of the challenge constantly posed by invading plant and animal parasites.

Changing Reproductive Strategies

Aphids can switch their reproductive strategy to suit the prevailing conditions. During the summer, aphid populations comprise only females, which produce live nymphs (also all females), without the involvement of meiosis or male fertilisation. This asexual reproduction maximises population numbers to take advantage of the abundant food supply. In autumn, they switch to sexual reproduction, producing winged males and females. These produce gametes by meiosis and mate. The eggs produced survive winter in diapause. These "sexual" eggs increase variation through recombination of alleles.

Winged male

Wingless adult

Nymph

1. Explain how sexual reproduction confers an advantage over asexual reproduction in the long term:

   _____

   _____

2. (a) Explain why species that reproduce asexually can out-perform sexually reproducing organisms in some situations:

   _____

   _____

   (b) Describe a situation where species that reproduce asexually might not be favoured: _____

   _____

3. Contrast **sexual** and **asexual** reproduction with respect to the inheritance of favourable mutations: _____

   _____

   _____

   _____

4. Explain how a reproductive strategy involving **alternation** of sexual and asexual phases might benefit a species:

   _____

   _____

   _____

   _____

Related activities: For Harm or Benefit?, Alleles, Sexual Selection

A 2

Crossing Over

Crossing over refers to the mutual exchange of pieces of chromosome and involves the swapping of whole groups of genes between the **homologous** chromosomes. This process can occur only during the first division of **meiosis**. Errors in crossing over can result in **chromosome mutations** (see the activity on this on the TRC), which can be very damaging to development. Crossing over can upset expected frequencies of offspring in dihybrid crosses. The frequency of crossing over (COV) for different genes (as followed by inherited, observable traits) can be used to determine the relative positions of genes on a chromosome and provide a **genetic map**. There has been a recent suggestion that crossing over may be necessary to ensure accurate cell division.

Pairing of Homologous Chromosomes

Every somatic cell has a pair of each type of chromosome in its nucleus. These chromosome pairs, one from each parent, are called **homologous** pairs or **homologues**. In prophase of the first division of **meiosis**, the homologues pair up to form **bivalents** in a process called **synapsis**. This allows the chromatids of the homologous chromosomes to come in very close contact.

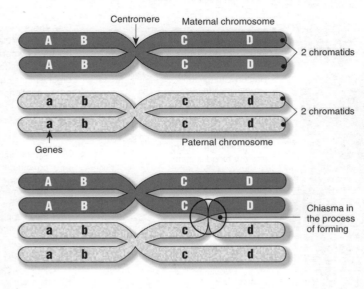

Chiasma Formation and Crossing Over

The pairing of the homologues allows **chiasmata** to form between the chromatids of homologous chromosomes. These are places where the chromatids become criss-crossed and the chromosomes exchange segments. In the diagram, the chiasma are in the process of forming and the exchange of pieces of chromosome have not yet taken place. Every point where the chromatids have crossed is a **chiasma**.

Separation

New combinations of genes arise from crossing over, resulting in what is called **recombination**. When the homologues separate at anaphase of meiosis I, each of the chromosomes pictured will have new genetic material (mixed types) that will be passed into the gametes soon to be formed. This process of recombination is an important source of variation for the gene pool of a population.

Gamete Formation

Once the final division of meiosis is complete, the two chromatids that made up each replicated chromosome become separated and are now referred to as chromosomes. Because chromatid segments were exchanged, **four** chromosomes that are quite different (genetically) are produced. If no crossing over had occurred, there would have been only two types (two copies of each). Each of these chromsomes will end up in a different gamete (sperm or egg).

1. Briefly explain how the process of crossing over is going to alter the genotype of gametes: _____

_____

2. Describe the importance of crossing over in the process of evolution: _____

_____

Related activities: Crossing Over Problems, Chromosome Mapping

Crossing Over Problems

The diagram below shows a pair of homologous chromosomes about to undergo chiasma formation during the first division of meiosis. There are known crossover points along the length of the chromatids (same on all four chromatids shown in the diagram). In the prepared spaces below, draw the gene sequences after crossing over has occurred on three unrelated and separate occasions (it would be useful to use different coloured pens to represent the genes from the two different chromosomes). See the diagrams on the previous page as a guide.

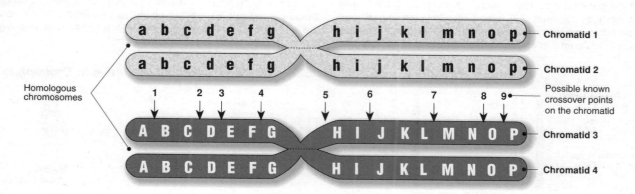

Sources of Variation

1. Crossing over occurs at a **single** point between the chromosomes above.

   (a) Draw the gene sequences for the four chromatids (on the right), after crossing over has occurred at crossover point: **2**

   (b) List which genes have been exchanged with those on its homologue (neighbour chromosome):

   _____

2. Crossing over occurs at **two** points between the chromosomes above.

   (a) Draw the gene sequences for the four chromatids (on the right), after crossing over has occurred between crossover points: **6** and **7**.

   (b) List which genes have been exchanged with those on its homologue (neighbour chromosome):

   _____

3. Crossing over occurs at **four** points between the chromosomes above.

   (a) Draw the gene sequences for the four chromatids (on the right), after crossing over has occurred between crossover points: **1** and **3**, and **5** and **7**.

   (b) List which genes have been exchanged with those on its homologue (neighbour chromosome):

   _____

4. Explain the genetic significance of **crossing over**: _____

_____

_____

_____

Linkage

Linkage refers to genes that are located on the same chromosome. Linked genes tend to be inherited together and fewer genetic combinations of their alleles are possible. Linkage reduces the variety of offspring that can be produced (contrast this with recombination). In genetic crosses, linkage is indicated when a greater proportion of the progeny resulting from a cross are of the parental type (than would be expected if the alleles were assorting independently). If the genes in question had been on separate chromosomes, there would have been more genetic variation in the gametes and therefore in the offspring. Note that

in the example below, wild type alleles are dominant and are denoted by an upper case symbol of the mutant phenotype (Cu or Eb). This symbology used for *Drosophila* departs from the convention of using the dominant gene to provide the symbol. This is necessary because there are many mutant alternative phenotypes to the wild type (e.g. curled and vestigial wings). A lower case symbol of the wild type (e.g. ss for straight wing), would not indicate the mutant phenotype involved. Alternatively, the wild type is sometimes denoted with a raised plus sign e.g. cu⁺cu⁺ and all symbols are in lower case.

Overview of Linkage

An Example of Linked Genes in *Drosophila*

The genes for wing shape and body colour are linked (they are on the same chromosome).

Sex of offspring is irrelevant in this case

Contact **Newbyte Educational Software** for details of their superb *Drosophila Genetics* software package which includes coverage of linkage and recombination. *Drosophila* images © Newbyte Educational Software.

Possible offspring
Only two kinds of genotype combinations are possible

1. Describe the effect of **linkage** on the inheritance of genes: _____

_____

2. (a) List the possible genotypes in the offspring (above, left) if genes A and B had been on **separate chromosomes**:

_____

(b) If the female *Drosophila* had been homozygous for the dominant wild type alleles (CuCu EbEb), state:

The genotype(s) of the F₁: _____ The phenotype(s) of the F₁: _____

3. Explain how linkage decreases the amount of genetic variation in the offspring: _____

_____

Recombination

Genetic recombination refers to the exchange of alleles between homologous chromosomes as a result of **crossing over**. The alleles of parental linkage groups separate and new associations of alleles are formed in the gametes. Offspring formed from these gametes show new combinations of characteristics and are known as **recombinants** (offspring with genotypes unlike either parent). The proportion of recombinants in the offspring can be used to calculate the frequency of recombination (crossover value). These values are fairly constant for any given pair of alleles and can be used to produce gene maps indicating the relative positions of genes on a chromosome. In contrast to linkage, recombination increases genetic variation. Recombination between the alleles of parental linkage groups is indicated by the appearance of recombinants in the offspring, although not in the numbers that would be expected had the alleles been on separate chromosomes (independent assortment). The example below uses the same genotypes as the previous activity, *Linkage*, but in this case crossing over occurs between the alleles in a linkage group in one parent. The symbology is the same for both activities.

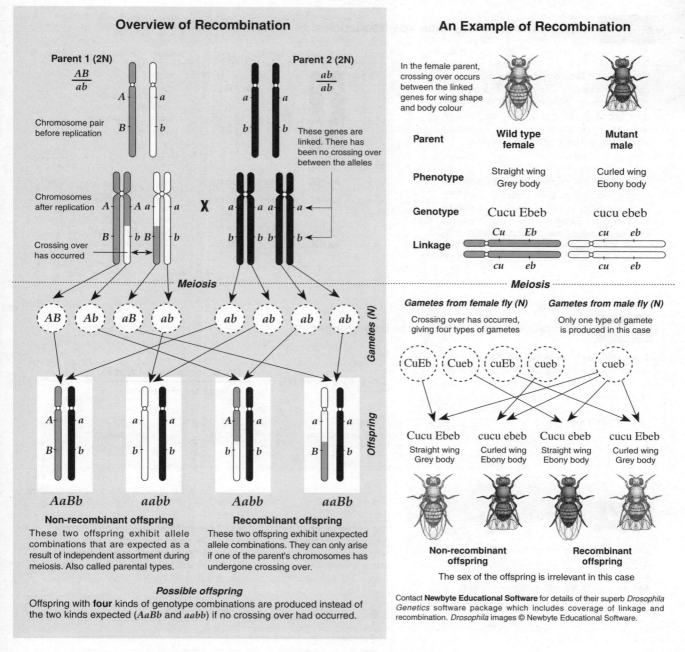

Overview of Recombination

Possible offspring

Offspring with **four** kinds of genotype combinations are produced instead of the two kinds expected (*AaBb* and *aabb*) if no crossing over had occurred.

An Example of Recombination

The sex of the offspring is irrelevant in this case

Contact **Newbyte Educational Software** for details of their superb *Drosophila Genetics* software package which includes coverage of linkage and recombination. *Drosophila* images © Newbyte Educational Software.

---

1. Describe the effect of **recombination** on the inheritance of genes: _____

_____

2. Explain how recombination increases the amount of genetic variation in offspring: _____

_____

3. Explain why it is not possible to have a recombination frequency of greater than 50% (half recombinant progeny):

_____

Chromosome Mapping

Crossover frequencies (crossover values or COV) can be used to map the relative positions of genes on a chromosome. Linked genes are genes that are located on the same chromosome, each found at a different **locus**. The amount of crossing over between linked genes is a direct measure of the distance between the genes on the chromosome. This is because genes that are further apart on the same chromosome have more potential crossing over points between them. Crossover values have traditionally been obtained as a result of test crosses for the genes concerned. A crossover value of 1% is equivalent to one unit of map distance (a *centimorgan cM*). To create a genetic map of a chromosome, two pieces of information are required: the **order** of the genes and the **relative** distances between them.

Mapping Genes Using Crossover Values

Step 1   Identify the offspring genotype frequencies

Parents with known genotypes: AaBb and aabb. *Without crossing over*, these parents can produce offspring of only two **genotypes**: AaBb and aabb. These in turn will produce only the two parental **phenotypes** (see the page on linkage).

However, if the progeny of these parents have 4 phenotypes, then crossing over must have occurred in the formation of some of the gametes. Progeny where such crossing over has occurred are called **recombinants**.

These two genes are linked (on the same chromosome)

AaBb	108	Parental genotype
aabb	97	Parental genotype

These genotypes will produce the parental phenotypes

Aabb	33	Recombinant
aaBb	42	Recombinant

These genotypes will produce non-parental phenotypes. Crossing over allows new allele combinations to come together.

Total   280

Step 2   Calculate the crossover values between genes

The frequency of recombination between the two genes (the number of offspring showing recombination) provides an indication of the relative distance between those genes. Expressed as a crossover value, it can be calculated as follows:

$$\text{Crossover Value (\%)} = \frac{\text{No. of recombinant types}}{\text{Total no. of offspring}} \times 100$$

Using the results from the offspring above: $\dfrac{33 + 42}{280} \times 100 = 26.8\%$ Crossover value between genes A and B

Step 3   Map the genes

Using the crossover value determined above, a chromosome map can be created, with each percent of crossover equivalent to one genetic map unit (a **centimorgan** or **cM**).

The **centimorgan** is so named after **Thomas Hunt Morgan**, an American geneticist who confirmed the ideas of earlier geneticists such as Sutton, and established the *chromosome theory of inheritance*. He worked with *Drosophila* and, in a series of breeding experiments, established the phenomenon of sex linkage. He also discovered crossover and devised the first chromosome map.

B ◄— 26.8 map units —► A

Known map distances for pairs of genes may be combined to produce a more complete map for many genes. In the example below, there are three genes for which the crossing over values are known. By trial and error, the genes could be laid out in different sequences to see if the genetic mapping distances (crossover frequencies) fit, but only one combination will work:

Crossover values:

B - A	22%
A - C	10%
B - C	12%

1. The following exercise involves some simple mapping problems. Use the procedure outlined on the previous page and the crossover values given to calculate the sequence and distance apart of the genes indicated. HINT: Always place on the map the two genes with the greatest mapping distance first, then plot the other genes relative to them.

Gene Map 1   Genes have the following crossover values:

X – G	18 %
B – G	3 %
B – X	15 %

Gene Map 2   Genes have the following crossover values:

P – m	7 %
K – m	28 %
P – K	21 %
m – C	5 %

Gene Map 3   Genes have the following crossover values:

Q – s	18 %
X – Q	18 %
X – s	36 %
Z – X	11 %
s – Z	25 %

2. In a cross between wild type female Drosophila flies (normal wing, grey body: *Vgvg Ebeb*) and male flies with vestigial wings and ebony bodies (*vgvgebeb*), the offspring were: *VgvgEbeb* = 585, *vgvgebeb* = 466, *vgvgEbeb* = 104, *Vgvgebeb* = 113.

(a) State the phenotypes of each genotype: _____

_____

(b) Calculate the crossover frequency (value): _____

3. Gene maps have traditionally been based on the recombinant frequencies obtained from breeding experiments. The mapping of the human genome is based on the presence of polymorphic markers (DNA with alternative forms, whose inheritance can be followed). The entire genomes of many organisms have already been mapped e.g. yeast, *E. coli,* and *Drosophila*). Describe some of the benefits to be gained from:

(a) Mapping the genome of non-human organisms: _____

_____

_____

(b) Mapping the human genome: _____

_____

_____

_____

Mutagens

Mutagens are chemicals or radiation that increase the likelihood of a mutation occurring. The rate of mutation induced by a mutagen is directly proportional to the dose received. Mutagens have a cumulative effect on an individual (i.e. small doses over a long period may be just as harmful as a single, larger dose). The four main classes of mutagens are outlined below.

Mutagen and Effect	Those Most at Risk
Ionising radiation Nuclear radiation from nuclear fallout, ultraviolet radiation from the sun and tanning lamps, X-rays and gamma rays from medical diagnosis and treatment. Ionising radiation is associated with the development of cancers, e.g. thyroid cancers and leukaemia (nuclear fallout), and skin cancer (high UV exposure).	• Nuclear radiation: Those working with radioisotopes, or living near nuclear plants, waste dumps, or testing sites. • UV radiation: Fair skinned people in tropical and sub-tropical regions. Those using tanning beds excessively. • X-rays, gamma rays: Early workers in radiology. Today, better protection and safer equipment has considerably lowered the risks to technicians and patients.
Viruses and microorganisms Some viruses integrate into the human chromosome, upsetting genes and triggering cancers. Examples include hepatitis B virus (liver cancer), HIV (Kaposi's sarcoma), and Epstein-Barr virus (Burkitt's lymphoma, Hodgkin's disease). Aflatoxins produced by the fungus *Aspergillus flavus* are potent inducers of liver cancer.	• Hepatitis B: Intravenous drug users. • HIV: Intravenous drug users, those with unsafe sexual activity (i.e. unprotected sex with new partners). • The development of Burkitt's lymphoma in response to infection with Epstein-Barr virus is triggered only after infection with malaria. Children in low-lying tropical regions are most at risk.
Alcohol and dietary components Diets high in fat, especially those containing burned and/or fatty, highly preserved meat, slow the passage of food through the gut giving time for mutagenic irritants to form in the bowel. High alcohol intake increases the risk of some cancers and also increases susceptibility to tobacco-smoking related cancers.	• Those with a diet high in total fats (particularly saturated fat) are at higher risk of developing bowel or rectal cancers. The risks may be increased by obesity. • The risks associated with specific dietary factors may be compounded by familial (inherited) susceptibility. • The risks associated with specific dietary factors may be compounded by other lifestyle choices, e.g. smoking.
Environmental poisons and irritants Many chemicals are mutagenic. Synthetic and natural examples include organic solvents such as benzene, asbestos, formaldehyde, tobacco tar, vinyl chlorides, coal tars, some dyes, and nitrites.	• Those working in the chemicals industries, including the glue, paint, rubber, resin, and leather industries. • Those in coal and other mining industries. • Petrol pump attendants exposed to petroleum volatiles and vehicle exhaust emissions. • Smokers and asbestos workers.

Tobacco smoking: Tobacco tar is one of the most damaging constituents of tobacco smoke. Tobacco tars contain at least 17 known carcinogens (cancer inducing mutagens) that cause chronic irritation of the respiratory system and are a major cause of cancer in smokers. In the last 5 years, there has been acknowledgement by the tobacco-producing companies that tobacco smoke is not only addictive but also carcinogenic.

Nuclear reactor failures: The most famous nuclear accident is that of Chornobyl, in the Ukraine, which suffered a catastrophic explosion and fire in 1986 during routine tests. Two explosions ejected some 8 tonnes of plutonium and other highly radioactive materials from the reactor. This led to widespread contamination of food sources, an increase in radiation-linked diseases, and ongoing environmental problems throughout northern parts of western Europe.

Ultraviolet rays from the sun: The ultraviolet rays (UV-A and UV-B) from the sun are particularly damaging to the skin and the retina of the eyes. The development of the ozone depleted region over Antarctica each year has increased the mutagenic effect from this source. One of the main diseases caused by this radiation is a form of skin cancer called melanoma. This is often an aggressive cancer that may be fatal if left untreated.

1. Discuss the role of **mutagens** and **carcinogens** in causing mutations: _____

_____

_____

_____

_____

A 2   **Related activities:** The Effect of Mutations

The Effect of Mutations

Mutations add, delete, or rearrange genetic material. They can happen spontaneously, as a result of errors during DNA replication, or they can be induced by **mutagens**. Not all mutations are inherited. Only those taking place in the cells that produce gametes will be inherited. If they occur in a body cell after the organism has begun to develop beyond the zygote stage, then they may give rise to a **chimera** (an organism with a mixture of genetically different cells). In some cases, mutations trigger the onset of **cancer** through the disruption of the normal controls regulating cell division.

Gametic Mutations

Sperm

Egg

Mutation

Fertilisation

Cleavage. Prior to implantation

Cells of tissues affected by the mutation

Foetus

Baby

Gametic mutations are *inherited* and occur in the testes of males and the ovaries of females.

Somatic Mutations

Sperm

Egg

Mutation

Somatic mutations occur in body cells. They are *not inherited* but may affect the person during their lifetime.

The Effect of Mutagens on DNA

UV Light

Thymine dimer

DNA of tumour suppressor gene

| G | C | T | T | G | C | A | G | T | A |
| C | G | A | A | C | G | T | C | A | T |

An example of how mutagens cause damage to the genes controlling the normal cell cycle is shown above. After exposure to UV light (a potent mutagen), adjacent thymine bases in DNA become cross-linked to form a 'thymine dimer'. This disrupts the normal base pairing and throws the controlling gene's instructions into chaos.

Mutant phenotype (gold colour)

Normal phenotype (red colour)

The photo above shows an example of a somatic mutation in a red delicious apple. A mutation occurred in the part of the flower that eventually developed into the fleshy part of the apple. The seeds would not be mutant.

1. Describe some of the general effects of **mutations**: _____

_____

_____

_____

2. Explain how **somatic mutations** differ from **gametic mutations** and comment on the significance of the difference:

_____

_____

_____

_____

3. Explain why the mutation seen in the red delicious apple (above right) will not be inherited: _____

_____

4. Explain why organisms such as *Drosophila* (fruit fly) and microorganisms are frequently used in the study of mutations:

_____

_____

_____

Related activities: Mutagens, For Harm or Benefit?

RA 2

For Harm or Benefit?

It is not correct to assume that all mutations are harmful. There are many documented cases where mutations conferring a survival advantage have arisen in a population. These **beneficial mutations** occur mostly amongst viruses and bacteria, but some (such as pesticide resistance) occur in multicellular organisms (especially those, like insects, with short **generation times**).

Mutations that are neither harmful nor beneficial are called **neutral** (silent) mutations. Neutral mutations can be virtually impossible to detect because they are not associated with any observable change in phenotype. If there is no selective pressure against a mutation, it may be carried silently in the population and prove to be beneficial (or harmful) at some time in the future.

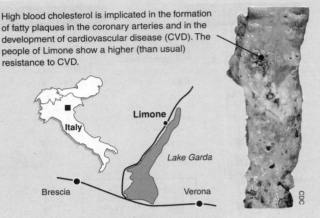

High blood cholesterol is implicated in the formation of fatty plaques in the coronary arteries and in the development of cardiovascular disease (CVD). The people of Limone show a higher (than usual) resistance to CVD.

Harmful Mutations

There are many well documented examples of mutations with harmful effects. These include the mutations giving rise to **cystic fibrosis** (CF) and **sickle cell disease**. The sickle cell mutation involves a change to only one base in a DNA sequence, whereas a common CF mutation involves the loss of a single triplet. The malformed proteins resulting from these mutations cannot carry out their normal functions. **Albinism** is caused by a mutation in the gene producing an enzyme in the metabolic pathway to **melanin**. It occurs in a large number of animals. The example above shows an albino python. Albinos are uncommon in the wild because they tend to be more vulnerable to predation and damaging UV radiation.

Beneficial Mutations

A well documented example of a beneficial mutation is the **tolerance to high cholesterol levels** in humans. In the small village of **Limone**, about 40 villagers have extraordinarily high levels of blood cholesterol, with no apparent harm to their coronary arteries. The village has 980 inhabitants and was largely isolated from the rest of the world until recently. The 40 villagers possess a mutation that alters a protein by **one amino acid** and makes it ten times more effective at mopping up excess cholesterol. Excess cholesterol is always disposed of, no matter what the dietary intake. All carriers of the mutation are related and are descended from one couple who arrived in Limone in 1636.

Neutral Mutations

Neutral mutations may be important in an evolutionary sense because they can be passed from generation to generation and may directly (or indirectly through subsequent mutation) be subject to selection pressure in the future. The example below shows how a change to the DNA sequence can be silenced if there is no change to the resulting amino acid sequence.

Amino acid sequence from the non-mutated DNA forms a normal polypeptide chain

Despite the change in the last base of a triplet, the amino acid sequence is unchanged

1. (a) Explain the difference between neutral (silent), beneficial, and harmful mutations: _____

_____

_____

(b) Identify which of these mutations is the most common and suggest why: _____

_____

2. Explain how the mutation that 40 of the villagers of Limone possess is beneficial under current environmental conditions:

_____

_____

Related activities: Metabolic Pathways, Antibiotic Resistance, Cystic Fibrosis Mutation, Sickle Cell Mutation

Gene Mutations

Gene mutations are small, localised changes in the structure of a DNA strand. These mutations may involve change in a single nucleotide (often called **point mutations**), or they may involve changes to a triplet (e.g. deletion or triplet repeat). If one amino acid in a protein is wrong, the biological function of the entire protein can be disrupted. Not all mutations may result in altered proteins. Because of the degeneracy of the genetic code, a substitution of the 3rd base in a codon may code for the same amino acid. The diagrams below and following show how various point mutations can occur. These alterations in the DNA are at the **nucleotide** level where individual **codons** are affected. Alteration of the precise nucleotide sequence of a coded gene in turn alters the mRNA transcribed from the mutated DNA and may affect the polypeptide chain that it normally creates.

Base Mismatching – Tautomerism

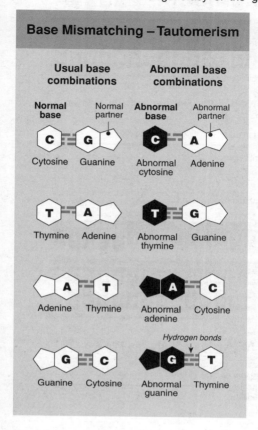

Usual base combinations

Normal base — Normal partner

C — G
Cytosine — Guanine

T — A
Thymine — Adenine

A — T
Adenine — Thymine

G — C
Guanine — Cytosine

Abnormal base combinations

Abnormal base — Abnormal partner

C — A
Abnormal cytosine — Adenine

T — G
Abnormal thymine — Guanine

A — C
Abnormal adenine — Cytosine

Hydrogen bonds

G — T
Abnormal guanine — Thymine

Base Mismatching

Watson and Crick proposed a theory of how base mismatching could occur. The diagram on the left shows suggested changes in bases and the resulting mismatch of complementary bases. On rare occasions some bases may have altered hydrogen-bond positions. As a result, during DNA replication, such abnormal bases pair with incorrect complementary bases. This gives rise to mutations in DNA molecules, which in turn are expressed as altered forms of mRNA and often altered proteins. NOTE: The abnormal bases on the right hand side of the diagram have a different arrangement of hydrogen bonds than normal.

A normal sequence without mutations

Original DNA: A A A A T G C T T C T C C A A G A T

Transcription

mRNA: U U U U A C G A A G A G G U U C U A

Translation

Amino acids: Phe — Tyr — Glu — Glu — Val — Leu

Amino acid sequence forms a normal polypeptide chain

Missense substitution

A single base is substituted for another base which may result in a codon that codes for a different amino acid. Some substitutions, however, may still code for the same amino acid, because of the high degree of degeneracy in the genetic code (i.e. many amino acids have 4 or 6 codons coding for them). In the illustrated example, placing a T where a C should have been, results in the amino acid **lysine** appearing where **glutamic** acid should be. This could affect how this protein functions.

Nonsense substitution

Some amino acids can be coded for by 4 or 6 different codons and are therefore less affected by substitutions. In the example illustrated, a single base substitution in the first nucleotide of the third codon has a dramatic effect on the nature of the polypeptide chain it is coding for. The codon no longer codes for an amino acid, but instead is an instruction for the termination of the translation process of protein synthesis. This results in a very short polypeptide chain that is likely to have little or no function since the **STOP** codon is introduced near the **START** codon.

Mutation: Substitute **T** instead of **C**

Original DNA: A A A A T G C T T C T C C A A G A T

Mutant DNA: A A A A T G T T T C T C C A A G A T

mRNA: U U U U A C A A A G A G G U U C U A

Amino acids: Phe — Tyr — Lys — Glu — Val — Leu

Polypeptide chain with wrong amino acid

Mutation: Substitute **A** instead of **C**

Original DNA: A A A A T G C T T C T C C A A G A T

Mutant DNA: A A A A T G A T T C T C C A A G A T

mRNA: U U U U A C U A A G A G G U U C U A

Amino acids: Phe — Tyr

Mutated DNA creates a STOP codon which prematurely ends synthesis of polypeptide chain

Related activities: For Harm or Benefit?, Cystic Fibrosis Mutation, Sickle Cell Mutation, Inherited Metabolic Disorders

RA 3

Mutations involving change in only one nucleotide may have no observable effect on the phenotype of the organism; the subtle changes in the DNA sequence may still produce a chain of identical amino acids in the protein, or at least produce a protein that is unaffected by the change. Because of the degeneracy of the genetic code, many mutations of this type are unlikely to cause any change in the biological activity of the protein (there are exceptions, e.g. sickle cell disease).

Reading frame shift by insertion

A major upset can occur when a single extra base is inserted into the DNA sequence. This has the effect of displacing all the other bases along one position and thereby creating a whole new sequence of codons. Such mutations are almost always likely to lead to a non-functional protein, but this does depend on the distance of the insertion or deletion from the START codon (i.e. the closer the insertion is to the START codon, the more the protein will be affected).

NOTE: could also lead to nonsense

Reading frame shift by deletion

In the same way that an insertion of an extra base into the DNA sequence has a large scale damaging effect, a deletion may also cause a frame shift. Again the result is usually a polypeptide chain of doubtful biological activity.

NOTE: could also lead to nonsense

Partial reading frame shift

Both an insertion and a deletion of bases within a gene can cause a frame shift effect where each codon no longer has the correct triplet of three bases. In this example, three codons have been affected, along with the amino acids they code for. The error is limited to the codons including and between the insertion and deletion. There is no biological activity if the amino acids altered are important to the functioning of the resulting protein.

Large scale frame shift resulting in a completely new sequence of amino acids. The resulting protein is unlikely to have any biological activity.

Large scale frame shift resulting in a completely new sequence of amino acids. The resulting protein is unlikely to have any biological activity.

Altered chain which may or may not produce a protein with biological activity

1. Explain what is meant by a **reading frame shift**: _____

_____

2. Not all gene mutations have the same effect on the organism, some are more disruptive than others.

   (a) Identify which type of gene mutations are the most damaging to an organism: _____

   (b) Explain why they are the most disruptive: _____

_____

3. Explain why **biological activity** of a protein might be affected by a reading frame shift: _____

_____

Antibiotic Resistance

Antibiotics are drugs that fight bacterial infections. After being discovered in the 1940s, they rapidly transformed medical care and dramatically reduced illness and death from bacterial disease. However, with increased and often indiscriminate antibiotic use, many bacteria quickly developed drug resistance. The increasing number of **multi-drug resistant** strains is particularly worrying; resistant infections inhibit the treatment of patients and increase mortality. Antibiotic resistance also adds considerably to the costs of treating disease and, as resistance spreads, new drugs have an increasingly limited life span during which they are effective. Resistant bacteria include *Klebsiella, Enterococcus, E. coli, Staphylococcus aureus, Enterobacter, Pseudomonas,* and *Mycobacterium tuberculosis.* Their resistance has developed through three main mechanisms (below right).

Methods by which Bacteria Acquire Resistance

Spontaneous resistance

Spontaneous mutation caused by radiation, chemicals, or transcription error.

Mutated gene codes for antibiotic resistance.

Conjugation

Plasmid is transferred via a pilus between the bacteria.

Plasmid giving resistance to antibiotic 1.

Plasmid gives resistance to antibiotic 2.

This bacterium contains plasmids that give resistance to both antibiotics 1 and 2.

Transduction

A virus has picked up an antibiotic resistance gene from another bacterium.

Bacterial DNA from the virus integrates into this cell's DNA, providing antibiotic resistance.

Transformation

Naked DNA containing a gene for antibiotic resistance is engulfed by the bacterium.

The naked DNA is taken in and integrated into the bacterial DNA, providing resistance.

Mechanisms of Resistance

Inactivation

A mutated enzyme produced in the bacterium destroys the antibiotic. Many bacteria that are resistant to penicillin possess such an enzyme (called penicillinase). Penicillinase inactivates penicillin by catalysing the destruction of bonds within the penicillin molecule, thereby inactivating it.

Alteration of target

Some antibiotics (e.g. streptomycin) inhibit bacterial protein synthesis. However, if only one amino acid in either of two positions on a ribosome is replaced, a bacterium can develop streptomycin resistance.

Some antibiotics, such as penicillin, interfere with cell wall synthesis. Therefore mutations to the cell wall proteins can result in resistance.

Alteration of permeability

In order to be effective, antibiotics have to get into the bacterial cell and interfere with its cellular processes. Bacterial cells can acquire resistance by excluding the antibiotic or by slowing down its entry enough to render the antibiotic ineffective.

Bacteria can develop proteins that actively pump antibiotics out of their cell faster than the antibiotics can enter.

1. Explain how spontaneous resistance can occur in a bacterium:

_____

_____

2. Explain how the misuse of antibiotics by patients can lead to the development of antibiotic resistant bacteria:

_____

_____

_____

Related activities: Features of Microbial Groups. Evolution in Bacteria

RA 2

Cystic Fibrosis Mutation

Cystic fibrosis an inherited disorder caused by a mutation of the **CF gene**. It is one of the most common lethal autosomal recessive conditions affecting caucasians, with an incidence in the UK of 1 in 2500 live births and a **carrier frequency** of 4%. It is uncommon in Asians and Africans. The CF gene's protein product, **CFTR**, is a membrane-based protein with a function in regulating the transport of chloride across the membrane. A faulty gene in turn codes for a faulty CFTR. More than 500 mutations of the CF gene have been described, giving rise to disease symptoms of varying severity. One mutation is particularly common and accounts for more than 70% of all defective CF genes. This mutation, called δ(delta)F508, leads to the absence of CFTR from its proper position in the membrane. This mutation is described below. Another CF mutation, R117H, which is also relatively common, produces a partially functional CFTR protein. The DNA sequence below is part of the transcribing sequence for the **normal** CF gene.

Normal CFTR *(1480 amino acids)*
Correctly controls chloride ion balance in the cell

Abnormal CFTR *(1479 amino acids)*
Unable to control chloride ion balance in the cell

The CF gene on chromosome 7

The CF gene is located on chromosome 7. The δF508 mutation of the CF gene describes a deletion of the 508th triplet, which in turn causes the loss of a single **amino acid** from the gene's protein product, the cystic fibrosis transmembrane conductance regulator (CFTR). This protein normally regulates the chloride channels in cell membranes, but the mutant form fails to achieve this. The portion of the DNA containing the mutation site is shown below:

The CFTR protein consists of 1480 amino acids

CFTR protein

The δF508 mutant form of CFTR fails to take up its position in the membrane. Its absence results in defective chloride transport and leads to a net increase in water absorption by the cell. This accounts for the symptoms of cystic fibrosis, where mucus-secreting glands, particularly in the lungs and pancreas, become fibrous and produce abnormally thick mucus. The widespread presence of CFTR throughout the body also explains why CF is a multisystem condition affecting many organs.

Base 1630
CFTR gene → DNA

C C G T G G T A A T T T C T T T T T A T A G T A G A A A C C A C C A

This triplet codes for the 500th amino acid

The 508th triplet is absent in the form with the δF508 mutation

1. (a) Write the mRNA sequence for the transcribing DNA strand above: _____

_____

   (b) Use a mRNA-amino acid table to determine the amino acid sequence coded by the mRNA for the fragment of the normal protein we are studying here (consult a textbook or *Advanced Biology AS*):

_____

2. (a) Rewrite the mRNA sequence for the mutant DNA strand: _____

_____

   (b) State what kind of mutation δF508 is: _____

   (c) Determine the amino acid sequence coded by the mRNA for the fragment of the δF508 mutant protein:

_____

   (d) Identify the amino acid that has been removed from the protein by this mutation: _____

3. Suggest why cystic fibrosis is a disease with varying degrees of severity: _____

_____

_____

Related activities: For Harm or Benefit?, Gene Mutations, Inherited Metabolic Disorders

Sickle Cell Mutation

Sickle cell disease (formerly called sickle cell anaemia) is an inherited disorder caused by a gene mutation which codes for a faulty beta (β) chain haemoglobin (Hb) protein. This in turn causes the red blood cells to deform causing a whole range of medical problems. The DNA sequence below is the beginning of the transcribing sequence for the **normal** β-chain Hb molecule.

Normal red blood cells

Each red blood cell contains about 270 million haemoglobin molecules. In their normal state, the red blood cells have a flattened disc shape.

Normal haemoglobin produces normal red blood cells

Mutant haemoglobin produces sickle-shaped red blood cells

Haemoglobin molecules are made up of 2 α-chains and 2 β-chains linked together

Sickle cells

The mutated form of haemoglobin is less soluble and precipitates when deprived of oxygen. This deforms the red blood cells to give them their sickle shape.

Location of the HBB gene on chromosome 11

HBB gene

p

q

The 438 nucleotides produce a protein made up of 146 amino acids

β-chain haemoglobin

Sickle cell disease

The effects of the sickle cells on human metabolism are considerable: sickle cells are rapidly removed from the circulation leading to jaundice and anaemia; a cascade of other ailments include heart defects, brain damage, kidney defects, skin lesions and enlargement of the spleen.

First base

Normal base: T
Substituted base: A

DNA

C A C G T G G A C T G A G G A C T C C T C

Codes for the 1st amino acid

The gene coding for the β-chain of haemoglobin protein is located on chromosome 11 and consists of 438 bases.

The sickle cell mutation involves the **substitution** of one base for another in the **HBB gene**, causing a single amino acid to be altered. This new amino acid has different properties and makes the haemoglobin behave in a different manner.

Sources of Variation

1. Identify how many of the following are exhibited or coded for in the DNA sequence above:

   (a) Bases: _____ (b) Triplets: _____ (c) Amino acids coded for: _____

2. Write the mRNA sequence for the transcribing DNA strand above.

   _____

3. Use a mRNA-amino acid table to determine the amino acids coded by the mRNA (in question 2 above) for the fragment of the **normal** protein we are studying here (consult a textbook or *Advanced Biology AS*):

   _____

   _____

4. Rewrite the transcribing DNA sequence above with the 17th nucleotide (base) changed from a **T** to **A.** This is the mutation that causes sickle cell disease.

   Mutant DNA: _____ Type of mutation: _____

5. Write the mRNA sequence for the **mutant** DNA strand above.

   _____

6. Use a mRNA-amino acid table to determine the amino acids coded by the mRNA (in question 5 above) for the fragment of the **mutant** protein we are studying here (consult a textbook or *Advanced Biology AS*):

   _____

   _____

7. Briefly explain what causes sickle cell disease: _____

   _____

   _____

Related activities: For Harm or Benefit?, Gene Mutations, Inherited Metabolic Disorders

RA 3

Inherited Metabolic Disorders

Humans have more than 6000 physiological diseases attributed to mutations in single genes and over one hundred syndromes known to be caused by chromosomal abnormality. The number of genetic disorders identified increases every year. The work of the Human Genome Project is enabling the identification of the genetic basis of these disorders. This will facilitate the development of new drug therapies and gene therapies. Four genetic disorders are summarised below.

Sickle Cell Disease	β-Thalassaemia	Cystic Fibrosis	Huntington Disease
Synonym: Sickle cell anaemia	**Synonyms**: Cooley anaemia, Mediterranean anaemia	**Synonyms**: Mucoviscidosis, CF	**Synonyms**: Huntington's chorea, HD (abbreviated)
Incidence: Occurs most commonly in people of African ancestry. West Africans: 1% (10-45% carriers) West Indians: 0.5%	**Incidence**: Most common type of thalassaemia affecting 1% of some populations. More common in Asia, Middle East and Mediterranean.	**Incidence**: Varies with populations: Northern Ireland: 1 in 1800 Asians in England: 1 in 10 000 Caucasians: 1 in 20-28 are carriers	**Incidence**: An uncommon genetic disease present in 1 in 20 000. The UK has about 8000 sufferers and some 48 000 at risk.
Gene type: Autosomal mutation which results in the substitution of a single nucleotide in the HBB gene that codes for the beta haemoglobin chain. The allele is codominant.	**Gene type**: Autosomal recessive mutation of the HBB gene coding for the haemoglobin beta chain. It may arise through a gene deletion or a nucleotide deletion or insertion.	**Gene type**: Autosomal recessive. Over 500 different recessive mutations (deletions, missense, nonsense, terminator codon) of the CFTR gene have been identified.	**Gene type**: An autosomal dominant mutation of the HD gene (IT15) caused by an increase in the length (36-125) of a CAG repeat region (normal range is 11-30 repeats).
Gene location: Chromosome 11 HBB	**Gene location**: Chromosome 11 HBB	**Gene location**: Chromosome 7 CFTR	**Gene location**: Chromosome 4 IT15
Symptoms: Include the following: pain, ranging from mild to severe, in the chest, joints, back, or abdomen; swollen hands and feet; jaundice; repeated infections, particularly pneumonia or meningitis; kidney failure; gallstones (at an early age); strokes (at an early age), anaemia.	**Symptoms**: The result of haemoglobin with few or no beta chains, causes a severe anaemia during the first few years of life. People with this condition are tired and pale because not enough oxygen reaches the cells.	**Symptoms**: Disruption of glands: the *pancreas*; *intestinal glands*; *biliary tree* (biliary cirrhosis); *bronchial glands* (chronic lung infections); and *sweat glands* (high salt content of which becomes depleted in a hot environment); *infertility* occurs in males/females.	**Symptoms**: Mutant gene forms defective protein: **Huntingtin**. Progressive, selective *nerve cell death* associated with chorea (jerky, involuntary movements), *psychiatric disorders,* and *dementia* (memory loss, disorientation, impaired ability to reason, and personality changes).
Treatment and outlook: Patients are given folic acid. Acute episodes may require oxygen therapy, intravenous infusions of fluid, and antibiotic drugs. Experimental therapies include bone marrow transplants and gene therapy.	**Treatment and outlook**: Patients require frequent blood transfusions which causes iron build-up in their heart, liver, and other organs. Bone marrow transplants and gene therapy hold promise and are probable future treatments.	**Treatment and outlook**: Conventional: chest physiotherapy, a modified diet, and the use of TOBI antibiotic to control lung infections. Outlook: Gene transfer therapy inserting normal CFTR gene using adenovirus vectors and liposomes.	**Treatment and outlook**: Surgical treatment may be possible. Research is underway to discover drugs that interfere with *Huntingtin* protein. Genetic counselling coupled with genetic screening of embryos may be developed into the future.

1. For each of the genetic disorder below, indicate the following:

    (a) Sickle cell disease: Gene name: __HBB__   Chromosome: __11__   Mutation type: __Substitution__

    (b) β-thalassaemia: Gene name: _____   Chromosome: _____   Mutation type: _____

    (c) Cystic fibrosis: Gene name: _____   Chromosome: _____   Mutation type: _____

    (d) Huntington disease: Gene name: _____   Chromosome: _____   Mutation type: _____

2. Explain the cause of the symptoms for people suffering from β-thalassaemia: _____

    _____

3. Suggest a reason for the differences in the country-specific incidence rates for some genetic disorders:

    _____

    _____

Related activities: Cystic Fibrosis Mutation, Sickle Cell Mutation, Implications of the HGP

Implications of the HGP

The **Human Genome Project** (HGP) is a publicly funded venture involving many different organisations throughout the world. In 1998, Celera Genomics in the USA began a competing project, as a commercial venture, in a race to be the first to determine the human genome sequence. In 2000, both organisations reached the first draft stage, and the entire genome is now available as a high quality (golden standard) sequence. In addition to determining the order of bases in the human genome, genes are being identified, sequenced, and mapped (their specific chromosomal location identified). The next challenge is to assign functions to the identified genes. By identifying and studying the protein products of genes (a field known as **proteomics**), scientists can develop a better understanding of genetic disorders. Long term benefits of the HGP are both medical and non-medical (see next page). Many biotechnology companies have taken out patents on gene sequences. This practice is controversial because it restricts the use of the sequence information to the patent holders. Other genome sequencing projects have arisen as a result of the initiative to sequence the human one. In 2002 the International HapMap Project was started with the aim of developing a haplotype map (HapMap) of the human genome. Initially data was gathered from four populations with African, Asian and European ancestry and additional populations may be included as analysis of human genetic variation continues.

Gene Mapping

This process involves determining the precise position of a gene on a chromosome. Once the position is known, it can be shown on a diagram.

One form of colour blindness

Production of a blood clotting factor

X Chromosome

Equipment used for DNA Sequencing

Banks of PCR machines prepare DNA for the sequencing gel stage. The DNA is amplified and chemically tagged (to make the DNA fluoresce and enable visualisation on a gel).

Banks of DNA sequencing gels and powerful computers are used to determine the base order in DNA.

Count of Mapped Genes

The length and number of mapped genes to date for each chromosome are tabulated below. The entire human genome contains approximately 20 000-25 000 genes.

Chromosome	Length (Mb)	No. of Mapped Genes
1	263	1873
2	255	1113
3	214	964
4	203	614
5	194	782
6	183	1217
7	171	995
8	155	591
9	145	803
10	144	872
11	144	1162
12	143	894
13	114	290
14	109	1013
15	106	510
16	98	658
17	92	1034
18	85	302
19	67	1129
20	72	599
21	50	386
22	56	501
X	164	1021
Y	59	122
Total:		**19 445**

As at: 3 June 2007 For an update see:
http://gdbwww.gdb.org/gdbreports/CountGeneByChromosome.html

Examples of Mapped Genes

The positions of an increasing number of genes have been mapped onto human chromosomes (see below). Sequence variations can cause or contribute to identifiable disorders. Note that chromosome 21 (the smallest human chromosome) has a relatively low gene density, while others are gene rich. This is possibly why trisomy 21 (Down syndrome) is one of the few viable human autosomal trisomies.

Key

☐ Variable regions (heterochromatin)

▨ Regions reflecting the unique patterns of light and dark bands seen on stained chromosomes

Rhesus blood type → Shape of red blood cells

ABO blood type

MN blood type

Production of amylase enzyme

Down syndrome, critical region

Structure of nails and kneecaps

Skin structure

Duffy blood type

Chromosome: 21   9   4   1

The aim of the HGP was to produce a continuous block of sequence information for each chromosome. Initially the sequence information was obtained to draft quality, with an error rate of 1 in 1000 bases. The **Gold Standard sequence**, with an error rate of <1 per 100 000 bases, was completed in October 2004. Key results of the research are:

- The analysis suggests that there are perhaps only 20 000-25 000 protein-coding genes in our human genome.
- The number of gaps has been reduced 400-fold to only 341
- It covers 99% of the gene containing parts of the genome and is 99.999% accurate.
- The new sequence correctly identifies almost all known genes (99.74%).
- Its accuracy and completeness allows systematic searches for causes of disease.

Related activities: Chromosome Mapping
Web links: Genetic Counselling

RA 3

Benefits and ethical issues arising from the Human Genome Project

Medical benefits

- Improved **diagnosis** of disease and predisposition to disease by genetic testing.
- Better identification of disease carriers, through genetic testing.
- Better **drugs** can be designed using knowledge of protein structure (from gene sequence information) rather than by trial and error.
- Greater possibility of successfully using **gene therapy** to correct genetic disorders.

Non-medical benefits

- Greater knowledge of **family relationships** through genetic testing, e.g. paternity testing in family courts.
- Advances **forensic science** through analysis of DNA at crime scenes.
- Improved knowledge of the evolutionary relationships between humans and other organisms, which will help to develop better, more accurate classification systems.

Possible ethical issues

- It is unclear whether third parties, e.g. health insurers, have rights to genetic test results.
- If treatment is unavailable for a disease, genetic knowledge about it may have no use.
- Genetic tests are costly, and there is no easy answer as to who should pay for them.
- Genetic information is hereditary so knowledge of an individual's own genome has implications for members of their family.

Couples can already have a limited range of genetic tests to determine the risk of having offspring with some disease-causing mutations.

When DNA sequences are available for humans and their ancestors, comparative analysis may provide clues about human evolution.

Legislation is needed to ensure that there is no discrimination on the basis of genetic information, e.g. at work or for health insurance.

1. Briefly describe the objectives of the Human Genome Project (HGP): _____

_____

_____

_____

2. Suggest a reason for developing a HapMap of the human genome: _____

_____

_____

3. Describe two possible **benefits** of Human Genome Project (HGP):

(a) Medical: _____

_____

(b) Non-medical: _____

_____

4. Explain what is meant by **proteomics** and explain its significance to the HGP and the ongoing benefits arising from it:

_____

_____

_____

_____

5. Suggest two possible points of view for one of the **ethical issues** described in the list above (top right):

(a) _____

_____

(b) _____

_____

Inheritance

AQA-A	AQA-B	SNAB	Edexcel	OCR
Complete: 1-14, 16-21, 25-27 **Extension:** 15	**Complete:** 1-14, 16, 19-21, 24, 27 **Extension:** 15, 17-18, 25-26	**Complete:** 1-14, 16-21, 25-27 **Extension:** 15	**Complete:** 1-14, 16, 19, 22-25 **Extension:** 15, 17-18, 26-27	**Complete:** 1-14, 16-21 **Extension:** 15, 25-27

Learning Objectives

☐ 1. Compile your own glossary from the **KEY WORDS** displayed in **bold type** in the learning objectives below.

Meiosis and Variation *(pages 103-104, 109-115)*

☐ 2. Describe the role of meiosis and fertilisation in contributing to genetic variation in the offspring. Appreciate the role of **crossing over** and **independent assortment** in meiosis, and explain when these occur. Coverage of this material, including the learning objectives is provided in *Sources of Variation*.

☐ 3. Appreciate that the **phenotype** is the result of the expression of the **genotype** (the genetic constitution of an organism) and the influence of the environment. Coverage of this material, including the learning objectives is provided in *Sources of Variation*.

The Study of Inheritance
(Required knowledge: pages 131, 133, 141-142)

☐ 4. Recall Mendel's principles of inheritance, stating their importance to the understanding of heredity: the *Theory of particulate inheritance*, the *Law of segregation*, and the *Law of independent assortment*.

☐ 5. Appreciate that Mendelian principles were important in providing a mechanism through which natural selection could act upon the genetic variation in populations.

☐ 6. Define the terms: **allele** and **locus**. Distinguish between **dominant**, **recessive**, and **codominant** alleles. Appreciate how new alleles are formed.

☐ 7. Define the term **trait** with respect to the study of genetics.

☐ 8. Describe how genes, and their dominant and recessive alleles, are represented symbolically by letters of the alphabet. Explain clearly what is meant by the terms: **heterozygous** and **homozygous**.

☐ 9. Distinguish between **genotype** and **phenotype** and use these terms appropriately when discussing the outcomes of genetic crosses.

☐ 10. Demonstrate an ability to use a **Punnett square** in the solution of different inheritance problems. Define the terms: **monohybrid**, and **dihybrid cross**.

☐ 11. Define and demonstrate an understanding of the terms commonly used in inheritance studies: **cross**, **carrier**, **selfing**, **pure-breeding**, **test-cross**, **back-cross**, **offspring** (progeny), F_1 **generation**, F_2 **generation**.

Inheritance Patterns *(pages 134-144, 155-157)*

For each of the cases below, use a Punnett square through to the F_2 generation and determine the probability of the occurrence of a particular genetic outcome (genotype and phenotype ratios).

☐ 12. Describe **monohybrid inheritance** involving a single trait or gene locus. Solve problems involving the inheritance of phenotypic traits that follow a simple **dominant-recessive** pattern. Define the term: **carrier**.

☐ 13. Distinguish between **codominance** and **incomplete dominance**, explaining how you would recognise these dominance patterns in genetic crosses.

☐ 14. Recognise that **multiple alleles** (alternative alleles) may exist for a single gene, e.g. ABO blood groups. Solve problems involving the inheritance of phenotypic traits involving **codominance** of multiple alleles.

☐ 15. Describe examples of recessive and/or dominant **lethal alleles**. Recognise how lethal alleles modify the ratio of progeny resulting from a cross. If required, explain the mechanisms by which the lethal allele operates.

☐ 16. Describe **dihybrid inheritance** involving two unlinked genes for two independent traits, where the genes described are carried on different chromosomes (**unlinked, autosomal genes**). Solve problems involving dihybrid inheritance of unlinked genes.

☐ 17. Demonstrate an ability to use the **chi-squared test** to test the significance of differences between the observed and expected results of genetic crosses.

☐ 18. Explain why experimental results may be expected only to approximate Mendelian ratios.

Sex Determination in Humans *(page 132)*

☐ 19. Understand the basis of **sex determination** in humans. Recognise humans as being of the **XX / XY** type. Distinguish **sex chromosomes** from **autosomes**.

Sex Linkage *(pages 145-147)*

☐ 20. Define the term: **sex linked**. Describe examples and solve problems involving different patterns of sex linked inheritance involving sex linked genes (e.g. inheritance of red-green colour-blindness, haemophilia, and the sex linked form of rickets).

☐ 21. Contrast the pattern of inheritance of **sex linked recessive traits** (e.g. haemophilia) and **sex linked dominant traits** (e.g. rickets).

Gene Interactions *(pages 148-157)*

☐ 22. Recognise and describe examples of a simple interaction between two genes: e.g. **collaboration** in the determination of **comb shape** of domestic hens. Show how genes that show collaboration influence the same trait, but produce a phenotype that could not result from the action of either gene independently. State the possible phenotypes for this type of interaction. Solve problems involving the inheritance of phenotypic traits involving collaboration.

☐ 23. Explain what is meant by **pleiotropy** and describe an example of a gene with pleiotropic effects, e.g. sickle cell gene mutation. Describe how the phenotype is affected by the presence of the sickle cell mutation.

☐ 24. Explain what is meant by **epistasis**, **epistatic gene**, and **hypostatic gene**. Solve problems involving the inheritance of phenotypic traits involving epistasis.

Epistasis case study 1: coat colour in mice. Several genes determine coat colour in mice. Mice that are homozygous recessive for the albino gene are white regardless of their genotype for coat colour.

Epistasis case study 2: where one gene can only be expressed in the presence of another. Such **complementary genes** control some flower colours (e.g. in sweet peas): one gene controls the production of a colourless intermediate (for a pigment) and another gene controls the transformation of that intermediate into the pigment. Both genes need to have a dominant allele present for the pigmented phenotype to be expressed.

☐ 25. Describe the distribution pattern of phenotypic variation produced by **polygenic inheritance**. Understand why **polygenes** are also called multiple genes. Provide examples of traits that follow this type of **quantitative inheritance** pattern.

☐ 26. Demonstrate genetic crosses for the inheritance of a polygenic trait (e.g. skin colour) using a Punnett square through to the F₂ generation and determine the probability of the occurrence of a particular genetic outcome (i.e. genotype and phenotype ratios).

☐ 27. Distinguish between **continuous** and **discontinuous variation** in phenotypes and recognise the genetic basis for each pattern.

 See the 'Textbook Reference Grid' on pages 8-9 for textbook page references relating to material in this topic.

Supplementary Texts

See pages 5-7 for additional details of these texts:

■ Adds, J., *et al.,* 2004. **Genetics, Evolution and Biodiversity**, (NelsonThornes), chpt. 6 as reqd.

■ Clegg, CJ., 1999. **Genetics and Evolution**, (John Murray), pp. 4-39 (in part).

■ Jones, N., *et al.,* 2001. **The Essentials of Genetics**, pp. 29-56, 61-74 (review), 77-110.

See page 7 for details of publishers of periodicals:

STUDENT'S REFERENCE

■ **What is a Gene?** Biol. Sci. Rev., 15(2) Nov. 2002, pp. 9-11. *A good introduction to genes and their role in heredity. This article provides a useful overview and an historical perspective.*

■ **What is Variation?** Biol. Sci. Rev., 13(1) Sept. 2000, pp. 30-31. *The nature of continuous and discontinuous variation in particular characters. The distribution pattern of traits that show continuous variation as a result of polygeny is discussed.*

■ **It Takes Two** New Scientist, 5 October 2002, pp. 34-37. *The relative influence of genes and environment. Studies of twins may help to clarify the nature versus nurture debate.*

■ **Determinedly Male** Biol. Sci. Rev., 7(3) Jan. 1995, pp. 28-31. *The basis of gender determination and sexual differentiation in animals.*

■ **Pay Attention Rover** New Scientist, 10 May 1997, pp. 30-33. *Genes account for about 25% of the differences between dogs in their ability to learn.*

■ **Queen Victoria's Gene** Biol. Sci. Rev., 8(2) November 1995, pp. 18-22. *Inheritance of haemophilia using pedigree analysis of Queen Victoria's as an example.*

■ **Secrets of The Gene** National Geographic, 196(4) Oct. 1999, pp. 42-75. *Thorough coverage of the nature of genes and the inheritance of particular genetic traits through certain populations.*

■ **Stuttering Gene Goes GAA GAA** New Scientist, 16 March 1996, p. 19. *Genetic stutters underlying conditions like Huntington's and fragile X syndrome show recessive inheritance patterns.*

■ **Beetles Show Why it Pays to Have Sex** New Scientist, 28 October 1995, p. 20. *New experiments with flour beetles show how sex benefits a population within a few generations.*

■ **Upwardly Mobile** New Scientist, 7 Nov. 1998, p. 6. *Recent studies indicate that 70% of the variation in adult height is due to one gene.*

■ **Blood Group Antigens** Biol. Sci. Rev., 9(5) May 1997, pp. 10-13. *Blood group antigens illustrate human genetic diversity at the molecular level.*

■ **Mendel's Legacy** Biol. Sci. Rev., 18(4) April 2006, pp. 34-37. *Examining the accuracy of Mendel's laws of inheritance compared with the current principles of genetics.*

TEACHER'S REFERENCE

Gender determination

■ **Why the Y is so Weird** Scientific American, Feb. 2001, pp. 42-47. *A comparison of the features of the X and Y chromosomes: why they look so different and how they behave during division.*

■ **The Double Life of Women** New Scientist, 10 May 2003, pp. 42-45. *Two X chromosomes make you a girl, but why are women not overloaded by having two. A good account of the role of X inactivation in embryonic development.*

■ **Decoding the Ys and Wherefores of Males** New Scientist, 21 June 2003, p. 15. *The publication of the genetic sequence of the Y chromosome has unveiled surprising information; the Y chromosome contains about 78 genes, and has mechanisms by which it can repair itself, generation to generation.*

Inheritance patterns

■ **Skin Deep** Scientific American, October 2002, pp. 50-57. *This article examines the evolution of skin colour in humans and presents powerful evidence for skin colour ("race") being the end result of opposing selection forces (the need for protection of folate from UV vs the need to absorb vitamin D). Clearly written and of high interest, this is a must for student discussion and a perfect vehicle for examining natural selection.*

■ **Does Race Exist?** Scientific American, December 2003, pp. 50-57. *The outward signs on which most definitions of race are based, such as skin colour and hair texture, are dictated by only a handful of genes. But the other genes of two people of the same 'race' can be very different, and two people of different race can share more genetic similarity than two of the 'same race'. The medical implications of this are still under debate.*

■ **Fair Enough** New Scientist, 12 October 2002, pp. 34-37. *The inheritance of skin colour in humans. This article examines why humans have such varied skin pigmentation and looks at the argument for there being a selective benefit to being dark or pale in different environments.*

■ **You Are What Your Mother Ate** New Scientist, 9 August 2003, pp. 14-15. *A few genes vital for development might be controlled by what your mother ate. This finding has intrigued scientists looking for the mechanisms behind epigenetic inheritance (the inheritance of phenotypic changes that are not the result of genomic changes).*

■ **The Unseen Genome: Beyond DNA** Scientific American, December 2003, pp. 78-85. *Most traits are transmitted by genes in DNA that encode proteins, but a separate code, written in chemical signatures outside DNA sequences, also influences health and appearance. This epigenetic code may explain why some diseases skip generations.*

■ **Where Did You Get Your Brains?** New Scientist, 3 May 1997, pp. 34-39. *What features do we inherit from our mothers and which from our fathers? The study of imprinted genes can make maternal/paternal inheritance clearer.*

■ **Spot Marks the X** Scientific American, April 1993, pp. 12-13. *The black and yellow patches in the calico cat's fur are the outward manifestations of a more subtle genetic quirk.*

See pages 10-11 for details of how to access **Bio Links** from our web site: **www.biozone.co.uk**. From Bio Links, access sites under the topics:

GENERAL BIOLOGY ONLINE RESOURCES
> **Online Textbooks and Lecture Notes**: • An on-line biology book • Kimball's biology pages • Learn.co.uk • Mark Rothery's biology web site ... *and others* > **General Online Biology resources**: • Ken's Bioweb resources • HowStuffWorks ... *and others* > **Glossaries**: • DNA glossary • Genetics glossary... *and others*

GENETICS: • MIT biology hypertextbook • Genetest • Gene Almanac • Virtual library on genetics ... *and others* > **Molecular Genetics**: • DNA from the beginning ... *and others* > **Inheritance**: • BioLogica • Coat color and pattern genetics of the domestic cat • Cat color genetics • Drag and drop genetics • Online Mendelian inheritance in man • Patterns of inheritance • The Biology project: Mendelian genetics • The role of genes ... *and others*

Presentation MEDIA to support this topic:
GENES & INHERITANCE
• Inheritance
• Gene Interactions

 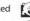

Alleles

Sexually reproducing organisms in nearly all cases have paired sets of chromosomes, one set coming from each parent. The equivalent chromosomes that form a pair are termed **homologues**. They contain equivalent sets of genes on them. But there is the potential for different versions of a gene to exist in a population and these are termed **alleles**.

Homologous Chromosomes

In sexually reproducing organisms, most cells have a homologous pair of chromosomes (one coming from each parent). The diagram below shows the position of three different genes on the same chromosome that control three different traits (A, B and C).

These two different versions of *gene A* create a condition known as **heterozygous**. Only the dominant allele (A) will be expressed.

When both chromosomes have identical copies of the dominant allele for *gene B* the organism is said to be **homozygous dominant** for that gene

When both chromosomes have identical copies of the recessive allele for *gene C* the organism is said to be **homozygous recessive** for that gene

Maternal chromosome that originated from the egg of this person's mother

Paternal chromosome that originated from the sperm of this person's father

Genes occupying the same **locus** or position on a chromosome code for the same trait

The diagram above shows the complete chromosome complement for a hypothetical organism. It has a total of ten chromosomes, comprising five, nearly identical pairs (each pair is numbered). One chromosome of each pair was supplied by each parent (mother and father). These pairs are called **homologues** or **homologous pairs**. Each homologue carries an identical assortment of genes, but the version (allele) of the gene from each parent may differ.

Inheritance

1. Define the following terms used to describe the allele combinations in the genotype for a given gene:

   (a) Heterozygous: _____

   (b) Homozygous dominant: _____

   (c) Homozygous recessive: _____

2. For a gene given the symbol 'A', name the alleles present in an organism that is identified as:

   (a) Heterozygous: _____ (b) Homozygous dominant: _____ (c) Homozygous recessive: _____

3. Explain what a homologous pair of chromosomes is: _____

   _____

   _____

4. Discuss the significance of genes existing as **alleles**: _____

   _____

   _____

   _____

   _____

Related activities: Meiosis

A 1

Sex Determination

The determination of the sex (gender) of an organism is controlled in most cases by the sex chromosomes provided by each parent. These have evolved to regulate the ratios of males and females produced and preserve the genetic differences between the sexes. In humans, males are referred to as the **heterogametic sex** because each somatic cell has one X chromosome and one Y chromosome. The determination of sex is based on the presence or absence of the Y chromosome. Without the Y chromosome, an individual will develop into a **homogametic** female (each somatic cell with two X chromosomes). In mammals, the male is always the heterogametic sex, but this is not necessarily the case in other taxa. In birds and butterflies, the female is the heterogametic sex, and in some insects the male is simply X whereas the female is XX.

Sex Determination in Humans

Y-bearing sperm cell

X-bearing sperm cell

X-bearing oocyte

X-bearing oocyte

XY: male

XX: female

The Sex Determining Region of the Y Chromosome

Scientists have known since 1959 that the Y chromosome is associated with being male. However, it was not until 1990 that a group of researchers, working for the Medical Research Council in London, discovered the gene on the Y chromosome that determines maleness. It was named **SRY**, for **Sex Determining Region of the Y**. The SRY gene produces a type of protein called a **transcription factor**. This transcription factor switches on the genes that direct the development of male structures in the embryo.

Y chromosome

SYR gene

X chromosome

Smaller and darker female half

Larger and lighter male half

Gynandromorphism occurs when an animal is a genetic mosaic and possesses both male and female characteristics (i.e. some of its cells are genetically male and others are female). This phenomenon is found particularly in insects, but also appears in birds and mammals. Gynandromorphism occurs due to the loss of an X chromosome in a stem cell of a female (XX), so that all the tissues derived from that cell are phenotypically male.

In the pill woodlouse, *Armadillium vulgare*, sex determination is characterised by female heterogamety (ZW) and male homogamety (ZZ). However, in several wild populations this system is overridden by an infectious bacterium. This bacterium causes genetically male woodlice to change into females. The bacteria are transmitted through the egg cytoplasm of the woodlouse. Therefore the conversion of males to females increases the propagation of the bacterium.

1. Explain what determines the sex of the offspring at the moment of conception in humans: _____

_____

2. Explain why human males are called the heterogametic sex: _____

_____

Basic Genetic Crosses

For revision purposes, examine the diagrams below on monohybrid crosses and complete the exercise for dihybrid (two gene) inheritance. A **test cross** is also provided to show how the genotype of a dominant phenotype can be determined. A test cross will yield one of two different results, depending on the genotype of the dominant individual. A **back cross** (not shown) refers to any cross between an offspring and one of its parents (or an individual genetically identical to one of its parents).

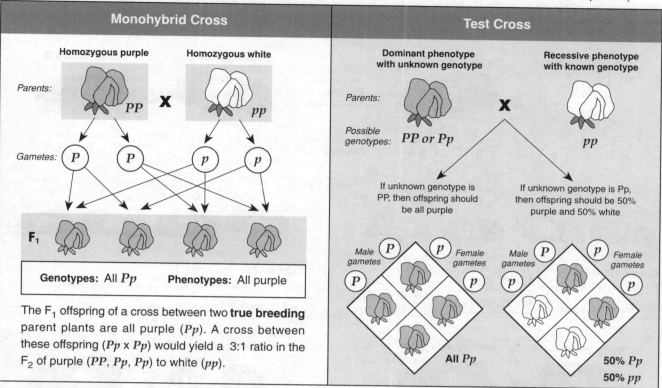

Monohybrid Cross

Homozygous purple **X** Homozygous white

PP pp

Parents:

Gametes: P P p p

F_1

Genotypes: All Pp **Phenotypes:** All purple

The F_1 offspring of a cross between two **true breeding** parent plants are all purple (Pp). A cross between these offspring (Pp x Pp) would yield a 3:1 ratio in the F_2 of purple (PP, Pp, Pp) to white (pp).

Test Cross

Dominant phenotype with unknown genotype **X** Recessive phenotype with known genotype

Parents:

Possible genotypes: PP or Pp pp

If unknown genotype is PP, then offspring should be all purple

If unknown genotype is Pp, then offspring should be 50% purple and 50% white

Male gametes P p Female gametes
P p

All Pp

Male gametes P p Female gametes
p p

50% Pp
50% pp

Inheritance

Dihybrid Cross

In pea seeds, yellow colour (Y) is dominant to green (y) and round shape (R) is dominant to wrinkled (r). Each **true breeding** parental plant has matching alleles for each of these characters ($YYRR$ or $yyrr$). F_1 offspring will all have the same genotype and phenotype (yellow-round: $YyRr$).

Parents: Homozygous yellow-round **X** Homozygous green-wrinkled

Gametes: YR yr

F_1 all yellow-round $YyRr$ **X** $YyRr$ for the F_2

1. Fill in the Punnett square (below right) to show the genotypes of the F_2 generation.

2. In the boxes below, use fractions to indicate the numbers of each phenotype produced from this cross.

   Yellow-round [ ]

   Green-round [ ]

   Yellow-wrinkled [ ]

   Green-wrinkled [ ]

3. Express these numbers as a ratio:

   _____

Offspring (F_2)   **Female gametes**

Possible fertilisations YR Yr yR yr

Male gametes: YR Yr yR yr

Related activities: Monohybrid Cross, Dihybrid Cross

A 2

Monohybrid Cross

The study of **single-gene inheritance** is achieved by performing **monohybrid crosses**. The six basic types of matings possible among the three genotypes can be observed by studying a pair of alleles that govern coat colour in the guinea pig. A dominant allele: given the symbol **B** produces **black** hair, and its recessive allele: **b**, produces white. Each of the parents can produce two types of gamete by the process of **meiosis** (in reality there are four, but you get identical pairs). Determine the **genotype** and **phenotype frequencies** for the crosses below (enter the frequencies in the spaces provided). For crosses 3 to 6, you must also determine gametes produced by each parent (write these in the circles), and offspring (F₁) genotypes and phenotypes (write in the genotype inside the offspring and state if black or white).

Cross 1:
(a) Genotype frequency: **100% Bb**
(b) Phenotype frequency: **100% black**

Parents / Gametes / Possible fertilisations / Offspring (F₁)

Cross 2:
(a) Genotype frequency: _____
(b) Phenotype frequency: _____

Cross 3:
(a) Genotype frequency: _____
(b) Phenotype frequency: _____

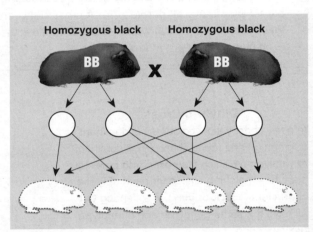

Cross 4:
(a) Genotype frequency: _____
(b) Phenotype frequency: _____

Cross 5:
(a) Genotype frequency: _____
(b) Phenotype frequency: _____

Cross 6:
(a) Genotype frequency: _____
(b) Phenotype frequency: _____

Related activities: Basic Genetic Crosses

Dominance of Alleles

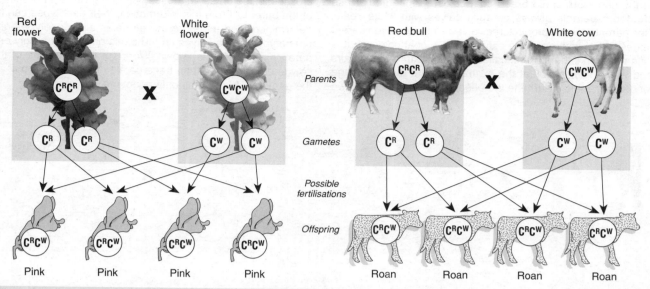

Incomplete Dominance

Incomplete dominance refers to the situation where the action of one allele does not completely mask the action of the other and neither allele has dominant control over the trait. The heterozygous offspring is **intermediate** in phenotype between the contrasting homozygous parental phenotypes. In crosses involving incomplete dominance the phenotype and genotype ratios are identical. Examples include snapdragons (*Antirrhinum*), where red and white-flowered parent plants are crossed to produce pink-flowered offspring. In this type of inheritance the phenotype of the offspring results from the partial influence of both alleles.

Codominance

Codominance refers to inheritance patterns when both alleles in a heterozygous organism contribute to the phenotype. Both alleles are **independently** and **equally expressed**. One example includes the human blood group AB which is the result of two alleles: A and B, both being equally expressed. Other examples include certain coat colours in horses and cattle. Reddish coat colour is not completely dominant to white. Animals that have both alleles have coats that are **roan**-coloured (coats with a mix of red and white hairs). The red hairs and white hairs are expressed equally and independently (not blended to produce pink).

Inheritance

1. In incomplete and codominance, two parents of differing phenotype produce offspring different from either parent. Explain the mechanism by which this occurs in:

   (a) Incomplete dominance: _____

   _____

   (b) Codominance: _____

   _____

2. For each situation below, explain how the heterozygous individuals differ in their phenotype from homozygous ones:

   (a) Incomplete dominance: _____

   (b) Codominance: _____

3. Describe the classical phenotypic ratio for a codominant gene resulting from the cross of two heterozygous parents (in the case of the cattle described above, this would be a cross between two roan cattle). Use the Punnett square (provided right) to help you:

   _____

   _____

   _____

Gametes from male

Gametes from female

4. A plant breeder wanted to produce flowers for sale that were only pink or white (i.e. no red). Determine the phenotypes of the two parents necessary to produce these desired offspring. Use the Punnett square (provided right) to help you:

   _____

   _____

Gametes from male

Gametes from female

Related activities: Multiple Alleles In Blood Groups

A 2

In the shorthorn cattle breed coat colour is inherited. White shorthorn parents always produce calves with white coats. Red parents always produce red calves. But when a red parent mates with a white one the calves have a coat colour that is different from either parent, called roan (a mixture of red hairs and white hairs). Look at the example on the previous page for guidance and determine the offspring for the following two crosses. In the cross on the left, you are given the phenotype of the parents. From this information, their genotypes can be determined, and therefore the gametes and genotypes and phenotypes of the calves. In the cross on the right, only one parent's phenotype is known. Work out the genotype of the cow and calves first, then trace back to the unknown bull via the gametes, to determine its genotype.

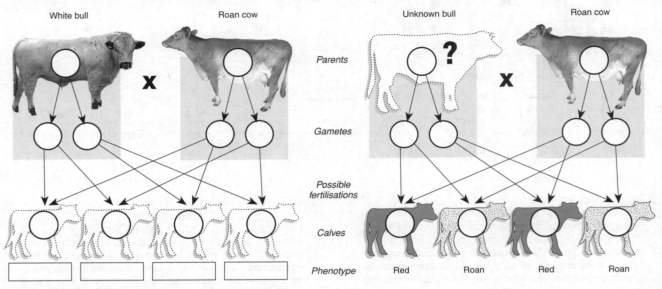

White bull   Roan cow   Unknown bull   Roan cow

Parents — Gametes — Possible fertilisations — Calves — Phenotype: Red   Roan   Red   Roan

5. A white bull is mated with a roan cow (above, left).

(a) Fill in the spaces on the diagram (above, left) to show the genotype and phenotype for parents and calves.

(b) State the phenotype ratio for this cross: _____

(c) Suggest how the farmer who owns these cattle could control the breeding so that the herd ultimately consisted of red coloured cattle only:

_____

_____

_____

6. A unknown bull is mated with a roan cow (above, right). A farmer has only roan shorthorn cows on his farm. He suspects that one of the bulls from his next door neighbours may have jumped the fence to mate with his cows earlier in the year. All the calves born were either red or roan. One neighbour has a red bull, the other has a roan bull.

(a) Fill in the spaces on the diagram (above, right) to show the genotype and phenotype for parents and calves.

(b) State which of the neighbour's bulls must have mated with the cows: **red** or **white** (*delete one*)

7. A plant breeder crossed two plants of the plant variety known as Japanese four o'clock. This plant is known to have its flower colour controlled by a gene which possesses incomplete dominant alleles. Pollen from a pink flowered plant was placed on the stigma of a red flowered plant.

Pink flower   Red flower

Parents

(a) Fill in the spaces on the diagram on the right to show the genotype and phenotype for parents and offspring.

Gametes

(b) State the phenotype ratio:

Possible fertilisations

_____

_____

Offspring

_____

Phenotype

Multiple Alleles in Blood Groups

The four common blood groups of the human 'ABO blood group system' are determined by three alleles: **A**, **B**, and **O** (also represented in some textbooks as: I^A, I^B, and i^O or just **i**). This is an example of a **multiple allele** system for a gene. The ABO antigens consist of sugars attached to the surface of red blood cells. The alleles code for enzymes (proteins) that join together these sugars. The allele **O** produces a non-functioning enzyme that is unable to make any changes to the basic antigen (sugar) molecule. The other two alleles *(A, B)* are **codominant** and are expressed equally. They each produce a different functional enzyme that adds a different, specific sugar to the basic sugar molecule. The blood group A and B antigens are able to react with antibodies present in the blood from other people and must be matched for transfusion.

Recessive allele: **O** produces a non-functioning protein
Dominant allele: **A** produces an enzyme which forms **A antigen**
Dominant allele: **B** produces an enzyme which forms **B antigen**

Blood group (phenotype)	Possible genotypes	Frequency in the UK
O	OO	45%
A	AA AO	43%
B		9%
AB		3%

If a person has the **AO** allele combination then their blood group will be group **A**. The presence of the recessive allele has no effect on the blood group in the presence of a dominant allele. Another possible allele combination that can create the same blood group is **AA**.

1. Use the information above to complete the table for the possible genotypes for blood group B and group AB.

2. Below are six crosses possible between couples of various blood group types. The first example has been completed for you. Complete the genotype and phenotype for the other five crosses below:

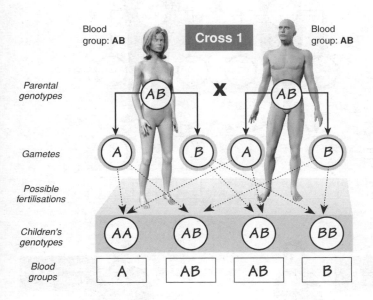

Cross 1

Blood group: **AB**   Blood group: **AB**

Parental genotypes: AB X AB

Gametes: A B A B

Possible fertilisations

Children's genotypes: AA AB AB BB

Blood groups: A AB AB B

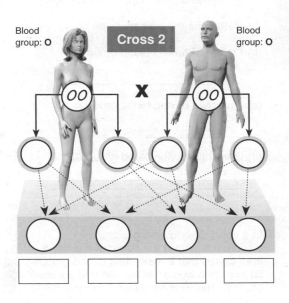

Cross 2

Blood group: **O**   Blood group: **O**

Parental genotypes: OO X OO

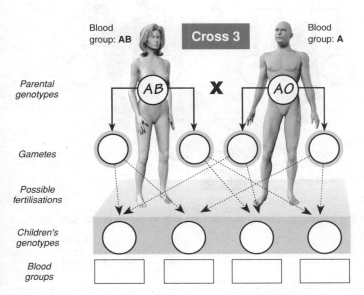

Cross 3

Blood group: **AB**   Blood group: **A**

Parental genotypes: AB X AO

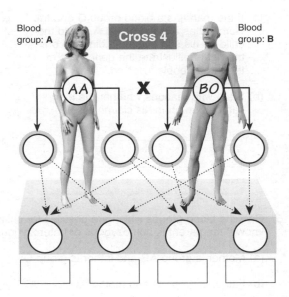

Cross 4

Blood group: **A**   Blood group: **B**

Parental genotypes: AA X BO

Parental genotypes
Gametes
Possible fertilisations
Children's genotypes
Blood groups

Inheritance

Related activities: Dominance of Alleles

A 2

Blood group: A — **Cross 5** — Blood group: O

Parental genotypes — AO X OO

Gametes

Possible fertilisations

Children's genotypes

Blood groups

Blood group: B — **Cross 6** — Blood group: O

Parental genotypes — BO X OO

Gametes

Possible fertilisations

Children's genotypes

Blood groups

3. A wife is heterozygous for blood group **A** and the husband has blood group **O**.

   (a) Give the genotypes of each parent (fill in spaces on the diagram on the right).

   Determine the probability of:

   (b) One child having blood group **O**:

   _____

   (c) One child having blood group **A**:

   _____

   (d) One child having blood group **AB**:

   _____

Blood group **A** X Blood group **O**

Parental genotypes

Gametes

Possible fertilisations

Children's genotypes

Blood groups

4. In a court case involving a paternity dispute (i.e. who is the father of a child) a man claims that a male child (blood group **B**) born to a woman is his son and wants custody. The woman claims that he is not the father.

   (a) If the man has a blood group **O** and the woman has a blood group **A**, could the child be his son? Use the diagram on the right to illustrate the genotypes of the three people involved.

   (b) State with reasons whether the man can be correct in his claim:

   _____

   _____

   _____

Blood group **A** X Blood group **O**

Parental genotypes

Gametes

Child's genotype

Blood group **B**

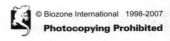

5. Give the blood groups which are possible for children of the following parents (remember that in some cases you don't know if the parent is homozygous or heterozygous).

   (a) Mother is group **AB** and father is group **O**: _____

   (b) Father is group **B** and mother is group **A**: _____

Dihybrid Cross

A cross (or mating) between two organisms where the inheritance patterns of **two genes** are studied is called a **dihybrid cross** (compared with the study of one gene in a monohybrid cross). There are a greater number of gamete types (four) produced when two genes are considered. Remember that the genes described are being carried by separate chromosomes and are sorted independently of each other during meiosis (that is why you get four kinds of gamete). The two genes below control two unrelated characteristics **hair colour** and **coat length**. Black and short are dominant.

Homozygous black, short hair

Homozygous white, long hair

Parents (P)   BBLL   X   bbll

Parents: The notation **P**, is only used for a cross between **true breeding** (homozygous) parents.

Gametes   BL BL BL BL   bl bl bl bl

Possible fertilisations

Gametes: Only one type of gamete is produced from each parent (although they will produce four gametes from each oocyte or spermatocyte). This is because each parent is homozygous for both traits.

Offspring (F1)   BbLl   X   BbLl

F1 offspring: There is only one **kind** of gamete from each parent, therefore only one kind of offspring produced in the first generation. The notation **F1** is only used to denote the heterozygous offspring of a cross between two true breeding parents.

F2 offspring: The F1 were mated with each other (**selfed**). Each individual from the F1 is able to produce four different kinds of gamete. Using a grid called a **Punnett square** (left), it is possible to determine the expected genotype and phenotype ratios in the F2 offspring. The notation **F2** is only used to denote the offspring produced by crossing F1 heterozygotes.

Offspring (F2)

Female gametes

Possible fertilisations

Male gametes	BL	Bl	bL	bl
BL	BBLL	BBLl	BbLL	BbLl
Bl	BBLl	BBll	BbLl	Bbll
bL	BbLL	BbLl	bbLL	bbLl
bl	BbLl	Bbll	bbLl	bbll

Each of the 16 animals shown here represents the possible zygotes formed by different combinations of gametes coming together at fertilisation.

The offspring can be arranged in groups with similar phenotypes:

Genotype

Phenotype

1	BBLL
2	BbLL
2	BBLl
4	BbLl

A total of 9 offspring with one of 4 different genotypes can produce black, short hair

9 black, short hair

| 1 | BBll |
| 2 | Bbll |

A total of 3 offspring with one of 2 different genotypes can produce black, long hair

3 black, long hair

| 1 | bbLL |
| 2 | bbLl |

A total of 3 offspring with one of 2 different genotypes can produce white, short hair

3 white, short hair

| 1 | bbll |

Only 1 offspring of a given genotype can produce white, long hair

1 white, long hair

Inheritance

Related activities: Basic Genetic Crosses

A 2

140

Cross Nº· 1

The dihybrid cross on the right has been partly worked out for you. You must determine:

1. The genotype and phenotype for each animal (write your answers in its dotted outline).

2. Genotype **ratio** of the offspring:

   _____

   _____

   _____

   _____

   _____

3. Phenotype **ratio** of the offspring:

   _____

   _____

   _____

   _____

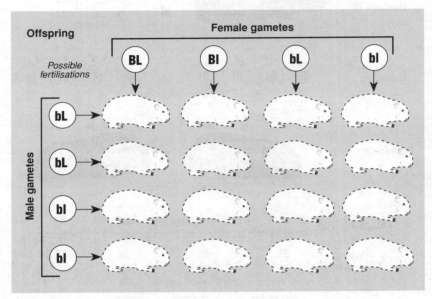

Cross Nº· 2

For the dihybrid cross on the right, determine:

1. Gametes produced by each parent (write these in the circles).

2. The genotype and phenotype for each animal (write your answers in its dotted outline).

3. Genotype **ratio** of the offspring:

   _____

   _____

   _____

4. Phenotype **ratio** of the offspring:

   _____

   _____

   _____

   _____

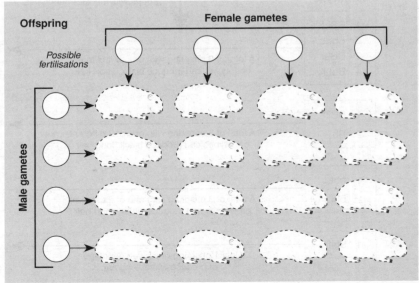

Problems in Mendelian Genetics

The following problems involve Mendelian crosses through to the F2 generation. The alleles involved are associated with various phenotypic traits in domestic breeds. See *Basic Genetic Crosses* if you need to review test crosses and back crosses.

1. The Himalayan colour-pointed, long-haired cat is a breed developed by crossing a pedigree (true-breeding), uniform-coloured, long-haired Persian with a pedigree colour-pointed (darker face, ears, paws, and tail) short-haired Siamese.
   The genes controlling hair colouring and length are on separate chromosomes: uniform colour **U**, colour pointed **u**, short hair **S**, long hair **s**.

 Persian   Siamese   Himalayan

   (a) Using the symbols above, indicate the genotype of each breed below its photograph (above, right). _____ _____ _____

   (b) State the genotype of the **F₁** (Siamese X Persian): _____

   (c) State the phenotype of the **F₁**: _____

   (d) Use the Punnett square to show the outcome of a cross between the F₁ (the F₂):

   (e) State the ratio of the F₂ that would be Himalayan: _____

   (f) State whether the Himalayan would be true breeding: _____

   (g) State the ratio of the F₂ that would be colour-point, short-haired cats: _____

   (h) Explain how two F₂ cats of the same phenotype could have different genotypes:

   _____

   _____

   (i) Explain how you could determine which of the F₂ colour-point, short-hairs were true breeding for these characters:

   _____

   _____

2. In rabbits, spotted coat **S** is dominant to solid colour **s,** while for coat colour: black **B** is dominant to brown **b**. A brown spotted rabbit is mated with a solid black one and all the offspring are black spotted (the genes are not linked).

   (a) State the genotypes: Male parent: _____ Female parent: _____ Offspring: _____

   (b) Use the Punnett square to show the outcome of a cross between the F₁ (the F₂):

   (c) Using ratios, state the phenotypes of the F₂ generation: _____

   _____

   (d) State the name given to this type of cross: _____

3. In guinea pigs, rough coat **R** is dominant over smooth coat **r** and black coat **B** is dominant over white **b**. The genes for coat texture and colour are not linked. In a cross of a homozygous rough black animal with a homozygous smooth white:

   (a) State the genotype of the **F₁**: _____

   (b) State the phenotype of the **F₁**: _____

   (c) Use the Punnett square to show the outcome of a cross between the F₁ (the F₂):

   (d) Using ratios, state the phenotypes of the F₂ generation:

   _____

   _____

   _____

Related activities: Basic Genetic Crosses, Dominance of Alleles, Lethal Alleles
Web links: Drag and Drop Genetics

RA 3

Inheritance

(e) Use the Punnett square (right) to show the outcome of a cross between the offspring of a **back cross** of the F$_1$ to the rough, black parent:

(f) Using ratios, state the phenotype of the F$_2$ generation:

_____

(g) Use the Punnett square to show the outcome of a cross between the offspring of a **test cross** of the F$_1$ to the smooth, white parent:

(h) Using ratios, state the phenotypes of the F$_2$ generation:

_____

_____

_____

(i) A rough black guinea pig was crossed with a rough white one produced the following offspring: 28 rough black, 31 rough white, 11 smooth black, and 10 smooth white. Determine the genotypes of the parents:

_____

_____

4. Chickens with shortened wings and legs are called creepers. When creepers are mated to normal birds, they produce creepers and normals with equal frequency. When creepers are mated to creepers they produce two creepers to one normal. Crosses between normal birds produce only normal progeny. Explain these results:

_____

_____

5. Black wool of sheep is due to a recessive allele (**b**), and white wool to its dominant allele (**B**). A white ram is crossed to a white ewe. Both animals carry the allele for black (b). They produce a white ram lamb, which is then back crossed to the female parent. Determine the probability of the **back cross** offspring being black:

_____

6. Mallard ducks have their plumage colour controlled by a gene with three alleles: **M^R** restricted mallard pattern, **M** mallard pattern, and **m** dusky mallard pattern. The dominance hierarchy is: **M^R** > **M** > **m** (i.e. **M^R** is more dominant than **M**, which is more dominant than **m**). Determine the genotypic and phenotypic ratios expected in the **F$_1$** of the following crosses:

(a) **M^{R}M^R X M^{R}M:**   Genotypes: _____

Phenotypes: _____

(b) **M^{R}M^R X M^{R}m:**   Genotypes: _____

Phenotypes: _____

(c) **M^{R}M   X M^{R}m:**   Genotypes: _____

Phenotypes: _____

(d) **M^{R}m   X Mm:**   Genotypes: _____

Phenotypes: _____

(e) **Mm   X mm:**   Genotypes: _____

Phenotypes: _____

7. A dominant gene (**W**) produces wire-haired texture in dogs; its recessive allele (**w**) produces smooth hair. A group of heterozygous wire-haired individuals are crossed and their F$_1$ progeny are then test-crossed. Determine the expected genotypic and phenotypic ratios among the **test cross** progeny:

_____

_____

Lethal Alleles

Lethal alleles are mutations of a gene that produce a gene product, which is not only nonfunctional, but may affect the organism's survival. Some lethal alleles are fully dominant and kill in one dose in the heterozygote. Others, such as in the **Manx** cat and yellow mice (below), produce viable offspring with a recognisable phenotype in the heterozygote. In some lethal alleles, the lethality is fully recessive and the alleles confer no detectable effect in the heterozygote at all. Furthermore, lethal alleles may take effect at different stages in development (e.g. in juveniles or, as in **Huntington disease**, in adults).

When **Lucien Cuenot** investigated inheritance of coat colour in yellow mice in 1905, he reported a peculiar pattern. When he mated two yellow mice, about $^2/_3$ of their offspring were yellow, and $^1/_3$ were non-yellow. This was a departure from the expected Mendelian ratio of 3:1. A test cross of the yellow offspring showed that they were all heterozygous.

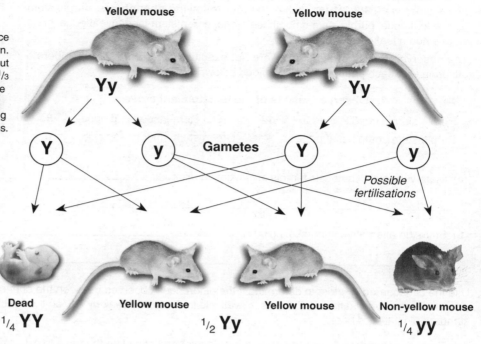

Yellow mouse   **Yellow mouse**

Yy   **Yy**

(Y) (y) **Gametes** (Y) (y)

Possible fertilisations

Dead   **Yellow mouse**   **Yellow mouse**   **Non-yellow mouse**

$^1/_4$ **YY** $^1/_2$ **Yy** $^1/_4$ **yy**

Inheritance

Cats possess a gene for producing a tail. The tailless **Manx** phenotype in cats is produced by an allele that is lethal in the homozygous state. The Manx allele **M^L** severely interferes with normal spinal development. In heterozygotes (**M^LM**), this results in the absence of a tail. In **M^LM^L** homozygotes, the double dose of the gene produces an extremely abnormal embryo, which does not survive.

Manx cats are born without a tail

Normal cats are born with a tail

Normal cat   **Manx cat**

1. Distinguish between recessive lethal alleles and dominant lethal alleles: _____

    _____

    _____

2. In Manx cats, the allele for taillessness (**M^L**) is incompletely dominant over the recessive allele for normal tail (**M**). Tailless Manx cats are heterozygous (**M^LM**) and carry a recessive allele for normal tail. Normal tailed cats are **MM**. A cross between two Manx (tailless) cats, produces two Manx to every one normal tailed cat (not a regular 3 to 1 Mendelian ratio).

    (a) State the genotypes arising from this type of cross: _____

    (b) State the phenotype ratio of Manx to normal cats and explain why it is not the expected 3:1 ratio: _____

    _____

3. Explain why Huntington disease persists in the human population when it is caused by a lethal, dominant allele:

    _____

    _____

Related activities: Dominance of Alleles, Inherited Metabolic Disorders

Using Chi-Squared in Genetics

The following problems examine the use of the chi-squared (χ^2) test in genetics. A worked example illustrating the use of the chi-squared test for a genetic cross is provided in *Advanced Biology AS*.

1. In a tomato plant experiment, two heterozygous individuals were crossed (the details of the cross are not relevant here). The predicted Mendelian ratios for the offspring of this cross were **9:3:3:1** for each of the **four following phenotypes**: purple stem-jagged leaf edge, purple stem-smooth leaf edge, green stem-jagged leaf edge, green stem-smooth leaf edge.

   The observed results of the cross were not exactly as predicted. The numbers of offspring with each phenotype are provided below:

Observed results of the tomato plant cross			
Purple stem-jagged leaf edge	**12**	Green stem-jagged leaf edge	**8**
Purple stem-smooth leaf edge	**9**	Green stem-smooth leaf edge	**0**

   (a) State your null hypothesis for this investigation (H0): _____

   _____

   (b) State the alternative hypothesis (HA): _____

   _____

2. Use the chi-squared (χ^2) test to determine if the differences observed between the phenotypes are significant. The table of critical values of χ^2 at different *P* values is available in *Advanced Biology AS*, on the Teacher Resource CD-ROM, or in any standard biostatistics book.

   (a) Enter the observed values (number of individuals) and complete the table to calculate the χ^2 value:

Category	O	E	O – E	(O – E)^2	$\dfrac{(O - E)^2}{E}$
Purple stem, jagged leaf					
Purple stem, smooth leaf					
Green stem, jagged leaf					
Green stem, smooth leaf					
Σ					Σ

   (b) Calculate χ^2 value using the equation:

   $$\chi^2 = \sum \frac{(O - E)^2}{E} \qquad \chi^2 = \underline{\hspace{2cm}}$$

   (c) Calculate the degrees of freedom: _____

   (d) Using a χ^2 table, state the *P* value corresponding to your calculated χ^2 value:

   _____

   (e) State your decision: reject H0 / do not reject H0

   (circle one)

3. Students carried out a pea plant experiment, where two heterozygous individuals were crossed. The predicted Mendelian ratios for the offspring were **9:3:3:1** for each of the **four following phenotypes**: round-yellow seed, round-green seed, wrinkled-yellow seed, wrinkled-green seed.

   The observed results were as follows:

Round-yellow seed	**441**	Wrinkled-yellow seed	**143**
Round-green seed	**159**	Wrinkled-green seed	**57**

   Use a separate piece of paper to complete the following:

   (a) State the null and alternative hypotheses (H0 and HA).

   (b) Calculate the χ^2 value.

   (c) Calculate the degrees of freedom and state the *P* value corresponding to your calculated χ^2 value.

   (d) State whether or not you reject your null hypothesis: reject H0 / do not reject H0 (circle one)

4. Comment on the whether the χ^2 values obtained above are similar. Suggest a reason for any difference:

   _____

   _____

Related activities: Using Chi-Squared in Ecology
Web link: Using the Chi-squared Test in Genetics

Sex Linked Genes

Sex linkage is a special case of linkage occurring when a gene is located on a sex chromosome (usually the X). The result of this is that the character encoded by the gene is usually seen only in one sex (the heterogametic sex) and occurs rarely in the homogametic sex. In humans, recessive sex linked genes are responsible for a number of heritable disorders in males, e.g. haemophilia. Women who have the recessive allele are said to be **carriers**. One of the loci controlling coat colour in cats is sex-linked. The two alleles, red (for orange colour) and non-red (for black colour), are found only on the X-chromosome.

Allele types

X_o = Non-red (=black)
X_O = Red

Genotypes

X_oX_o, X_oY = Black coated female, male
X_OX_O, X_OY = Orange coated female, male
X_oX_O = Tortoiseshell (intermingled black and orange in fur) in female cats only

Phenotypes

1. An owner of a cat is thinking of mating her black female cat with an orange male cat. Before she does this, she would like to know what possible coat colours could result from such a cross. Use the symbols above to fill in the diagram on the right. Summarise the possible genotypes and phenotypes of the kittens in the tables below.

	Genotypes	Phenotypes
Male kittens		

	Genotypes	Phenotypes
Female kittens		

Parent cats

Black female X **Orange male**

Gametes

Possible fertilisations (kittens)

2. A female tortoiseshell cat mated with an unknown male cat in the neighbourhood and has given birth to a litter of six kittens. The owner of this female cat wants to know what the appearance and the genotype of the father was of these kittens. Use the symbols above to fill in the diagram on the right. Also show the possible fertilisations by placing appropriate arrows.

Describe the father cat's:

(a) Genotype: _____

(b) Phenotype: _____

Tortoiseshell female Parent cats **Unknown male**

X ?

Gametes

Possible fertilisations (kittens)

2 orange females | 1 tortoiseshell female | 1 black male | 2 orange males

3. The owner of another cat, a black female, also wants to know which cat fathered her two tortoiseshell female and two black male kittens. Use the symbols above to fill in the diagram on the right. Show the possible fertilisations by placing appropriate arrows.

Describe the father cat's:

(a) Genotype: _____

(b) Phenotype: _____

(c) Was it the same male cat that fathered both this litter and the one above?
YES / NO *(delete one)*

Black female Parent cats **Unknown male**

X ?

Gametes

Possible fertilisations (kittens)

1 tortoiseshell female | 1 tortoiseshell female | 1 black male | 1 black male

Dominant allele in humans

A rare form of rickets in humans is determined by a **dominant** allele of a gene on the **X chromosome** (it is not found on the Y chromosome). This condition is not successfully treated with vitamin D therapy. The allele types, genotypes, and phenotypes are as follows:

Allele types		Genotypes	Phenotypes
X_R = affected by rickets		X_RX_R, X_RX =	Affected female
X = normal		X_RY =	Affected male
		XX, XY =	Normal female, male

As a genetic counsellor you are presented with a married couple where one of them has a family history of this disease. The husband is affected by this disease and the wife is normal. The couple, who are thinking of starting a family, would like to know what their chances are of having a child born with this condition. They would also like to know what the probabilities are of having an affected boy or affected girl. Use the symbols above to complete the diagram right and determine the probabilities stated below (expressed as a proportion or percentage).

4. Determine the probability of having:

   (a) Affected children: _____

   (b) An affected girl: _____

   (c) An affected boy: _____

Normal wife   Affected husband

Parents   X

Gametes

Possible fertilisations

Children

Another couple with a family history of the same disease also come in to see you to obtain genetic counselling. In this case the husband is normal and the wife is affected. The wife's father was not affected by this disease. Determine what their chances are of having a child born with this condition. They would also like to know what the probabilities are of having an affected boy or affected girl. Use the symbols above to complete the diagram right and determine the probabilities stated below (expressed as a proportion or percentage).

5. Determine the probability of having:

   (a) Affected children: _____

   (b) An affected girl: _____

   (c) An affected boy: _____

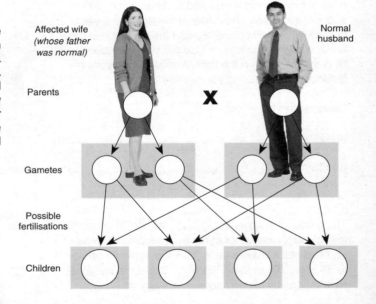

Affected wife *(whose father was normal)*   Normal husband

Parents   X

Gametes

Possible fertilisations

Children

6. Describing examples other than those above, discuss the role of **sex linkage** in the inheritance of genetic disorders:

_____

_____

_____

_____

_____

Inheritance Patterns

Complete the following monohybrid crosses for different types of inheritance patterns in humans: autosomal recessive, autosomal dominant, sex linked recessive, and sex linked dominant inheritance.

1. Inheritance of autosomal recessive traits
Example: *Albinism*

Albinism (lack of pigment in hair, eyes and skin) is inherited as an autosomal recessive allele (not sex linked).

Using the codes: **PP** (normal)
Pp (carrier)
pp (albino)

(a) Enter the parent phenotypes and complete the Punnett square for a cross between two carrier genotypes.

(b) Give the ratios for the phenotypes from this cross.

Phenotype ratios: _____

_____

2. Inheritance of autosomal dominant traits
Example: *Woolly hair*

Woolly hair is inherited as an autosomal dominant allele. Each affected individual will have at least one affected parent.

Using the codes: **WW** (woolly hair)
Ww (woolly hair, heterozygous)
ww (normal hair)

(a) Enter the parent phenotypes and complete the Punnett square for a cross between two heterozygous individuals.

(b) Give the ratios for the phenotypes from this cross.

Phenotype ratios: _____

_____

3. Inheritance of sex linked recessive traits
Example: *Haemophilia*

Inheritance of haemophilia is sex linked. Males with the recessive (haemophilia) allele, are affected. Females can be carriers.

Using the codes: **XX** (normal female)
XX$_h$ (carrier female)
X$_h$X$_h$ (haemophiliac female)
XY (normal male)
X$_h$Y (haemophiliac male)

(a) Enter the parent phenotypes and complete the Punnett square for a cross between a normal male and a carrier female.

(b) Give the ratios for the phenotypes from this cross:

Phenotype ratios: _____

_____

4. Inheritance of sex linked dominant traits
Example: *Sex linked form of rickets*

A rare form of rickets is inherited on the X chromosome.

Using the codes: **XX** (normal female); **XY** (normal male)
X$_R$X (affected heterozygote female)
X$_R$X$_R$ (affected female)
X$_R$Y (affected male)

(a) Enter the parent phenotypes and complete the Punnett square for a cross between an affected male and heterozygous female.

(b) Give the ratios for the phenotypes from this cross.

Phenotype ratios: _____

_____

Related activities: Monohybrid Cross, Sex Linkage

A 1

Inheritance

Interactions Between Genes

Genes   Product   Character

A
B   **Pleiotropy**
      Several characters are
C   affected by the single gene
D

Protein synthesis

E   **Polygeny**
      Several genes control the
      expression of the same
      character

F   **No interaction**
G   Some genes only control
      the expression of a single
H   character with no other
      input from other genes

Sickle Cell Disease as an Example of Pleiotropy

A person homozygous (**Hb^{S}Hb^S**) for the sickle cell allele produces mutant haemoglobin. This affects a number of organs.

Brain damage leading to seizures, stroke, or loss of consciousness.

Skin lesions

Heart and blood vessel damage

Spleen enlarges in childhood but atrophies later, leading to susceptibility to infection.

Kidney damage leading to blood in the urine.

Types of Gene Interaction

Some genes may not just control a single characteristic or trait in the phenotype of an organism. It is thought that most genes probably have an effect on one or more phenotypic traits, a phenomenon known as **pleiotropy**. In some cases a single characteristic may be controlled by more than one gene; a situation known as **polygeny**. A further possible type of gene interaction, called **epistasis**, involves two non-allelic genes (at different loci), where the action of one gene masks or otherwise alters the expression of other genes. An example is albinism, which appears in rodents that are homozygous recessive for colour even if they have the alleles for agouti or black fur (the gene for colour is epistatic and the hypostatic gene determines the nature of the colour).

Pleiotropy

A single gene may produce a product that can influence a number of traits in the phenotype of an organism. Such a gene is said to be **pleiotropic**. The gene Hb codes for production of haemoglobin, an important oxygen-carrying molecule in the blood. A point mutation to this gene produces sickle cell disease. The phenotype has poor oxygen-carrying capability and deformed red blood cells leading to haemolytic anaemia. A range of other organ abnormalities (above) also occur as a result of the mutant haemoglobin. In the diagram above, **Hb^S** stands for the single gene that codes for the mutated haemoglobin molecule. A person with sickle cell disease is homozygous for the mutated allele **Hb^{S}Hb^S**. This condition is eventually fatal. The normal genotype is HbHb.

1. Discuss the basic differences between **polygeny**, **pleiotropy**, and **epistasis**, giving examples to illustrate your answer:

_____

_____

_____

_____

_____

_____

2. (a) Describe the cause of sickle cell disease: _____

_____

(b) State the genotype of an affected individual: _____

(c) Describe the phenotype of an individual who is homozygous recessive for the sickle cell mutation: _____

_____

(d) Explain why the sickle cell gene is regarded as pleiotropic: _____

_____

Related activities: Sickle Cell Mutation, Polygenes, Epistasis
Summary of Gene Interactions

Epistasis

In its narrowest definition, **epistatic genes** are those that mask the effect of other genes. In this definition, they are also known as supplementary genes. Typically there are **three possible phenotypes** for a dihybrid cross involving this type of gene interaction. One example of this type of epistasis occurs between the genes controlling coat colour in rodents and other mammals. Skin and hair colour is the result of melanin, a pigment which may be either black/brown (eumelanin) or reddish/yellow (phaeomelanin). Melanin itself is synthesised via several biochemical steps from the amino acid tyrosine. The control of coat colour and patterning in mammals is complex and involves at least five major interacting genes. One of these genes (gene C), controls the production of the pigment melanin, while another gene (gene B), is responsible for whether the colour is black or brown (this interaction is illustrated for mice, below). Epistasis literally means "standing upon". In albinism, the homozygous recessive condition, cc, "stands upon" the other coat colour genes, blocking their expression.

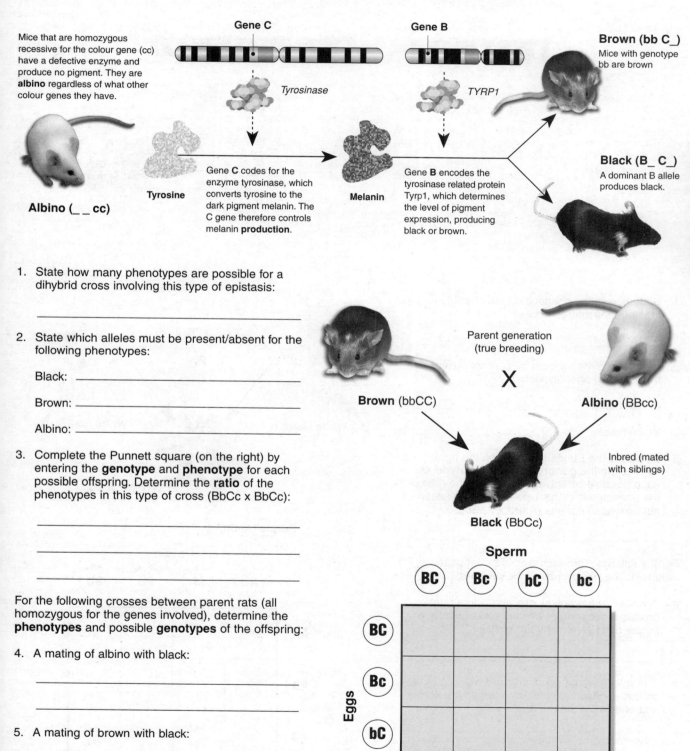

Mice that are homozygous recessive for the colour gene (cc) have a defective enzyme and produce no pigment. They are **albino** regardless of what other colour genes they have.

Gene C

Tyrosinase

Albino (_ _ cc)

Tyrosine

Gene **C** codes for the enzyme tyrosinase, which converts tyrosine to the dark pigment melanin. The C gene therefore controls melanin **production**.

Melanin

Gene B

TYRP1

Gene **B** encodes the tyrosinase related protein Tyrp1, which determines the level of pigment expression, producing black or brown.

Brown (bb C_)
Mice with genotype bb are brown

Black (B_ C_)
A dominant B allele produces black.

Inheritance

1. State how many phenotypes are possible for a dihybrid cross involving this type of epistasis:

   _____

2. State which alleles must be present/absent for the following phenotypes:

   Black: _____

   Brown: _____

   Albino: _____

3. Complete the Punnett square (on the right) by entering the **genotype** and **phenotype** for each possible offspring. Determine the **ratio** of the phenotypes in this type of cross (BbCc x BbCc):

   _____

   _____

   _____

For the following crosses between parent rats (all homozygous for the genes involved), determine the **phenotypes** and possible **genotypes** of the offspring:

4. A mating of albino with black:

   _____

   _____

5. A mating of brown with black:

   _____

   _____

   _____

Brown (bbCC)

X

Albino (BBcc)

Parent generation (true breeding)

Inbred (mated with siblings)

Black (BbCc)

Sperm

BC   Bc   bC   bc

Eggs

BC

Bc

bC

bc

Related activities: Dihybrid Cross, Summary of Gene Interactions

A 3

Complementary Genes

In some cases of epistasis, some genes can only be expressed in the presence of other genes: the genes are **complementary**. Both genes must have a dominant allele present for the final end product of the phenotype to be expressed. Typically, there are two possible phenotypes for this condition. An example of such an interaction would be if one gene controls the production of a pigment (intermediate) and another gene controls the

transformation of that intermediate into the pigment (by producing a controlling enzyme). Such genes have been found to control some flower colours. The diagram below right illustrates how one kind of flower colour in sweet peas is controlled by two complementary genes. The purple pigment is produced only in the presence of the dominant allele for each of the two genes. If a dominant is absent for either gene, then the flower is white.

Gene A — Protein synthesis — Enzyme A
Gene B — Protein synthesis — Enzyme B

Precursor substance (colourless) → Intermediate substance (colourless) → End product (purple)

1. State how many phenotypes are possible for this type of gene interaction:

   _____

2. State which alleles must be present/absent for the following phenotypes:

   Purple flower: _____

   White flower: _____

3. Complete the Punnett square (on the right) by entering the **genotype** and **phenotype** for each possible **offspring**. Determine the ratio of the phenotypes in this type of cross, between heterozygous parents (AaBb x AaBb):

   _____

For the following three crosses of sweet peas, determine the **genotypes** of the **parents**:

4. A white-flowered plant, crossed with a purple, produces offspring of which three-eighths are purple and five-eighths are white.

   _____

5. A purple-flowered plant, crossed with a purple, produces offspring of which one-half are purple and one-half white.

   _____

6. A white-flowered plant, crossed with another white, produces offspring of which three-fourths are white and one-fourth purple.

   _____

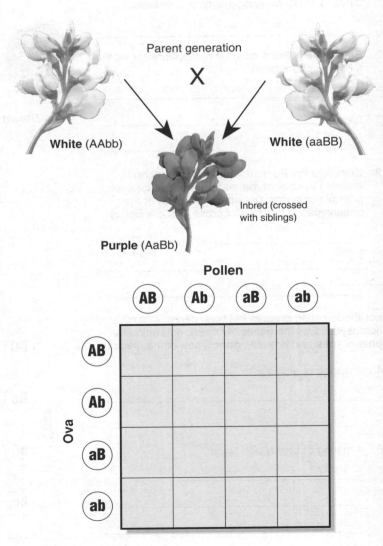

Parent generation

White (AAbb)   X   White (aaBB)

Purple (AaBb)

Inbred (crossed with siblings)

Pollen

	AB	Ab	aB	ab
AB				
Ab				
aB				
ab				

Ova

Related activities: Control of Metabolic Pathways, Dihybrid Cross, Summary of Gene Interactions

Polygenes

A single phenotype may be influenced, or determined, by more than one gene. Such phenotypes exhibit continuous variation in a population. Examples are skin colour and height, although the latter has been found to be influenced mostly by a single gene.

A light-skinned person   A dark-skinned person

In the diagram (right), the five possible phenotypes for skin colour are represented by nine genotypes. The production of the skin pigment melanin is controlled by two genes. The amount of pigment produced is directly proportional to the number of dominant alleles for either gene. No dominant allele results in an **albino** (aabb). Full pigmentation (black skin) requires four dominant alleles (AABB).

White	Light	Medium	Dark	Black
aabb	**Aabb**	**AAbb**	**AaBB**	**AABB**
	aaBb	**AaBb**	**AABb**	
		aaBB		

1. State how many phenotypes are possible for this type of gene interaction:

   _____

2. State which alleles must be present/absent for the following phenotypes:

   Black: _____

   Medium: _____

   White: _____

3. Complete the Punnett square (on the right) by entering the **genotype** and **phenotype** for each possible offspring. Determine the ratio of the phenotypes in this type of cross, between heterozygous parents (AaBb x AaBb):

   _____

   _____

   _____

   _____

For the following two crosses between humans, determine the phenotypes of the offspring:

4. A mating of white with black:

   _____

5. A mating between two individuals of medium skin colour:

   _____

6. In a polygenic inheritance illustrated above, two genes (A and B) are able to produce 5 phenotypes. Determine how many possible phenotypes could be produced if three genes were involved (i.e. genes A, B and C produce genotypes aabbcc, Aabbcc, etc.):

Parent generation

X

Black (AABB)   **White** (aabb)

F₁ generation

Breeding with an identical phenotype

Medium (AaBb)

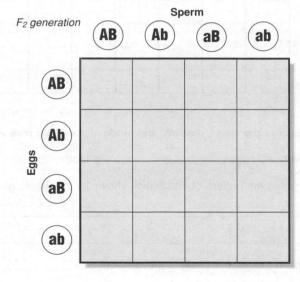

F₂ generation

Sperm

AB   Ab   aB   ab

Eggs

AB

Ab

aB

ab

   _____

   _____

   _____

7. Discuss the differences between **continuous** and **discontinuous** variation, giving examples to illustrate your answer:

_____

_____

_____

_____

_____

8. From a sample of no less than 30 adults, collect data (by request or measurement) for one continuous variable (e.g. height, weight, shoe size, or hand span). Record and tabulate your results in the space below, and then plot a frequency histogram of the data on the grid below:

Raw data

Tally Chart (frequency table)

Variable: _____

Frequency

(a) Calculate the mean, median, and mode of your data (see *Advanced Biology AS* if you need help):

Mean: _____ **Mode:** _____ **Median:** _____

(b) Describe the pattern of distribution shown by the graph, giving a reason for your answer: _____

_____

_____

(c) Explain the genetic basis of this distribution: _____

_____

(d) Explain the importance of a large sample size when gathering data relating to a continuous variable:

_____

Collaboration

There are genes that may influence the same trait, but produce a phenotype that could not result from the action of either gene independently. These are termed collaborative genes (they show collaboration). There are typically four possible phenotypes for this condition. An example of this type of interaction can be found in the comb shape of domestic hens.

The dash (missing allele) in the bold genotypes on the right indicates that the allele may be dominant or recessive, it will not affect the phenotype (i.e. the resulting character displayed).

Single comb

Genotypes:

rrpp

rrpp

Pea comb

Genotypes:

rrP_

rrPp, rrPP

Rose comb

Genotypes:

R_pp

Rrpp, RRpp

Walnut comb

Genotypes:

R_P_

RRPP, RrPP, RrPp, RRPp

1. State **how many phenotypes** are possible for this type of gene interaction:

_____

2. State which alleles must be present or absent for the following phenotypes:

   Pea comb: _____

   Rose comb: _____

   Single comb: _____

   Walnut comb: _____

3. Complete the Punnett square (on the right) by entering the **genotype** and **phenotype** for each possible **offspring**. Determine the ratio of the phenotypes in this type of cross, between heterozygous parents (RrPp x RrPp):

_____

4. Determine the **genotype** of the **parents** for each of the following crosses:

   (a) A rose crossed with a walnut produces offspring, 3/8 of which are walnut, 3/8 rose, 1/8 pea, and 1/8 single.

   _____

   (b) A walnut crossed with a single, produces in the F$_1$ generation, 1/4 walnut, 1/4 pea, 1/4 rose, and 1/4 single.

   _____

   (c) A rose crossed with a pea produces six walnut and five rose offspring.

   _____

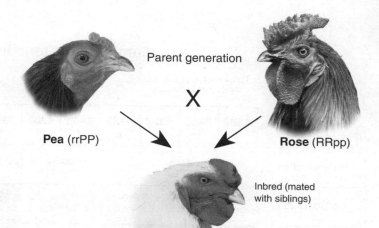

Parent generation

X

Pea (rrPP)   **Rose** (RRpp)

Inbred (mated with siblings)

Walnut (RrPp)

Sperm

	RP	Rp	rP	rp
RP				
Rp				
rP				
rp				

Eggs

Related activities: Dihybrid Cross, Summary of Gene Interactions

A 3

Summary of Gene Interactions

The table below presents various results for interactions involving two genes. Each gene has two alleles, dominant and recessive. The genes are unlinked and each one appears on a different chromosome. The first row shows no interaction, with an ordinary Mendelian F_2 ratio. In this case, genes **A** and **B** represent two genes for **two separate traits** as in the classical Mendelian experiments.

The second row shows a simple gene interaction producing a 9:3:3:1 ratio, but the genes at two independent loci are interacting to affect a **single character**. In the particular cross from which row 2 was taken, the F_1 (AaBb) had a walnut comb

shape, a phenotype not shown by any of the P generation (rose-combed hens: AAbb and pea-combed roosters: aaBB). The walnut comb is caused by the interaction between one or more dominant genes at each of the two loci. The double recessive (aabb) produces a single comb phenotype which did not appear in either the P or the F_1 generation.

All examples, apart from the first one, represent various types of gene interaction and the F_2 phenotypic ratios that result. In the **polygene** example, both A and B have the same effect on kernel colour, which is determined by the dosage of dominant alleles, ranging from four (for the deepest colour), to none (for white).

Various types of gene interaction are illustrated here for the offspring arising from a cross between two heterozygous parents (AaBb). Each of these gene interactions gives a characteristic ratio. For comparison, a dihybrid cross with no interaction is shown below:

Grandparents	**P**	**AAbb**	X	**aaBB**

Parents	**F₁**	**AaBb**	X	**AaBb** •

Parents chosen to produce the ratios in the table below (interbreeding of the F_1)

Offspring	**F₂**	**AABB** **AaBB** **AaBb** **AABb**	**AAbb** **Aabb**	**aaBB** **aaBb**	**aabb**

These genotypes are grouped into offspring with similar phenotypes following the classical 9:3:3:1 ratio. In this situation, there is **no interaction** between the two genes (see row 1 in the table below)

Summary of Gene Interactions

Number of offspring with this genotype

Possible genotypes from AaBb x AaBb cross

Ratio of offspring with each phenotype

Type of gene interaction	AABB 1	AABb 2	AaBB 2	AaBb 4	AAbb 1	Aabb 2	aaBB 1	aaBb 2	aabb 1	Dihybrid ratio	Organism	Trait
No interaction	Yellow round				Yellow angular		Green round		Green angular	9:3:3:1	Peas	Seed colour Seed shape
Collaboration	Walnut				Rose		Pea		Single	9:3:3:1	Chickens	Comb shape
Polygenes	Dark red	Medium dark red		Medium red		Light red	Medium red	Light red	White	1:4:6:4:1	Wheat	Kernel colour
Epistasis (complementary genes)	Purple				White					9:7	Sweet pea flowers	Colour
Epistasis (supplementary genes)	Black				Brown		White			9:3:4	Mice	Coat colour
Inhibitor genes (outside requirements)	White						Yellow		Green	12:3:1	Squash	Colour

Example

12 of the 16 offspring are likely to be **white**

3 of the 16 offspring are likely to be **yellow**

1 of the 16 offspring are likely to be **green**

Blocks with similar shading indicate genotypes that produce similar phenotypes. In this example: AABB, AABb, AaBB, AaBb, AAbb, Aabb; any with a dominant A.

1. Explain what is meant by a **gene interaction**: _____

   _____

2. Suggest what might lead you to suspect that there was an interaction between two genes: _____

   _____

Related activities: Interactions Between Genes, Epistasis, Complementary Genes, Polygenes, Collaboration

Inheritance in Domestic Cats

Cats have been domesticated for thousands of years. During this time, certain traits or characteristics have been considered fashionable or desirable in a cat by people in different parts of the world. In the domestic cat, the 'wild type' is the short-haired tabby. All other coat colours found in cats are modifications of this ancestral tabby pattern. Inheritance of coat characteristics and a few other features in cats is interesting because they exhibit the most common genetic phenomena. Some selected traits for domestic cats are identified below, together with a list of the kinds of genetic phenomena easily demonstrated in cats.

Inheritance Patterns in Domestic Cats

Dominance	The polydactylism gene with the dominant allele (Pd) produces a paw with extra digits.
Recessiveness	The dilution gene with the recessive allele (d) produces a diluted black to produce grey, or orange to cream.
Epistasis	The dominant agouti gene (A) must be present for the tabby gene (T) to be expressed.
Multiple alleles	The albino series (C) produces a range of phenotypes from full pigment intensity to true albino.
Incomplete dominance	The spotting gene (S) has three phenotypes ranging from extensive spotting to no spotting at all.
Lethal genes	The Manx gene (M) that produces a stubby or no tail is lethal when in the homozygous dominant condition (MM causes death in the womb).
Pleiotropy	The white gene (W) also affects eye colour and can cause congenital deafness (one gene with three effects).
Sex linkage	The orange gene (O) is sex (X) linked and can convert black pigment to orange. Since female cats have two X chromosomes they have three possible phenotypes (black, orange and tortoiseshell) whereas males can normally only exhibit two phenotypes (black and orange).
Environmental effects	The dark colour pointing in Siamese and Burmese cats where the gene (cs) is only active in the cooler extremities such as the paws, tail and face.

(NOTE: Some of these genetic phenomena are covered elsewhere)

Eyes
May have a range of colouring for the irises: blue, yellow, pink.

Ears
May be normal pointed ears, or the ears may be folded.

Coat colour
A wide range of coat colours are available, controlled by a variety of genes. Basic colours include black, white, orange and agouti. Colour patterns can range from solid, patched, spotted or tabby.

Coat length
Hair is usually either long or short. There is a breed with extremely short hair, so much so that it looks hairless (sphynx).

Coat texture
Smooth hair is the common phenotype, but there is an allele that causes curly hair.

Tail
Most cats have a long tail. An allele for short, stubby tails is almost completely restricted to the bobcat and Manx breeds.

Paws
Most cats have five digits on the front paw and four on the rear. The occurrence of polydactyly with as many as six or seven digits affects as many as one out of five cats (in some parts of the world it is even higher than this).

Genes controlling inherited traits in domestic cats

Wild forms		Mutant forms		Wild forms		Mutant forms	
Allele	*Phenotype*	*Allele*	*Phenotype*	*Allele*	*Phenotype*	*Allele*	*Phenotype*
A	Agouti	*a*	Black (non-agouti)	**m**	Normal tail	**M**	Manx tail, shorter than normal (stubby)
B	Black pigment	*b*	Brown pigment	**o**	Normal colours (no red, usually black)	**O**	Orange (sex linked)
C	Unicoloured	**cch**	Silver				
		cs	Siamese (pointing: dark at extremities)	**pd**	Normal number of toes	**Pd**	Polydactylism; has extra toes
		ca	Albino with blue eyes				
		c	Albino with pink eyes	**R**	Normal, smooth hair	**r**	Rex hair, curly
D	Dense pigment	*d*	Dilute pigment	**s**	Normal coat colour without white spots	**S**	Colour interspersed with white patches or spots (piebald white spotting)
fd	Normal, pointed ears	**Fd**	Folded ears				
Hr	Normal, full coat	*h*	Hairlessness	**T**	Tabby pattern (mackerel striped)	**Ta**	Abyssinian tabby
i	Fur coloured all over	*I*	Inhibitor: part of the hair is not coloured (silver)			**tb**	Blotched tabby, classic pattern of patches or stripes
L	Short hair	*l*	Long hair, longer than normal	**w**	Normal coat colour, not all white	**W**	All white coat colour (dominant white)
						Wh	Wirehair

Variation in Coat Colour in Domestic Cats

Non-agouti
A completely jet black cat has no markings on it whatsoever. It would have the genotype: *aaB–D–* since no dominant agouti allele must be present, and the black pigment is not diluted.

Siamese
The colour pointing of Siamese cats is caused by warm temperature deactivation of a gene that produces melanin pigment. Cooler parts of the body are not affected and appear dark.

Tortoiseshell
Because this is a sex linked trait, it is normally found only in female cats (*XO, Xo*). The coat is a mixture of orange and black fur irregularly blended together.

Agouti hair
Enlarged view of agouti hair. Note that the number of darkly pigmented stripes can vary on the same animal.

 Stripes of dark pigment — Lighter colour

Sex linked orange
The orange (*XO, XO*) cat has an orange coat with little or no patterns such as tabby showing.

Blotched tabby
Lacks stripes but has broad, irregular bands arranged in whorls (*tb*).

Wild type
Mackerel (striped) tabby (*A–B–T–*) with evenly spaced, well-defined, vertical stripes. The background colour is **agouti** with the stripes being areas of completely black hairs.

Calico
Similar to a tortoiseshell, but with substantial amounts of white fur present as well. Black, orange and white fur.

Orange, Black, White

Marmalade
The orange colour (*XO, XO*) is expressed, along with the alleles for the tabby pattern. The orange allele shows epistatic dominance and overrides the expression of the normal agouti colour so that the tabby pattern appears dark orange.

Golden yellow coat, Deeper colour stripes

Other Inherited Features in Domestic Cats

Manx tail (Mm)
The Manx breed of cat has little or no tail. This dominant allele is lethal if it occurs in the homozygous condition.

No tail

Polydactylism (Pd–)
This is a dominant mutation. The number of digits on the front paw should be five, with four digits on the rear paw.

6 digits on the paw

Ear fold (Fd–)
Most cats have normal pointed ears. A dominant mutation exists where the ear is permanently folded forwards.

Ears folded forwards

What Genotype Has That Cat?

Consult the table of genes listed on the previous pages and enter the allele symbols associated with each of the phenotypes in the column headed 'Allele'. For this exercise, study the appearance of real cats around your home or look at colour photographs of different cats. For each cat, complete the checklist of traits listed below by simply placing a tick in the appropriate spaces. These traits are listed in the same order as the genes for **wild forms** and **mutant forms** on page 155. On a piece of paper, write each of the cat's genotypes. Use a dash (-) for the second allele for characteristics that could be either heterozygous or homozygous dominant (see the sample at the bottom of the page).

NOTES:
1. Agouti fur colouring is used to describe black **hairs** with a light band of pigment close to its tip.
2. Patches of silver fur (called chinchilla) produces the silver tabby phenotype in agouti cats. Can also produce "smoke" phenotype in Persian long-haired cats, causing reduced intensity of the black.
3. Describes the dark extremities (face, tail and paws) with lighter body (e.g. Siamese).
4. The recessive allele makes black cats blue-grey in colour and yellow cats cream.
5. Spottiness involving less than half the surface area is likely to be heterozygous.

Inheritance

Phenotype Record Sheet for Domestic Cats

Gene	Phenotype	Allele	Sample	Cat 1	Cat 2	Cat 3	Cat 4
Agouti colour	Agouti[1]						
	Non-agouti		✔				
Pigment colour	Black		✔				
	Brown						
Colour present	Unicoloured		✔				
	Silver patches[2]						
	Pointed[3]						
	Albino with blue eyes						
	Albino with pink eyes						
Pigment density	Dense pigment						
	Dilute pigment[4]		✔				
Ear shape	Pointed ears		✔				
	Folded ears						
Hairiness	Normal, full coat		✔				
	Hairlessness						
Hair length	Short hair		✔				
	Long hair						
Tail length	Normal tail (long)		✔				
	Stubby tail or no tail at all						
Orange colour	Normal colours (non-orange)		✔				
	Orange						
Number of digits	Normal number of toes		✔				
	Polydactylism (extra toes)						
Hair curliness	Normal, smooth hair		✔				
	Curly hair (rex)						
Spottiness	No white spots						
	White spots (less than half)[5]		✔				
	White spots (more than half)						
Stripes	Mackerel striped (tabby)						
	Blotched stripes						
White coat	Not all white		✔				
	All white coat colour						

Sample cat (see ticks in chart above)

To give you an idea of how to read the chart you have created, here is an example genotype of the author's cat with the following features: *A smoky grey uniform-coloured cat, with short smooth hair, normal tail and ears, with 5 digits on the front paws and 4 on the rear paws, small patches of white on the feet and chest.* (Note that the stripe genotype is completely unknown since there is no agouti allele present).

GENOTYPE: | aa B– C– dd fdfd Hr– ii L– mm oo pdpd R– Ss ww |

Related activities: Inheritance in Domestic Cats

RA 3

Population Genetics and Speciation

AQA-A	AQA-B	SNAB	Edexcel	OCR
Complete:	Complete:	Complete:	Complete:	Complete:
1-2, 7-13, 15-16, 27-33	1-2, 4-5, 11-12, 15-16, 26-33	1-6, 7-9, 11-14, 16-19, 23-33	1-2, 7-8, 11-13, 16-19, 26, 29-33	1-9, 11-14, 16-27, 29-31

Learning Objectives

☐ 1. Compile your own glossary from the **KEY WORDS** displayed in **bold type** in the learning objectives below.

A Modern Synthesis of Evolution *(pages 160-162)*

☐ 2. Give a precise definition of the term **evolution**, explaining how evolution is a feature of **populations** and not of individuals.

☐ 3. Name some of the main contributors to the modern theory of evolution. List some of the important observations that **Darwin** made on his voyage on the Beagle that led him to formulate his theory.

☐ 4. Outline the fundamental ideas in Darwin's *"Theory of evolution by natural selection"* and describe in which way the theory was inadequate. Recognise that Alfred Russel Wallace was following a similar line of thought.

☐ 5. Appreciate how Darwin's original theory has been modified in the **new synthesis** to take into account our understanding of genetics and inheritance. Recognise examples of recently documented evolutionary change.

☐ 6. Understand the use of the term **fitness** and explain how evolution through **adaptation**, equips species for survival. Recognise structural and physiological adaptations of organisms to their environment.

The Gene Pool Concept *(pages 167-171, 173-176)*

☐ 7. Understand the concept of the **gene pool** and explain the term **deme**. Recognise that populations may be of various sizes and geographical extent.

☐ 8. Explain the term **allele frequency** and describe how allele frequencies are expressed for a population.

☐ 9. Understand the criteria (below) that must be satisfied in order to achieve **genetic equilibrium** in a population and explain why they are important. Identify the consequences of the fact that these criteria are rarely, perhaps never, met in reality.
(a) Large population size with **random mating**.
(b) No mutations (or mutations are balanced).
(c) No selective advantage for any genotype.
(d) No net migration (gene flow) between populations.
(e) Normal meiosis; chance operates in gametogenesis.

☐ 10. Explain how the **Hardy-Weinberg equation** provides a simple mathematical model of **genetic equilibrium** in a population. Demonstrate an ability to use the Hardy-Weinberg equation to calculate the allele, genotype, and phenotype frequencies from appropriate data.

Processes in Gene Pools *(pages 167-168)*

☐ 11. Recognise that changes within a gene pool occur when any or all of the above criteria for genetic equilibrium are not met. Identify **microevolution** as changes in the allele frequencies of gene pools. Recognise the four forces in microevolution that may alter allele frequencies: **natural selection**, **genetic drift**, **gene flow**, and **mutation**.

Identify which of these microevolutionary processes increase genetic diversity in the gene pool and which decrease it. Recognise the role of natural selection and genetic drift in sorting the genetic variability and establishing adaptive genotypes.

Natural selection *(pages 161, 163-66, 172, 177-79)*

☐ 12. Explain how **natural selection** is responsible for most evolutionary change by selectively reducing or changing genetic variation through differential survival and reproduction. Interpret data and/or use unfamiliar information to explain how natural selection produces change within a population.

☐ 13. Describe three types of natural selection: **stabilising**, **directional**, and **disruptive selection**. Describe the outcome of each type in a population exhibiting a normal curve in phenotypic variation.

☐ 14. With respect to natural selection, explain how environmental factors (e.g. climate) can act as a stabilising force or a force of change (evolution).

☐ 15. Describe examples of evolution by **natural selection**. Examples could include:
(a) **Transient polymorphism** as shown by **industrial melanism** in peppered moths (*Biston betularia*).
(b) The sickle cell trait as the basis for **balanced polymorphism** in malarial regions.
(c) The evolution of **antibiotic resistance** in bacteria.
(d) Changes to the size and shape of the beaks of Galapagos finches.
(e) The evolution of **pesticide resistance** in insects or **heavy metal tolerance** in plants.

Sexual selection *(page 180)*

☐ 16. EXTENSION: Discuss the role of **sexual selection** in affecting anatomy and behavior. Suggest how sexual selection may lead to the evolution of elaborate secondary sexual characteristics, particularly in males.

Genetic drift *(pages 169-170, 187)*

☐ 17. Explain what is meant by **genetic drift** and describe the conditions under which it is important. Distinguish between genetic drift and natural selection. Explain, using diagrams or a gene pool model, how genetic drift may lead to loss or **fixation of alleles** (where a gene is represented in the population by only one allele).

Mutation *(pages 103, 169-170)*

☐ 18. Recognise **mutations** as the ultimate source of all new alleles. Explain how mutations alter the genetic equilibrium of a population.

Gene flow *(pages 169-170)*

☐ 19. Explain, using diagrams or a gene pool model, how **migration** leads to **gene flow** between natural populations, and may affect allele frequencies.

Artificial Selection *(pages 181-185)*

☐ 20. Explain the basis of **artificial selection (selective breeding)**. Describe examples of artificial selection, e. g. in plants, livestock, and companion animals.

☐ 21. Using examples, explain what is meant by **F₁ hybrid vigour** and explain its genetic basis. Explain what is meant by the term **genetic gain**.

☐ 22. Recognise **genetic improvement** in livestock breeding as the gain towards a desired **phenotype**. Explain the contribution of technology to the rate of genetic improvement achieved in breeding programmes.

Special Events in Gene Pools *(pages 186-188)*

☐ 23. Explain the **founder effect**, including reference to its genetic and evolutionary consequences.

☐ 24. Explain the **population bottleneck effect**, including reference to its genetic and evolutionary consequences.

☐ 25. Appreciate how the founder effect and population (genetic) bottlenecks may accelerate the pace of evolutionary change. Explain the importance of **genetic drift** in populations that undergo these events.

Speciation *(pages 167, 189-196)*

☐ 26. Recall your definition of **evolution** and distinguish between **microevolution** and **macroevolution** (the formation of completely new species, genera etc.).

☐ 27. Provide a clear definition of the term **species**. Describe how the nature of some species can create problems for our definition. Give examples of organisms for which our definition of a species is problematic.

☐ 28. Providing examples, explain the concept of a **ring species** and closely related species where the **reproductive isolation** is no longer maintained.

☐ 29. Explain what is meant by **speciation**. Recognise the role of **natural selection** and **reproductive isolation** in the formation of new species.

☐ 30. Explain how populations may become **reproductively isolated** through changes in behaviour or physiology, polyploidy, or niche differentiation.

☐ 31. Distinguish between and describe **prezygotic** and **postzygotic reproductive isolating mechanisms**.

☐ 32. Explain the events occurring in **allopatric speciation**, identifying situations in which it is most likely to occur.

☐ 33. Explain the events occurring in **sympatric speciation** and describe the situations in which it is most likely to occur. Explain why reproductive isolating mechanisms tend to be much more pronounced between sympatric (as opposed to allopatric) species. Recognise the role of **polyploidy** in instant speciation events.

Textbooks

 See the 'Textbook Reference Grid' on pages 8-9 for textbook page references relating to material in this topic.

Supplementary Texts

See pages 6-7 for additional details of these texts:

■ Adds, J., *et al.*, 2004. **Genetics, Evolution and Biodiversity**, (NelsonThornes), chpt. 8.

■ Clegg, C.J., 1999. **Genetics and Evolution**, (John Murray), pp. 60-78.

■ Jones, N., *et al.*, 2001. **The Essentials of Genetics**, pp. 190-232.

Periodicals

See page 7 for details of publishers of periodicals:

STUDENT'S REFERENCE

■ **The Species Enigma** New Scientist, 13 June 1998 (Inside Science). *An excellent article on the nature of species, ring species, and the status of hybrids. Includes species life cycles and extinction.*

■ **Optimality** Biol. Sci. Rev., 17(4), April 2005, pp. 2-5. *Environmental stability and optimality of structure and function can explain evolutionary stasis in animals. Examples are described.*

■ **The Hardy-Weinberg Principle** Biol. Sci. Rev., 15(4), April 2003, pp. 7-9. *A succinct explanation of the basis of the Hardy-Weinberg principle, and its uses in estimating genotype frequencies and predicting change in populations.*

Presentation MEDIA to support this topic:
EVOLUTION
• **Population Genetics**
• **Evolution**

■ **Speciation** Biol. Sci. Rev., 16(2) Nov. 2003, pp. 24-28. *An excellent account of speciation. It covers the nature of species, reproductive isolation, how separated populations diverge, and sympatric speciation. Case examples include the cichlids of Lake Victoria and the founder effect in mynahs.*

■ **Polymorphism** Biol. Sci. Rev., 14(1), Sept. 2001, pp. 19-21. *An account of polymorphism in populations, with several case studies (including Biston moths) provided as illustrative examples.*

■ **Butterflies or Bitterflies** Biol. Sci. Rev., 11(3) January 1999, pp. 18-21. *Mimicry as the basis for natural selection in polymorphic populations.*

■ **How Life Branches Out** New Scientist, 24 June 2006, pp. 50-53. *The origin of Amazonia's spectacular biodiversity is attributable to sympatric speciation, not the Pleistocene refugia hypothesis.*

■ **Tails** (sic) **of Love and War** Biol. Sci. Rev., 11(1) Sept. 1998, pp. 30-33. *A discussion of the mechanisms of evolutionary change with examples of natural selection in animal populations.*

■ **The Cheetah: Losing the Race?** Biol. Sci. Rev., 14(2) Nov. 2001, pp. 7-10. *The evolutionary bottleneck experienced by cheetahs and its implications for the genetics of the species.*

TEACHER'S REFERENCE

■ **15 Answers to Creationist Nonsense** Scientific American, July 2002, pp. 62-69. *A synopsis of the common arguments presented by Creationists and the answers offered by science.*

■ **Evolution: Five Big Questions** New Scientist, 14 June 2003, pp. 32-39, 48-51. *A discussion of the most covered points regarding evolution and the mechanisms by which it occurs.*

■ **Live and Let Live** New Scientist, 3 July 1999, pp. 32-36. *Recent research suggests that hybrids are intact entities subject to the same evolutionary pressures as pure species.*

■ **Desert Island** Scientific American, June 2005, pp. 11-12. *A new study suggests that climate can effectively isolate organisms and foster endemism.*

■ **The Challenge of Antibiotic Resistance** Scientific American, March 1998, pp. 32-39. *An excellent article on the basis of antibiotic resistance in bacteria and its selective advantage.*

■ **Replaying Life** New Scientist, 13 February 1999, pp. 29-33. *Rapid evolution in bacteria driven by habitat diversity and niche differentiation.*

■ **Skin Deep** Scientific American, October 2002, pp. 50-57. *This article examines the evolution of skin colour in humans and presents powerful evidence for skin colour ("race") being the end result of opposing selection pressure. Clearly written and of high interest, this is a perfect vehicle for examining natural selection.*

■ **Together We're Stronger** New Scientist, 15 March 2003. (Inside Science). *The mechanisms behind the evolution of social behaviour in animals. The evolution of eusociality in hymenopteran insects is the case study provided.*

■ **Listen, We're Different** New Scientist, 17 July 1999, pp. 32-35. *An excellent account of speciation in periodic cicadas as a result of behavioural and temporal isolating mechanisms.*

■ **How the Species Became** New Scientist, 11 Oct. 2003, pp. 32-35. *Stability in species and new ideas on speciation. Species are stable if changes in form or behaviour are damped out, but unstable if the changes escalate as new generations shuffle parental genes and natural selection discards the allele combinations that do not work well.*

■ **Cichlids of the Rift Lakes** Scientific American, February 1999, pp. 44-49. *An excellent, thorough account of the recent speciation events documented in the cichlid fishes of Lake Victoria.*

Internet

See pages 10-11 for details of how to access **Bio Links** from our web site: **www.biozone.co.uk**. From Bio Links, access sites under the topics:

EVOLUTION: • A history of evolutionary thought • Evolution on the web for biology students • Harvard University biology links: evolution … *and others* > **Charles Darwin:** • Darwin and evolution overview • … *and others* > **Evolution Theory and Evidence:** • Introduction to evolutionary biology • Transitional vertebrate fossils FAQ … *and others* > **The Fossil Record:** • Geological time scale …*and others*

GENETICS > **Population Genetics:** • Introduction to evolutionary biology • Microevolution and population genetics • Random genetic drift • Population genetics: lecture notes … *and others*

Population Genetics and Speciation

The Modern Theory of Evolution

Although **Charles Darwin** is credited with the development of the theory of evolution by natural selection, there were many people that contributed ideas upon which he built his own. Since Darwin first proposed his theory, aspects that were problematic (such as the mechanism of inheritance) have now been explained. The development of the modern theory of evolution has a history going back at least two centuries. The diagram below illustrates the way in which some of the major contributors helped to form the currently accepted model, or **new synthesis**. Understanding of evolutionary processes continued to grow through the 1980s and 1990s as comparative molecular sequence data were amassed and understanding of the molecular basis of developmental mechanisms improved. Most recently, in the exciting new area of evolutionary developmental biology (**evo-devo**), biologists have been exploring how developmental gene expression patterns explain how groups of organisms evolved.

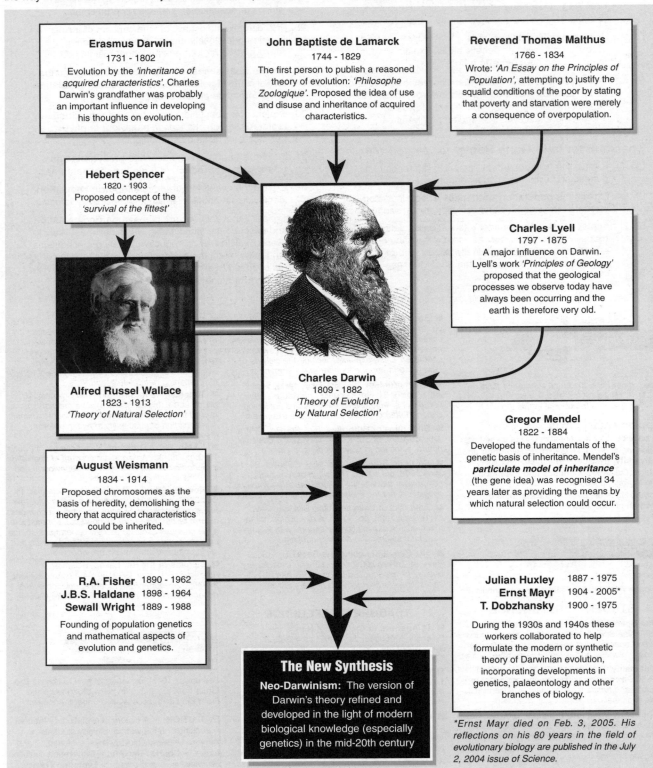

Erasmus Darwin
1731 - 1802
Evolution by the *'inheritance of acquired characteristics'*. Charles Darwin's grandfather was probably an important influence in developing his thoughts on evolution.

John Baptiste de Lamarck
1744 - 1829
The first person to publish a reasoned theory of evolution: *'Philosophe Zoologique'*. Proposed the idea of use and disuse and inheritance of acquired characteristics.

Reverend Thomas Malthus
1766 - 1834
Wrote: *'An Essay on the Principles of Population'*, attempting to justify the squalid conditions of the poor by stating that poverty and starvation were merely a consequence of overpopulation.

Hebert Spencer
1820 - 1903
Proposed concept of the *'survival of the fittest'*

Charles Lyell
1797 - 1875
A major influence on Darwin. Lyell's work *'Principles of Geology'* proposed that the geological processes we observe today have always been occurring and the earth is therefore very old.

Alfred Russel Wallace
1823 - 1913
'Theory of Natural Selection'

Charles Darwin
1809 - 1882
'Theory of Evolution by Natural Selection'

Gregor Mendel
1822 - 1884
Developed the fundamentals of the genetic basis of inheritance. Mendel's *particulate model of inheritance* (the gene idea) was recognised 34 years later as providing the means by which natural selection could occur.

August Weismann
1834 - 1914
Proposed chromosomes as the basis of heredity, demolishing the theory that acquired characteristics could be inherited.

R.A. Fisher 1890 - 1962
J.B.S. Haldane 1898 - 1964
Sewall Wright 1889 - 1988
Founding of population genetics and mathematical aspects of evolution and genetics.

Julian Huxley 1887 - 1975
Ernst Mayr 1904 - 2005*
T. Dobzhansky 1900 - 1975
During the 1930s and 1940s these workers collaborated to help formulate the modern or synthetic theory of Darwinian evolution, incorporating developments in genetics, palaeontology and other branches of biology.

The New Synthesis
Neo-Darwinism: The version of Darwin's theory refined and developed in the light of modern biological knowledge (especially genetics) in the mid-20th century

Ernst Mayr died on Feb. 3, 2005. His reflections on his 80 years in the field of evolutionary biology are published in the July 2, 2004 issue of Science.

1. From the diagram above, choose one of the contributors to the development of evolutionary theory (excluding Charles Darwin himself), and write a few paragraphs discussing their role in contributing to Darwin's ideas. You may need to consult an encyclopaedia or other reference to assist you.

Darwin's Theory

In 1859, Darwin and Wallace jointly proposed that new species could develop by a process of natural selection. Natural selection is the term given to the mechanism by which better adapted organisms survive to produce a greater number of viable offspring. This has the effect of increasing their proportion in the population so that they become more common. It is Darwin who is best remembered for the theory of evolution by natural selection through his famous book: 'On the origin of species by means of natural selection', written 23 years after returning from his voyage on the Beagle, from which much of the evidence for his theory was accumulated. Although Darwin could not explain the origin of variation nor the mechanism of its transmission (this was provided later by Mendel's work), his basic theory of evolution by natural selection (outlined below) is widely accepted today. The study of population genetics has greatly improved our understanding of evolutionary processes, which are now seen largely as a (frequently gradual) change in allele frequencies within a population. Students should be aware that scientific debate on the subject of evolution centres around the relative merits of various alternative hypotheses about the nature of evolutionary processes. The debate is not about the existence of the phenomenon of evolution itself.

Darwin's Theory of Evolution by Natural Selection

Overproduction
Populations produce too many young: many must die

Populations tend to produce more offspring than are needed to replace the parents. Natural populations normally maintain constant numbers. There must therefore be a certain number dying.

Variation
Individuals show variation: some are more favourable than others

Individuals in a population vary in their phenotype and therefore, their genotype. Some variants are better suited to the current conditions than others and find it easier to survive and reproduce.

Natural Selection
Natural selection favours the best suited at the time

The struggle for survival amongst overcrowded individuals will favour those variations which have the best advantage. This does not necessarily mean that those struggling die, but they will be in a poorer condition.

Inherited
Variations are Inherited. The best suited variants leave more offspring.

The variations (both favourable and unfavourable) are passed on to offspring. Each new generation will contain proportionally more descendents from individuals with favourable characters than those with unfavourable.

Population Genetics and Speciation

1. In your own words, describe how Darwin's theory of evolution by natural selection provides an explanation for the change in the appearance of a species over time:

_____

_____

_____

_____

Related activities: The Modern Theory of Evolution, Darwin's Finches

A 2

Adaptations and Fitness

An **adaptation**, is any heritable trait that suits an organism to its natural function in the environment (its niche). These traits may be structural, physiological, or behavioural. The idea is important for evolutionary theory because adaptive features promote fitness. **Fitness** is a measure of an organism's ability to maximise the numbers of offspring surviving to reproductive age. Adaptations are distinct from properties which, although they may be striking, cannot be described as adaptive unless they are shown to be functional in the organism's natural habitat. Genetic adaptation must not be confused with **physiological adjustment** (acclimatisation), which refers to an organism's ability to adapt during its lifetime to changing environmental conditions (e.g. a person's acclimatisation to altitude). Examples of adaptive features arising through evolution are illustrated below.

Ear Length in Rabbits and Hares

The external ears of many mammals are used as important organs to assist in thermoregulation (controlling loss and gain of body heat). The ears of rabbits and hares native to hot, dry climates, such as the jack rabbit of south-western USA and northern Mexico, are relatively very large. The Arctic hare lives in the tundra zone of Alaska, northern Canada and Greenland, and has ears that are relatively short. This reduction in the size of the extremities (ears, limbs, and noses) is typical of cold adapted species.

Arctic hare: *Lepus arcticus*

Black-tail jackrabbit: *Lepus californicus*

Body Size in Relation to Climate

Regulation of body temperature requires a large amount of energy and mammals exhibit a variety of structural and physiological adaptations to increase the effectiveness of this process. Heat production in any endotherm depends on body volume (heat generating metabolism), whereas the rate of heat loss depends on surface area. Increasing body size minimises heat loss to the environment by reducing the surface area to volume ratio. Animals in colder regions therefore tend to be larger overall than those living in hot climates. This relationship is know as **Bergman's rule** and it is well documented in many mammalian species. Cold adapted species also tend to have more compact bodies and shorter extremities than related species in hot climates.

Fennec fox

Arctic fox

The **fennec fox** of the Sahara illustrates the adaptations typical of mammals living in hot climates: a small body size and lightweight fur, and long ears, legs, and nose. These features facilitate heat dissipation and reduce heat gain.

The **Arctic fox** shows the physical characteristics typical of cold-adapted mammals: a stocky, compact body shape with small ears, short legs and nose, and dense fur. These features reduce heat loss to the environment.

Number of Horns in Rhinoceroses

Not all differences between species can be convincingly interpreted as adaptations to particular environments. Rhinoceroses charge rival males and predators, and the horn(s), when combined with the head-down posture, add effectiveness to this behaviour. Horns are obviously adaptive, but it is not clear that the possession of one (Indian rhino) or two (black rhino) horns is necessarily related directly to the environment in which those animals live.

Great Indian rhino

African black rhino

1. Distinguish between adaptive features (genetic) and acclimatisation: _____

_____

_____

2. Explain the nature of the relationship between the length of extremities (such as limbs and ears) and climate:

_____

_____

3. Explain the adaptive value of a larger body size at high latitude: _____

_____

Related activities: Darwin's Finches

Natural Selection

Natural selection operates on the phenotypes of individuals, produced by their particular combinations of alleles. In natural populations, the allele combinations of some individuals are perpetuated at the expense of other genotypes. This differential survival of some genotypes over others is called **natural selection**. The effect of natural selection can vary; it can act to maintain the genotype of a species or to change it.

Stabilising selection maintains the established favourable characteristics and is associated with stable environments. In contrast, **directional selection** favours phenotypes at one extreme of the phenotypic range and is associated with gradually changing environments. **Disruptive selection** is a much rarer form of selection favouring two phenotypic extremes, and is a feature of fluctuating environments.

Stabilising Selection

Extreme variations are culled from the population (there is selection against them). Those with the established (middle range) adaptive phenotype are retained in greater numbers. This reduces the variation for the phenotypic character. In the example right, light and dark snails are eliminated, leaving medium coloured snails. Stabilising selection can be seen in the selection pressures on human birth weights.

Directional Selection

Directional selection is associated with gradually changing conditions, where the adaptive phenotype is shifted in one direction and one aspect of a trait becomes emphasised (e.g. colouration). In the example right, light coloured snails are eliminated and the population becomes darker. Directional selection was observed in peppered moths in England during the Industrial Revolution. They responded to the air pollution of industrialisation by increasing the frequency of darker, melanic forms.

Disruptive or Diversifying Selection

Disruptive selection favours two extremes of a trait at the expense of intermediate forms. It is associated with a fluctuating environment and gives rise to **balanced polymorphism** in the population. In the example right, there is selection against medium coloured snails, which are eliminated. There is considerable evidence that predators, such as insectivorous birds, are more likely to find and eat common morphs and ignore rare morphs. This enables the rarer forms to persist in the population.

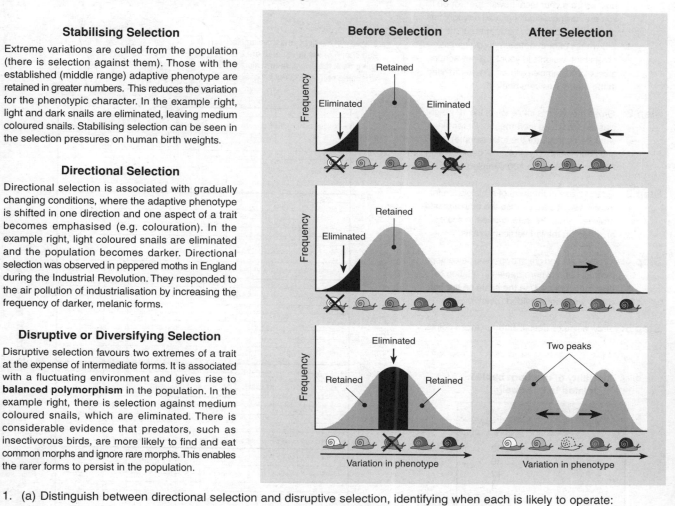

Population Genetics and Speciation

1. (a) Distinguish between directional selection and disruptive selection, identifying when each is likely to operate:

_____

_____

_____

(b) Identify which of the three types of selection described above will lead to evolution, and explain why:

_____

_____

2. Explain how a change in environment may result in selection becoming directional rather than stabilising:

_____

_____

3. Explain how, in a population of snails, through natural selection, shell colour could change from light to dark over time:

_____

_____

_____

Related activities: Industrial Melanism
Web links: Natural Selection in Populations, Changes in a Gene Pool

A 3

Selection for Human Birth Weight

Selection pressures operate on populations in such a way as to reduce mortality. For humans, giving birth is a special, but often traumatic, event. In a study of human birth weights it is possible to observe the effect of selection pressures operating to constrain human birth weight within certain limits. This is a good example of **stabilising selection**. This activity explores the selection pressures acting on the birth weight of human babies. Carry out the steps below:

Step 1: Collect the birth weights from 100 birth notices from your local newspaper (or 50 if you are having difficulty getting enough; this should involve looking back through the last 2-3 weeks of birth notices). If you cannot obtain birth weights in your local newspaper, a set of 100 sample birth weights is provided in the Model Answers booklet.

Step 2: Group the weights into each of the 12 weight classes (of 0.5 kg increments). Determine what percentage (of the total sample) fall into each weight class (e.g. 17 babies weigh 2.5-3.0 kg out of the 100 sampled = 17%)

Step 3: Graph these in the form of a histogram for the 12 weight classes (use the graphing grid provided right). Be sure to use the scale provided on the left vertical (y) axis.

Step 4: Create a second graph by plotting percentage mortality of newborn babies in relation to their birth weight. Use the scale on the right y axis and data provided (below).

Step 5: Draw a line of 'best fit' through these points.

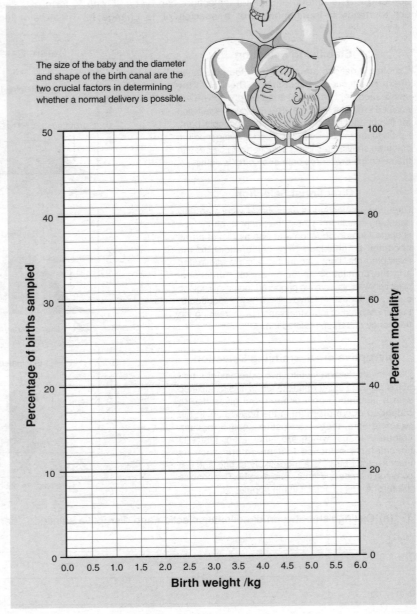

The size of the baby and the diameter and shape of the birth canal are the two crucial factors in determining whether a normal delivery is possible.

Mortality of newborn babies related to birth weight

Weight /kg	Mortality /%
1.0	80
1.5	30
2.0	12
2.5	4
3.0	3
3.5	2
4.0	3
4.5	7
5.0	15

Source: Biology: The Unity & Diversity of Life (4th ed), by Starr and Taggart

1. Describe the shape of the histogram for birth weights: _____

2. State the optimum birth weight in terms of the lowest newborn mortality: _____

3. Describe the relationship between newborn mortality and birth weight: _____

_____

4. Describe the selection pressures that are operating to control the range of birth weight: _____

_____

_____

5. Describe how medical intervention methods during pregnancy and childbirth may have altered these selection pressures:

_____

_____

Related activities: Natural Selection

Darwin's Finches

The Galapagos Islands, 920 km off the west coast of Ecuador, played a major role in shaping Darwin's thoughts about evolution. While exploring the islands in 1835, he was struck by the unique and peculiar species he found there, in particular, the island's finches. The Galapagos group is home to 13 species of finches in four genera. This variety has arisen as a result of evolution from one ancestral species. Initially, a number of small finches, probably grassquits, made their way from South America to the Galapagos Islands. In the new environment, which was relatively free of competitors, the colonisers underwent an adaptive radiation, producing a range of species each with its own unique feeding niche. Although similar in their plumage, nest building techniques, and calls, the different species can be distinguished by the size and shape of their beaks. Each species has a beak adapted for a different purpose, such as crushing seeds, pecking wood, or probing flowers. Between them, the 13 species of this endemic group fill the roles of seven different families of South American mainland birds. DNA analyses have confirmed Darwin's insight and have shown that all 13 species evolved from a flock of about 30 birds arriving a million years ago.

The Evolution of Darwin's Finches

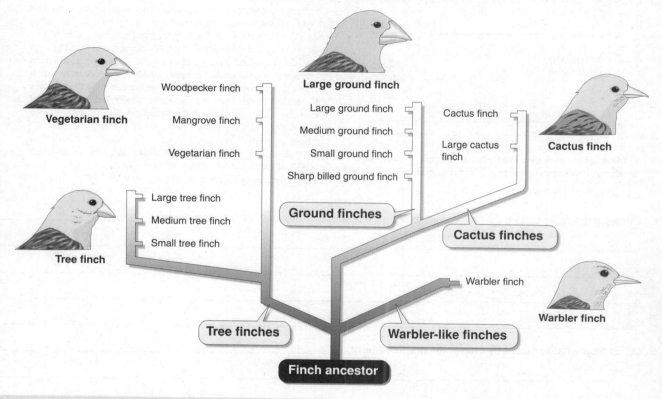

Vegetarian finch

Woodpecker finch
Mangrove finch
Vegetarian finch

Large ground finch

Large ground finch
Medium ground finch
Small ground finch
Sharp billed ground finch

Cactus finch
Large cactus finch

Cactus finch

Large tree finch
Medium tree finch
Small tree finch

Tree finch

Ground finches

Cactus finches

Warbler finch

Warbler finch

Tree finches

Warbler-like finches

Finch ancestor

Small tree finch   **Large tree finch**

As the name implies, **tree finches** are largely arboreal and feed mainly on insects. The bill is sharper than in ground finches and better suited to grasping insects. Paler than ground or cactus finches, they also have streaked breasts.

There are four species of **ground finches** with crushing-type bills used for seed eating. On Wolf Island, they are called vampire finches because they peck the skin of animals to draw blood, which they then drink (see left).

Cactus finches have most probably descended from ground finches. They have a probing beak and feed on insects on the cactus or the cactus itself. On islands where there are no ground finches, there is more variation in beak size than on the islands where the species coexist.

The **warbler finch** is named for its resemblance to the unrelated warblers. The beak of the warbler finch is the thinnest of the Galapagos finches. It is also the most widespread species, found throughout the archipelago. Warbler finches prey on flying and ground dwelling insects.

Related activities: Adaptation and Fitness, Darwin's Theory

A 2

Population Genetics and Speciation

166

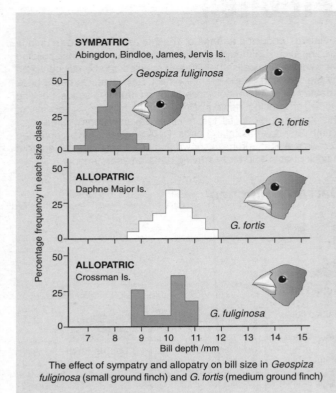

SYMPATRIC
Abingdon, Bindloe, James, Jervis Is.

Geospiza fuliginosa

G. fortis

ALLOPATRIC
Daphne Major Is.

G. fortis

ALLOPATRIC
Crossman Is.

G. fuliginosa

Percentage frequency in each size class

Bill depth /mm

The effect of sympatry and allopatry on bill size in *Geospiza fuliginosa* (small ground finch) and *G. fortis* (medium ground finch)

Sympatry and Character Displacement

There is good evidence that finch evolution appears to be driven by a combination of allopatric and sympatric events. Coexisting species of ground finches on four islands (top graph) show large differences in bill sizes, enabling each species to feed on different sized seeds. However when either species exists in the absence of the other on different islands (lower graphs), it possesses intermediate bill sizes (about 10 mm) enabling it to feed without partitioning seed resources. This phenomenon, whereby competition causes two closely related species to become more different in regions where their ranges overlap, is referred to as **character displacement**.

Character displacement is evident in other populations of finches as well. There are well-studied populations of the large cactus finch (*G. conirostris*) on Genovesa and Espanola Islands, but their bill sizes are quite different. On Genovesa, the large ground finch (*G. magnirostris*) coexists with the large cactus finch. In these sympatric populations, the variability in bill size *within* each species is minimal but there is little overlap *between* the species with respect to this trait. On Espanola, where the large ground finch either never arrived, or became extinct, the situation is quite different. With no competition on Espanola, the large cactus finch displays a greater variability in bill size. Its bill is somewhat intermediate between the two finches on Genovesa, and it can feed equally well in both niches all year round.

Data based on an adaptation by Strickberger (2000)

1. Describe the main factors that have contributed to the adaptive radiation of Darwin's finches: _____

_____

_____

_____

_____

2. (a) Explain what is meant by **character displacement**: _____

_____

_____

(b) Discuss how the incidence of character displacement observed in the Galapagos finches supports the view that their adaptive radiation from a common ancestor has been driven by a combination of allopatric and sympatric events:

_____

_____

_____

_____

3. The range of variability shown by a phenotype in response to environmental variation is called **phenotypic plasticity**.

(a) Discuss the evidence for phenotypic plasticity in Galapagos finches: _____

_____

_____

_____

(b) Explain what this suggests about the biology of the original finch ancestor: _____

_____

_____

Gene Pool Exercise

Cut out each of the beetles on this page and use them to reenact different events within a gene pool as described in this topic (see *Gene Pools and Evolution, Changes in a Gene Pool, Founder Effect, Population Bottlenecks, Genetic Drift*).

Population Genetics and Speciation

Related activities: Gene Pools and Evolution, Changes in a Gene Pool,
The Founder Effect, Population Bottlenecks, Genetic Drift

P

This page has deliberately been left blank

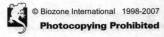

Gene Pools and Evolution

The diagram below illustrates the dynamic nature of **gene pools**. It portrays two imaginary populations of one beetle species. Each beetle is a 'carrier' of genetic information, represented here by the alleles (A and a) for a single **codominant gene** that controls the beetle's colour. Normally, there are three versions of the phenotype: black, dark, and pale. Mutations may create other versions of the phenotype. Some of the **microevolutionary processes** that can affect the genetic composition (**allele frequencies**) of the gene pool are illustrated. See the activity *Gene Pool Exercise* for cut-out beetles to simulate this activity.

Immigration: Populations can gain alleles when they are introduced from other gene pools. Immigration is one aspect of gene flow.

Mutations: Spontaneous mutations can develop that alter the allele frequencies of the gene pool, and even create new alleles. Mutation is very important to evolution, because it is the original source of genetic variation that provides new material for natural selection.

Emigration: Genes may be lost to other gene pools.

Deme 1

The term deme describes a local population that is genetically isolated from other populations in the species. Demes usually have some clearly definable genetic or other character that sets them apart from other populations.

Natural selection: Selection pressure against certain allele combinations may reduce reproductive success or lead to death. Natural selection sorts genetic variability, and accumulates and maintains favourable genotypes in a population. It tends to reduce genetic diversity within the gene pool and increase differences between populations.

Geographical barriers: Isolate the gene pool and prevent *regular* gene flow between populations.

Gene flow: Genes are exchanged with other gene pools as individuals move between them. Gene flow is a source of new genetic variation and tends to reduce differences between populations that have accumulated because of natural selection or genetic drift.

Key to genotypes and phenotypes

Black
Homozygous dominant

Dark
Heterozygous

Pale
Homozygous recessive

Mottled
Homozygous dominant (mutant)

Deme 2

Boundary of gene pool

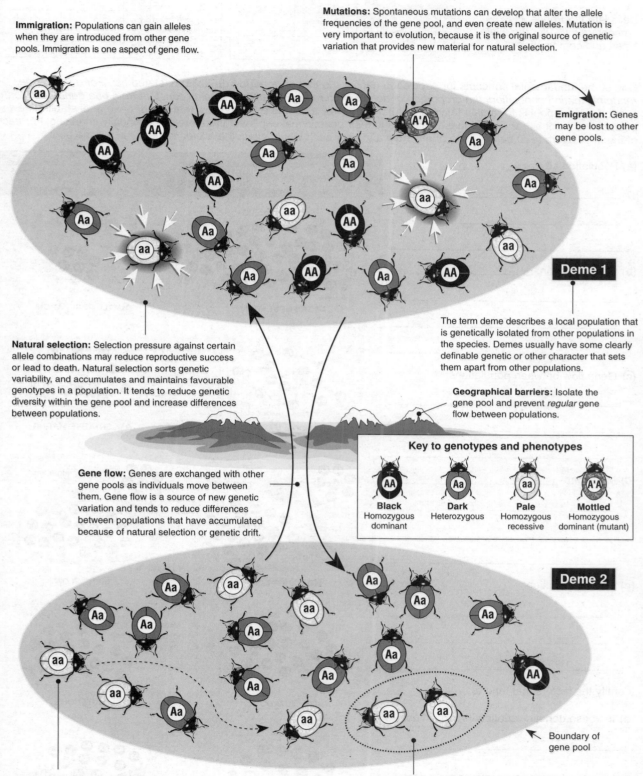

Mate selection (non-random mating): Individuals may not select their mate randomly and may seek out particular phenotypes, increasing the frequency of these "favoured" alleles in the population.

Genetic drift: Chance events can cause the allele frequencies of small populations to "drift" (change) randomly from generation to generation. Genetic drift can play a significant role in the microevolution of very small populations. The two situations most often leading to populations small enough for genetic drift to be significant are the **bottleneck effect** (where the population size is dramatically reduced by a catastrophic event) and the **founder effect** (where a small number of individuals colonise a new area).

Population Genetics and Speciation

Related activities: Gene Pool Exercise
Web links: Genes and Evolution, Allopatric Speciation

PA 2

1. For each of the 2 demes shown on the previous page (treating the mutant in deme 1 as a AA):

   (a) Count up the numbers of **allele types** (**A** and **a**).

   (b) Count up the numbers of **allele combinations** (**AA, Aa, aa**).

2. Calculate the frequencies as percentages (%) for the allele types and combinations:

Deme 1		Number counted	%
Allele types	A		
	a		
Allele combinations	AA		
	Aa		
	aa		

Deme 2		Number counted	%
Allele types	A		
	a		
Allele combinations	AA		
	Aa		
	aa		

3. One of the fundamental concepts for population genetics is that of **genetic equilibrium**, stated as: *"For a very large, randomly mating population, the proportion of dominant to recessive alleles remains constant from one generation to the next"*. If a gene pool is to remain unchanged, it must satisfy all of the criteria below that favour gene pool stability. Few populations meet all (or any) of these criteria and their genetic makeup must therefore by continually changing. For each of the five factors (a-e) below, state briefly **how** and **why** each would affect the allele frequency in a gene pool:

   (a) Population size: _____

   _____

   _____

   _____

   (b) Mate selection: _____

   _____

   _____

   _____

   (c) Gene flow between populations:

   _____

   _____

   _____

   (d) Mutations: _____

   _____

   _____

   _____

   (e) Natural selection: _____

   _____

   _____

   _____

   _____

4. Identify the factors that tend to:

   (a) Increase genetic variation in populations:

   _____

   _____

   (b) Decrease genetic variation in populations:

   _____

   _____

Factors Favouring Gene Pool Stability — **Factors Favouring Gene Pool Change**

LARGE POPULATION — SMALL POPULATION

RANDOM MATING — ASSORTATIVE MATING

NO GENE FLOW (Barrier to gene flow) — GENE FLOW (Immigration, Emigration)

NO MUTATION — MUTATIONS (New recessive allele)

NO NATURAL SELECTION — NATURAL SELECTION

Changes in a Gene Pool

The diagram below shows an imaginary population of beetles undergoing changes as it is subjected to two 'events'. The three phases represent a progression in time (i.e. the same gene pool, undergoing change). The beetles have three phenotypes determined by the amount of pigment deposited in the cuticle. Three versions of this trait exist: black, dark, and pale. The gene controlling this character is represented by two alleles **A** and **a**. Your task is to analyse the gene pool as it undergoes changes.

Phase 1: Initial gene pool

Calculate the frequencies of the *allele types* and *allele combinations* by counting the actual numbers, then working them out as percentages.

Black   **Dark**   **Pale**

	A	a	AA	Aa	aa
No.	27		7		
%	54		28		

Allele types   *Allele combinations*

Phase 2: Natural selection

In the same gene pool at a later time there was a change in the allele frequencies. This was due to the loss of certain allele combinations due to natural selection. Some of those with a genotype of aa were eliminated (poor fitness).

Calculate as for above. Do not include the individuals surrounded by small white arrows in your calculations; they are dead!

	A	a	AA	Aa	aa
No.					
%					

Phase 3: Immigration and emigration

This particular kind of beetle exhibits wandering behaviour. The allele frequencies change again due to the introduction and departure of individual beetles, each carrying certain allele combinations.

Calculate as above. In your calculations, include the individual coming into the gene pool (AA), but remove the one leaving (aa).

	A	a	AA	Aa	aa
No.					
%					

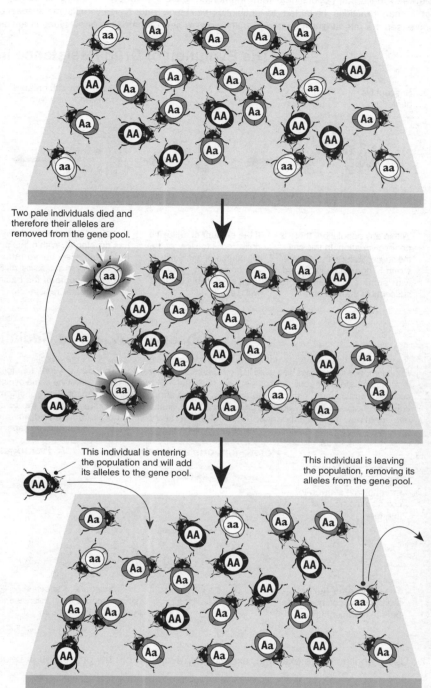

Two pale individuals died and therefore their alleles are removed from the gene pool.

This individual is entering the population and will add its alleles to the gene pool.

This individual is leaving the population, removing its alleles from the gene pool.

Population Genetics and Speciation

1. Explain how the number of dominant alleles (A) in the genotype of a beetle affects its phenotype:

_____

_____

2. For each phase in the gene pool above (place your answers in the tables provided; some have been done for you):

 (a) Determine the relative frequencies of the two alleles: A and a. Simply total the **A** alleles and **a** alleles separately.
 (b) Determine the frequency of how the alleles come together as allele pair combinations in the gene pool (AA, Aa, and aa). Count the number of each type of combination.
 (c) For each of the above, work out the frequencies as percentages:

Allele frequency = Number of counted alleles ÷ Total number of alleles x 100

Related activities: Natural Selection, Gene Pool Exercise
Web links: Natural Selection in Populations

Evolution in Bacteria

As a result of their short **generation times**, bacterial populations can show significant evolutionary change within relatively short periods of time. The development of **antibiotic resistance** is one such evolutionary change and it arises and spreads within and between bacterial populations with frightening ease. A variety of human practices have led to antibiotic resistance and have increased the rate at which bacterial strains acquire new properties.

These practices include the overuse and misuse of antibiotics by physicians, the use of antibiotics by immunosuppressed patients to prevent infection, the use of antibiotics in animal feed, and the spread of resistant bacteria to new areas because of air travel. For many strains of pathogenic bacteria, resistant mutants are increasingly replacing susceptible normal populations. This makes the search for new types of antibiotics increasingly urgent.

The Evolution of Drug Resistance in Bacteria

Susceptible bacterium | Slightly insensitive bacterium | Bacterium with greater resistance survives | Drug resistance genes can be transferred to non resistant strains

Within any population, there is genetic variation. In this case, the susceptibility of the bacterial strain is normally distributed, with some cells being more susceptible than others.

If the amount of antibiotic delivered is too low, or the full course of antibiotics is not completed, only the most susceptible bacteria will die.

Now a population of insensitive bacteria has developed. Within this population there will also be variation in the susceptibility to antibiotics. As treatment continues, some of the bacteria may acquire greater resistance.

A highly resistant population has evolved. The resistant cells can exchange genetic material with other bacteria, passing on the resistance genes. The antibiotic that was initially used against this bacterial strain will now be ineffective against it.

Observing Adaptive Radiation

Recently, scientists have demonstrated rapid evolution in bacteria. *Pseudomonas fluorescens* was used in the experiment and propagated in a simple heterogeneous environment consisting of a 25 cm^3 glass container containing 6 cm^3 of broth medium. Over a short period of time, the bacteria underwent morphological diversification, with a number of new morphs appearing. These morphs were shown to be genetically distinct. A striking feature of the evolved species is their niche specificity, with each new morph occupying a distinct habitat (below, left). In a follow up experiment (below, right), the researchers grew the same original bacterial strain in the same broth under identical incubation conditions, but in a homogeneous environment (achieved by shaking the broth). Without the different habitats offered by an undisturbed environment, no morphs emerged. The experiment illustrated the capacity of bacteria to evolve to utilise available niches.

Heterogeneous environment   Homogeneous environment

WS bacteria (wrinkly morphology) evolved to colonise the air-broth interface.

The FS species (fuzzy morphology) colonised the bottom of the container.

The ancestral SM species (smooth morphology) colonised the surface of the broth.

Because there is only one niche, no adaptive radiation occurs.

1. Using an illustrative example, explain why evolution of new properties in bacteria can be very rapid: _____

_____

_____

2. (a) In the example above, suggest why the bacteria evolved when grown in a heterogeneous environment:

_____

_____

(b) Predict what would happen if the FS morph was cultured in the homogeneous environment: _____

_____

_____

Related activities: Antibiotic Resistance

Population Genetics Calculations

The **Hardy-Weinberg equation** provides a simple mathematical model of genetic equilibrium in a gene pool, but its main application in population genetics is in calculating allele and genotype frequencies in populations, particularly as a means of studying changes and measuring their rate. The use of the Hardy-Weinberg equation is described below.

Punnett square

A   a

A   **AA**   **Aa**

a   **aA**   **aa**

Frequency of allele combination **AA** in the population is represented as p^2

Frequency of allele combination **aa** in the population is represented as q^2

Frequency of allele combination **Aa** in the population (add these together to get **2pq**)

$$(p + q)^2 \ = \ p^2 + 2pq + q^2 \ = \ 1$$

Frequency of allele types

p = Frequency of allele A

q = Frequency of allele a

Frequency of allele combinations

p^2 = Frequency of AA (homozygous dominant)

2pq = Frequency of Aa (heterozygous)

q^2 = Frequency of aa (homozygous recessive)

The Hardy-Weinberg equation is applied to populations with a simple genetic situation: dominant and recessive alleles controlling a single trait. The frequency of all of the dominant (A) and recessive alleles (a) equals the total genetic complement, and adds up to 1 or 100% of the alleles present.

How To Solve Hardy-Weinberg Problems

In most populations, the frequency of two alleles of interest is calculated from the proportion of homozygous recessives (q^2), as this is the only genotype identifiable directly from its phenotype. If only the dominant phenotype is known, q^2 may be calculated (1 – the frequency of the dominant phenotype). The following steps outline the procedure for solving a Hardy-Weinberg problem:

Remember that all calculations must be carried out using proportions, NOT PERCENTAGES!

1. Examine the question to determine what piece of information you have been given about the population. In most cases, this is the percentage or frequency of the homozygous recessive phenotype q^2, or the dominant phenotype p^2 + 2pq (see note above).

2. The first objective is to find out the value of p or q, If this is achieved, then every other value in the equation can be determined by simple calculation.

3. Take the square root of q^2 to find q.

4. Determine p by subtracting q from 1 (i.e. p = 1 – q).

5. Determine p^2 by multiplying p by itself (i.e. p^2 = p x p).

6. Determine 2pq by multiplying p times q times 2.

7. Check that your calculations are correct by adding up the values for $p^2 + q^2 + 2pq$ (the sum should equal 1 or 100%).

Worked example

In the American white population approximately 70% of people can taste the chemical phenylthiocarbamide (PTC) (the dominant phenotype), while 30% are non-tasters (the recessive phenotype).

Determine the frequency of:	Answers
(a) Homozygous recessive phenotype(q^2).	30% - provided
(b) The dominant allele (**p**).	45.2%
(c) Homozygous tasters (p^2).	20.5%
(d) Heterozygous tasters (**2pq**).	49.5%

Data: The frequency of the dominant phenotype (70% tasters) and recessive phenotype (30% non-tasters) are provided.

Working:

Recessive phenotype: q^2 =   30%
use 0.30 for calculation

therefore: **q**   =   0.5477
square root of 0.30

therefore: **p**   =   0.4523
1 – q = p
1 – 0.5477 = 0.4523

Use p and q in the equation (top) to solve any unknown:

Homozygous dominant p^2 =   0.2046
(p x p = 0.4523 x 0.4523)

Heterozygous: **2pq**   =   0.4953

1. A population of hamsters has a gene consisting of 90% M alleles (black) and 10% m alleles (grey). Mating is random.
   Data: Frequency of recessive allele (10% m) and dominant allele (90% M).

   Determine the proportion of offspring that will be black and the proportion that will be grey (show your working).

   _____

   _____

Recessive allele:	q =	
Dominant allele:	p =	
Recessive phenotype:	q^2 =	
Homozygous dominant:	p^2 =	
Heterozygous:	2pq =	

Population Genetics and Speciation

Related activities: Analysis of a Squirrel Gene Pool

RDA 3

2. You are working with pea plants and found 36 plants out of 400 were dwarf.
   Data: Frequency of recessive phenotype (36 out of 400 = 9%)

   (a) Calculate the frequency of the tall gene: _____

   (b) Determine the number of heterozygous pea plants:

   _____

   _____

Recessive allele:	q =	
Dominant allele:	p =	
Recessive phenotype:	q^2 =	
Homozygous dominant:	p^2 =	
Heterozygous:	2pq =	

3. In humans, the ability to taste the chemical phenylthiocarbaminde (PTC) is inherited as a simple dominant characteristic. Suppose you found out that 360 out of 1000 college students could not taste the chemical.
   Data: Frequency of recessive phenotype (360 out of 1000).

   (a) State the frequency of the gene for tasting PTC:

   _____

   (b) Determine the number of heterozygous students in this population:

   _____

Recessive allele:	q =	
Dominant allele:	p =	
Recessive phenotype:	q^2 =	
Homozygous dominant:	p^2 =	
Heterozygous:	2pq =	

4. A type of deformity appears in 4% of a large herd of cattle. Assume the deformity was caused by a recessive gene.
   Data: Frequency of recessive phenotype (4% deformity).

   (a) Calculate the percentage of the herd that are carriers of the gene:

   _____

   (b) Determine the frequency of the dominant gene in this case:

   _____

Recessive allele:	q =	
Dominant allele:	p =	
Recessive phenotype:	q^2 =	
Homozygous dominant:	p^2 =	
Heterozygous:	2pq =	

5. Assume you placed 50 pure bred black guinea pigs (dominant allele) with 50 albino guinea pigs (recessive allele) and allowed the population to attain genetic equilibrium (several generations have passed).
   Data: Frequency of recessive allele (50%) and dominant allele (50%).

   Determine the proportion (%) of the population that becomes white:

   _____

Recessive allele:	q =	
Dominant allele:	p =	
Recessive phenotype:	q^2 =	
Homozygous dominant:	p^2 =	
Heterozygous:	2pq =	

6. It is known that 64% of a large population exhibit the recessive trait of a characteristic controlled by two alleles (one is dominant over the other).
   Data: Frequency of recessive phenotype (64%). Determine the following:

   (a) The frequency of the recessive allele: _____

   (b) The percentage that are heterozygous for this trait: _____

   (c) The percentage that exhibit the dominant trait: _____

   (d) The percentage that are homozygous for the dominant trait: _____

   (e) The percentage that has one or more recessive alleles: _____

7. Albinism is recessive to normal pigmentation in humans. The frequency of the albino allele was 10% in a population.
   Data: Frequency of recessive allele (10% albino allele).

   Determine the proportion of people that you would expect to be albino:

   _____

   _____

Recessive allele:	q =	
Dominant allele:	p =	
Recessive phenotype:	q^2 =	
Homozygous dominant:	p^2 =	
Heterozygous:	2pq =	

Analysis of a Squirrel Gene Pool

In Olney, Illinois, in the United States, there is a unique population of albino (white) and grey squirrels. Between 1977 and 1990, students at Olney Central College carried out a study of this population. They recorded the frequency of grey and albino squirrels. The albinos displayed a mutant allele expressed as an albino phenotype only in the homozygous recessive condition. The data they collected are provided in the table below. Using the **Hardy-Weinberg equation** for calculating genotype frequencies, it was possible to estimate the frequency of the normal 'wild' allele (G) providing grey fur colouring, and the frequency of the mutant albino allele (g) producing white squirrels. This study provided real, first hand data that students could use to see how genotype frequencies can change in a real population.

Thanks to **Dr. John Stencel**, Olney Central College, Olney, Illinois, US, for providing the data for this exercise.

Grey squirrel, usual colour form

Albino form of grey squirrel

Population of grey and white squirrels in Olney, Illinois (1977-1990)

Year	Grey	White	Total	GG	Gg	gg	Freq. of g	Freq. of G
1977	602	182	784	26.85	49.93	23.21	48.18	51.82
1978	511	172	683	24.82	50.00	25.18	50.18	49.82
1979	482	134	616	28.47	49.77	21.75	46.64	53.36
1980	489	133	622	28.90	49.72	21.38	46.24	53.76
1981	536	163	699	26.74	49.94	23.32	48.29	51.71
1982	618	151	769	31.01	49.35	19.64	44.31	55.69
1983	419	141	560	24.82	50.00	25.18	50.18	49.82
1984	378	106	484	28.30	49.79	21.90	46.80	53.20
1985	448	125	573	28.40	49.78	21.82	46.71	53.29
1986	536	155	691	27.71	49.86	22.43	47.36	52.64
1987	No data collected this year							
1988	652	122	774	36.36	47.88	15.76	39.70	60.30
1989	552	146	698	29.45	49.64	20.92	45.74	54.26
1990	603	111	714	36.69	47.76	15.55	39.43	60.57

1. **Graph population changes**: Use the data in the first three columns of the table above to plot a line graph. This will show changes in the phenotypes: numbers of grey and white (albino) squirrels, as well as changes in the total population. Plot: **grey**, **white**, and **total** for each year:

(a) Determine by how much (as a %) total population numbers have fluctuated over the sampling period:

_____

(b) Describe the overall trend in total population numbers and any pattern that may exist:

_____

_____

_____

_____

_____

Related activities: Population Genetics Calculations, Sampling Populations

DA 3

Population Genetics and Speciation

2. **Graph genotype changes**: Use the data in the genotype columns of the table on the previous page to plot a line graph. This will show changes in the allele combinations (**GG**, **Gg**, **gg**). Plot: **GG**, **Gg**, and **gg** for each year:

Describe the overall trend in the frequency of:

(a) Homozygous dominant (**GG**) genotype:

_____

_____

(b) Heterozygous (**Gg**) genotype:

_____

_____

(c) Homozygous recessive (**gg**) genotype:

_____

_____

_____

3. **Graph allele changes**: Use the data in the last two columns of the table on the previous page to plot a line graph. This will show changes in the *allele frequencies* for each of the dominant (**G**) and recessive (**g**) alleles.
Plot: the frequency of **G** and the frequency of **g**:

(a) Describe the overall trend in the frequency of the dominant allele (**G**):

_____

_____

_____

(b) Describe the overall trend in the frequency of the recessive allele (**g**):

_____

_____

_____

_____

_____

4. (a) State which of the three graphs best indicates that a significant change may be taking place in the gene pool of this population of squirrels:

_____

(b) Give a reason for your answer: _____

_____

_____

5. Describe a possible cause of the changes in allele frequencies over the sampling period: _____

_____

_____

_____

Industrial Melanism

Natural selection may act on the frequencies of phenotypes (and hence genotypes) in populations in one of three different ways (through stabilising, directional, or disruptive selection). Over time, natural selection may lead to a permanent change in the genetic makeup of a population. The increased prevalence of melanic forms of the peppered moth, *Biston betularia*, during the Industrial Revolution, is one of the best known examples of directional selection following a change in environmental conditions. Although the protocols used in the central experiments on *Biston*, and the conclusions drawn from them, have been queried, it remains one of the clearest documented examples of phenotypic change in a polymorphic population.

Industrial melanism in peppered moths, *Biston betularia*

The **peppered moth**, *Biston betularia*, occurs in two forms (morphs): the grey mottled form, and a dark melanic form. Changes in the relative abundance of these two forms was hypothesised to be the result of selective predation by birds, with pale forms suffering higher mortality in industrial areas because they are more visible. The results of experiments by H.D. Kettlewell supported this hypothesis but did not confirm it, since selective predation by birds was observed but not quantified. Other research indicates that predation by birds is not the only factor determining the relative abundance of the different colour morphs.

Grey or mottled morph: vulnerable to predation in industrial areas where the trees are dark.

Melanic or carbonaria morph: dark colour makes it less vulnerable to predation in industrial areas.

Museum collections of the peppered moth made over the last 150 years show a marked change in the frequency of the melanic form. Moths collected in 1850 (above left), prior to the major onset of the industrial revolution in England. Fifty years later (above right) the frequency of the darker melanic forms had greatly increased. Even as late as the mid 20th century, coal-based industries predominated in some centres, and the melanic form occurred in greater frequency in these areas (see map, right).

Frequency of peppered moth forms in 1950

This map shows the relative frequencies of the two forms of peppered moth in the UK in 1950; a time when coal-based industries still predominated in some major centres.

Key to frequency graphs

- Grey or speckled form
- Melanic or carbonaria form

☐ Industrial areas
☐ Non-industrial areas

Scale 60 km / 60 miles

A grey (mottled) form of *Biston*, camouflaged against a lichen covered bark surface. In the absence of soot pollution, mottled forms appear to have the selective advantage.

A melanic form of *Biston*, resting on a dark branch, so that it appears as part of the branch. Note that the background has been faded out so that the moth can be seen.

Related activities: Natural Selection, Adaptations and Fitness

RDA 2

Population Genetics and Speciation

Changes in frequency of melanic peppered moths

In the 1940s and 1950s, coal burning was still at intense levels around the industrial centres of Manchester and Liverpool. During this time, the melanic form of the moth was still very dominant. In the rural areas further south and west of these industrial centres, the grey or speckled forms increased dramatically. With the decline of coal burning factories and the Clean Air Acts in cities, the air quality improved between 1960 and 1980. Sulfur dioxide and smoke levels dropped to a fraction of their previous levels. This coincided with a sharp fall in the relative numbers of melanic moths.

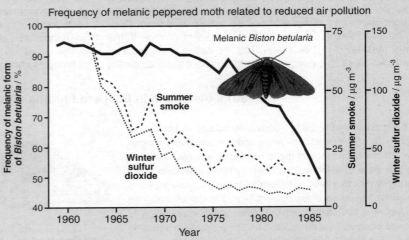

Frequency of melanic peppered moth related to reduced air pollution

1. The populations of peppered moth in England have undergone changes in the frequency of an obvious phenotypic character over the last 150 years. Describe the phenotypic character that changed in its frequency:

   _____

2. (a) Identify the (proposed) selective agent for phenotypic change in *Biston*: _____

   (b) Describe how the selection pressure on the light coloured morph has changed with changing environmental conditions over the last 150 years:

   _____

   _____

3. The industrial centres for England in 1950 were located around London, Birmingham, Liverpool, Manchester, and Leeds. Glasgow in Scotland also had a large industrial base. Comment on how the relative frequencies of the two forms of peppered moth were affected by the geographic location of industrial regions:

   _____

   _____

   _____

   _____

4. The level of pollution dropped around Manchester and Liverpool between 1960 and 1985.

   (a) State how much the pollution dropped by: _____

   (b) Describe how the frequency of the darker melanic form changed during the period of reduced pollution:

   _____

   _____

5. In the example of the peppered moths, state whether the selection pressure is disruptive, stabilising, or directional:

   _____

6. Outline the key difference between natural and artificial selection: _____

   _____

7. Discuss the statement "the environment directs natural selection": _____

   _____

   _____

   _____

Heterozygous Advantage

There are two mechanisms by which natural selection can affect allele frequencies. Firstly, there may be selection against one of the homozygotes. When one homozygous type (for example, aa), has a lower fitness than the other two genotypes (in this case, Aa or AA), the frequency of the deleterious allele will tend to decrease until it is completely eliminated. In some situations, both homozygous conditions (aa and AA) have lower fitness than the heterozygote; a situation that leads to **heterozygous advantage** and may result in the stable coexistence of both alleles in the population (**balanced polymorphism**). There are remarkably few well-documented examples in which the evidence for heterozygous advantage is conclusive. The maintenance of the sickle cell mutation (a codominant allele) in malaria-prone regions is one such example.

The Sickle Cell Allele (Hb^S)

Sickle cell disease is caused by a mutation to a gene that directs the production of the human blood protein called haemoglobin. The mutant allele is known as **Hb^S** and produces a form of haemoglobin that differs from the normal form by just one amino acid in the β-chain. This minute change however causes a cascade of physiological problems in people with the allele. Some of the red blood cells containing mutated haemoglobin alter their shape to become irregular and spiky; the so-called **sickle cells**.

Sickle cells have a tendency to clump together and work less efficiently. In people with just one sickle cell allele plus a normal allele (the heterozygote condition **Hb^{S}Hb**), there is a mixture of both red blood cell types and they are said to have the sickle cell trait. They are generally unaffected by the disease except in low oxygen environments (e.g. climbing at altitude). People with two Hb^S genes (**Hb^{S}Hb^S**) suffer severe illness and even death. For this reason Hb^S is considered **a lethal gene**.

Heterozygous Advantage in Malarial Regions

Falciparum malaria is widely distributed throughout central Africa, the Mediterranean, Middle East, and tropical and semi-tropical Asia (Fig. 1). It is transmitted by the *Anopheles* mosquito, which spreads the protozoan *Plasmodium falciparum* from person to person as it feeds on blood.

SYMPTOMS: These appear 1-2 weeks after being bitten, and include headache, shaking, chills, and fever. Falciparum malaria is more severe than other forms of malaria, with high fever, convulsions, and coma. It can be fatal within days of the first symptoms appearing.

THE PARADOX: The Hb^S allele offers considerable protection against malaria. Sickle cells have low potassium levels, which causes plasmodium parasites inside these cells to die. Those with a normal phenotype are very susceptible to malaria, but heterozygotes (**Hb^{S}Hb**) are much less so. This situation, called **heterozygous advantage**, has resulted in the Hb^S allele being present in moderately high frequencies in parts of Africa and Asia despite its harmful effects (Fig. 2). This is a special case of balanced polymorphism, called a **balanced lethal system** because neither of the homozygotes produces a phenotype that survives, but the heterozygote is viable.

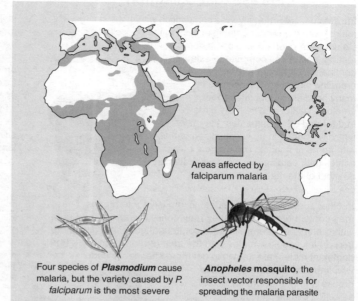

Areas affected by falciparum malaria

Four species of *Plasmodium* cause malaria, but the variety caused by *P. falciparum* is the most severe

Anopheles **mosquito**, the insect vector responsible for spreading the malaria parasite

Fig. 1: Incidence of falciparum malaria

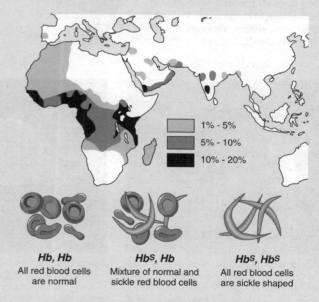

1% - 5%
5% - 10%
10% - 20%

Hb, Hb
All red blood cells are normal

Hb^S, Hb
Mixture of normal and sickle red blood cells

Hb^S, Hb^S
All red blood cells are sickle shaped

Fig. 2: Frequency of the sickle cell allele

1. With respect to the sickle cell allele, explain how **heterozygous advantage** can lead to **balanced polymorphism**:

_____

_____

_____

_____

_____

Related activities: Sickle Cell Mutation, Natural Selection
Inherited Metabolic Disorders

EA 3

Sexual Selection

The success of an individual is measured not only by the number of offspring it leaves, but also by the quality or likely reproductive success of those offspring. This means that it becomes important who its mate will be. It was Darwin (1871) who first introduced the concept of sexual selection; a special type of natural selection that produces anatomical and behavioural traits that affect an individual's ability to acquire mates. Biologists today recognise two types: **intrasexual selection** (usually male-male competition) and **intersexual selection** or mate selection. One result of either type is the evolution of **sexual dimorphism**.

Intrasexual Selection

Intrasexual selection involves competition within one sex (usually males) with the winner gaining access to the opposite sex. Competition often takes place before mating, and males compete to establish dominance or secure a territory for breeding or mating. This occurs in many species of ungulates (**deer**, antelope, cattle) and in many birds. In deer and other ungulates, the males typically engage in highly ritualised battles with horns or antlers. The winners of these battles gain dominance over rival males and do most of the mating.

In other species, males compete vigorously for territories. These may contain resources or they may consist of an isolated area within a special arena used for communal courtship display (a **lek**). In lek species, males with the best territories on a lek (the dominant males) are known to get more chances to mate with females. In some species of grouse (right), this form of sexual selection can be difficult to distinguish from intersexual selection, because once males establish their positions on the lek the females then choose among them. In species where there is limited access to females and females are promiscuous, **sperm competition** (below, centre) may also be a feature of male-male competition.

Intersexual Selection

In intersexual selection (or **mate choice**), individuals of one sex (usually the males) advertise themselves as potential mates and members of the other sex (usually the females), choose among them. Intersexual selection results in development of exaggerated ornamentation, such as elaborate plumages. Female preference for elaborate male ornaments is well supported by both anecdotal and experimental evidence. For example, in the **long-tailed widow bird** (*Euplectes progne*), females prefer males with long tails. When tails are artificially shortened or lengthened, females still prefer males with the longest tails; they therefore select for long tails, not another trait correlated with long tails.

As shown above, there was no significant difference in breeding success between the groups before the tails were altered. When the tails were cut and lengthened, breeding success went down and up respectively.

In male-male competition for mates, ornamentation is used primarily to advertise superiority to rival males, and not to mortally wound opponents. However, injuries do occur, most often between closely matched rivals, where dominance must be tested and established through the aggressive use of their weaponry rather than mere ritual duels.

Sperm competition occurs when females remate within a relatively short space of time. The outcome of sperm competition may be determined by mating order. In some species, including those that guard their mates, the first male has the advantage, but in many the advantage accrues to the sperm of the second or subsequent males.

How do male features, such as the extravagant plumage of the peacock, persist when increasingly elaborate plumage must become detrimental to survival at some point? At first, preference for such traits must confer a survival advantage. Male adornment and female preference then advance together until a stable strategy is achieved.

1. Explain the difference between **intrasexual selection** and **mate selection**, identifying the features associated with each:

_____

_____

_____

2. Suggest how sexual selection results in marked **sexual dimorphism**: _____

_____

_____

Artificial Selection

The ability of people to control the breeding of domesticated animals and crop plants has resulted in an astounding range of phenotypic variation over relatively short time periods. Most agricultural plants and animals, as well as pets, have undergone **artificial selection** (selective breeding). The dog is a striking example of this, as there are now over 400 different breeds. Artificial selection involves breeding from individuals with the most desirable phenotypes. The aim of this is to alter the average phenotype within the species. As well as selecting for physical characteristics, desirable behavioural characteristics (e.g. the ability to 'read' the body language of humans) has also been selected for in dogs. All breeds of dog are members of the same species, *Canis familiaris*. This species descended from a single wild species, the grey wolf *Canis lupus*, over 15 000 years ago. Five ancient dog breeds are recognised, from which all other breeds are thought to have descended by artificial selection.

Grey wolf *Canis lupus pallipes*

The grey wolf is distributed throughout Europe, North America, and Asia. Amongst members of this species, there is a lot of variation in coat coloration. This accounts for the large variation in coat colours of dogs today.

The Ancestor of Domestic Dogs

Until recently, it was unclear whether the ancestor to the modern domestic dogs was the desert wolf of the Middle East, the woolly wolf of central Asia, or the grey wolf of Northern Hemisphere. Recent genetic studies (mitochondrial DNA comparisons) now provide strong evidence that the ancestor of domestic dogs throughout the world is the grey wolf. It seems likely that this evolutionary change took place in a single region, most probably China.

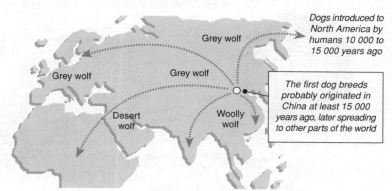

Dogs introduced to North America by humans 10 000 to 15 000 years ago

Grey wolf

Grey wolf

Grey wolf

Desert wolf

Woolly wolf

The first dog breeds probably originated in China at least 15 000 years ago, later spreading to other parts of the world

Mastiff-type
Canis familiaris inostranzevi
Originally from Tibet, the first records of this breed of dog go back to the Stoneage.

Greyhound
Canis familiaris leineri
Drawings of this breed on pottery dated from 8000 years ago in the Middle East make it one of the oldest.

Pointer-type
Canis familiaris intermedius
Probably derived from the greyhound breed for the purpose of hunting small game.

Sheepdog
Canis familiaris metris optimae
Originating in Europe, this breed has been used to guard flocks from predators for thousands of years.

Wolf-like
Canis familiaris palustris
Found in snow covered habitats in northern Europe, Asia (Siberia), and North America (Alaska).

Population Genetics and Speciation

1. Explain how artificial selection can result in changes in a gene pool over time: _____
   _____
   _____
   _____

2. Describe the behavioural tendency of wolves that predisposed them to becoming a domesticated animal:
   _____

3. List the physical and behavioural traits that would be desirable (selected for) in the following uses of a dog:

   (a) Hunting large game (e.g. boar and deer): _____

   (b) Game fowl dog: _____

   (c) Stock control (sheep/cattle dog): _____

   (d) Family pet (house dog): _____

   (e) Guard dog: _____

Related activities: The Domestication of Wheat, Livestock Improvement
Web links: Dogs and More Dogs

RA 2

182

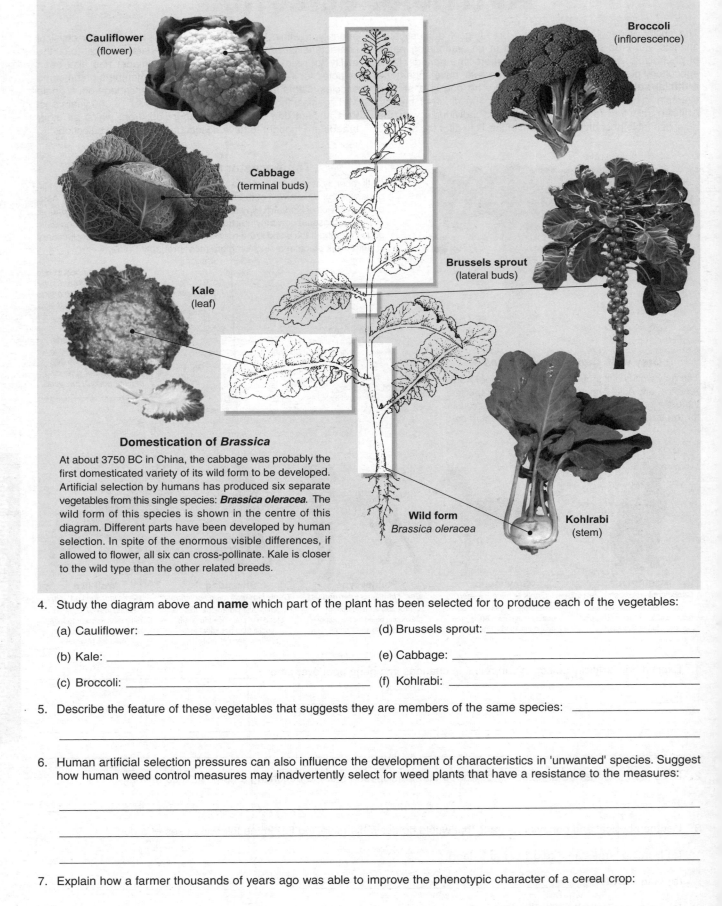

Domestication of *Brassica*

At about 3750 BC in China, the cabbage was probably the first domesticated variety of its wild form to be developed. Artificial selection by humans has produced six separate vegetables from this single species: ***Brassica oleracea***. The wild form of this species is shown in the centre of this diagram. Different parts have been developed by human selection. In spite of the enormous visible differences, if allowed to flower, all six can cross-pollinate. Kale is closer to the wild type than the other related breeds.

4. Study the diagram above and **name** which part of the plant has been selected for to produce each of the vegetables:

(a) Cauliflower: _____   (d) Brussels sprout: _____

(b) Kale: _____   (e) Cabbage: _____

(c) Broccoli: _____   (f) Kohlrabi: _____

5. Describe the feature of these vegetables that suggests they are members of the same species: _____
_____

6. Human artificial selection pressures can also influence the development of characteristics in 'unwanted' species. Suggest how human weed control measures may inadvertently select for weed plants that have a resistance to the measures:

_____
_____
_____

7. Explain how a farmer thousands of years ago was able to improve the phenotypic character of a cereal crop:

_____
_____
_____

The Domestication of Wheat

Wheat has been cultivated for more than 9000 years and has undergone many changes during the process of its domestication. The evolution of wheat involved two natural hybridisation events, accompanied by **polyploidy**. **Hybrids** are the offspring of genetically dissimilar parents. They are important because they recombine the genetic characteristics of (often inbred) parental lines and show increased **heterozygosity**. This is associated with greater adaptability, survival, growth, and fertility in the offspring; a phenomenon known as **hybrid vigour** or heterosis. There is evidence to show that **interspecific hybridisation** (i.e. between species) was an important evolutionary mechanism in the domestication of wheat. **Polyploidy** has also played a major role in the evolution of crop plants. Most higher organisms are

diploid, i.e. two sets of chromosomes (2N), one set derived from each parent. If there are more than two sets, the organism is said to be **polyploid**. Diploids formed from hybridisation of genetically very dissimilar parents, e.g. from different species, are often infertile because the two sets of chromosomes are not able to pair properly at meiosis. In such hybrids, there are no gametes produced or the gametes are abnormal. In some cases of **allopolyploidy**, the chromosomes can be doubled and a tetraploid is formed from the diploid. This restores fertility to a hybrid, because each of the original chromosome sets can pair properly with each other during meiosis. These processes are outlined in the diagram below showing the history of domestication in wheat.

Polyploidy Events in the Evolution of Wheat

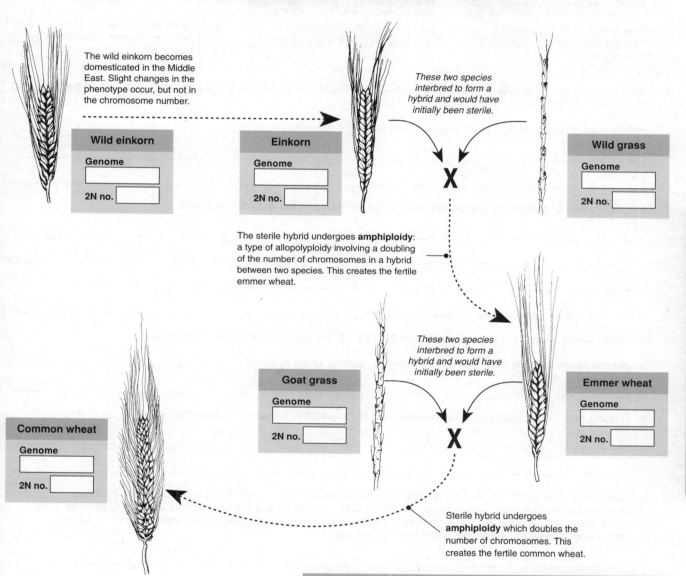

The wild einkorn becomes domesticated in the Middle East. Slight changes in the phenotype occur, but not in the chromosome number.

Wild einkorn
Genome
2N no.

Einkorn
Genome
2N no.

These two species interbred to form a hybrid and would have initially been sterile.

Wild grass
Genome
2N no.

The sterile hybrid undergoes **amphiploidy**: a type of allopolyploidy involving a doubling of the number of chromosomes in a hybrid between two species. This creates the fertile emmer wheat.

These two species interbred to form a hybrid and would have initially been sterile.

Goat grass
Genome
2N no.

Emmer wheat
Genome
2N no.

Common wheat
Genome
2N no.

Sterile hybrid undergoes **amphiploidy** which doubles the number of chromosomes. This creates the fertile common wheat.

Population Genetics and Speciation

The table on the right and the diagram above show the evolution of the common wheat. Common wheat is thought to have resulted from two sets of crossings between different species to produce hybrids. Wild einkorn (14 chromosomes, genome AA) evolved into einkorn, which crossed with a wild grass (14 chromosomes, genome BB) and gave rise to emmer wheat (28 chromosomes, genome AABB). Common wheat arose when emmer wheat was crossed with another type of grass (goat grass).

Common name	Species	Genome	Chromosomes N
Wild einkorn	*Triticum aegilopiodes*	**AA**	7
Einkorn	*Triticum monococcum*	**AA**	7
Wild grass	*Aegilops speltoides*	**BB**	7
Emmer wheat	*Triticum dicoccum*	**AABB**	14
Goat grass	*Aegilops squarrosa*	**DD**	7
Common wheat	*Triticum aestivum*	**AABBDD**	21

Related activities: Artificial Selection

A 3

184

Ancient cereal grasses had heads which shattered readily so that seeds would be scattered widely.

Modern wheat has been selected for its non shattering heads, high yield, and high gluten content.

Teosinte

Modern corn

Corn has also evolved during its domestication. Teosinte is thought to be the ancestor to both corn and maize.

1. Using the table on the previous page, label each of the wheats and grasses in the diagram with the correct **genome** and **2N** chromosome number for each plant.

2. Explain what is meant by F_1 **hybrid vigour** (heterosis): _____

_____

_____

3. Discuss the role of **polyploidy** and **interspecific hybridisation** in the evolution of wheat: _____

_____

_____

_____

_____

4. Cultivated wheat arose from wild, weedy ancestors through the selection of certain characters.

   (a) Identify the phenotypic traits that are desirable in modern wheat cultivars: _____

   _____

   (b) Suggest how ancient farmers would have carried out a selective breeding programme: _____

   _____

   _____

5. Cultivated American cotton plants have a total of 52 chromosomes (2N = 52). In each cell there are 26 large chromosomes and 26 small chromosomes. Old World cotton plants have 26 chromosomes (2N = 26), all large. Wild American cotton plants have 26 chromosomes, all small. Briefly explain how cultivated American cotton may have originated from Old World cotton and wild American cotton:

_____

_____

_____

_____

6. Discuss the need to maintain the biodiversity of wild plants and ancient farm breeds: _____

_____

_____

_____

_____

Livestock Improvement

The domestication of livestock has a long history dating back at least 8000 years. Modern sheep and cattle breeds have been developed over centuries of breeding for particular qualities, e.g. milk or meat production, tolerance to climate or terrain, or as draught animals. Furthermore, different countries have different criteria for selection, based on their local environments and consumer preferences. Cattle breeding in the UK today centres around selection for milk or meat production, with particular breeds being suited to different uses. Sheep are bred for meat, wool, and/or milk production. Traditional **selective breeding** involves choosing breeding stock with desirable qualities. This method produces a steady but slow gain in the desired qualities of the line. This **genetic gain** has accelerated in recent times with the advent of more reliable ways in which to assess genetic value and assist reproduction. New technologies refine the selection process and increase the rate at which stock improvements are made. Rates are predicted to accelerate further, as new technologies (e.g. genomic selection) available and less costly.

Beef breeds: Simmental, Aberdeen-Angus, Hereford (above), Galloway, Charolais. Consumer demand has led to the shift towards continental breeds such as Charolais because they are large, with a high proportion of lean muscle. **Desirable traits**: high muscle to bone ratio, rapid growth and weight gain, hardy, easy calving, docile temperament.

Dairy breeds: Jersey, Friesian (above), Holstein, Aryshire. **Desirable traits**: high yield of milk with high butterfat, milking speed, docile temperament, and udder characteristics such as teat placement.

Special breeds: Some cattle are bred for their suitability for climate or terrain. Scottish highland cattle (above) are a hardy, long coated breed and produce well where other breeds cannot thrive.

Artificial Selection and Genetic Gain in Cattle

Cattle are selected on the basis of particular desirable traits (e.g. milk fat or muscle mass). Most of the genetic improvement in dairy cattle has relied on selection of high quality progeny from proven stock and extensive use of superior sires through artificial insemination (AI). In beef cattle, AI is useful for introducing new breeds.

Predicted genetic gains based on different breeding methods

A **breed** is defined as a group of animals that, through selection and breeding, have come to resemble one another and pass on their traits reliably to their offspring. Improved breeding techniques accelerate what is called the **genetic progress**, i.e. the gain toward a desirable phenotype.

The graph (above) illustrates the **predicted** gains based on artificial insemination and standard selection techniques (based on criteria such as production or temperament). These are compared with the predicted gains using breeding values and various reproductive technologies such as embryo multiplication and transfer (EMT) of standard and transgenic stock, marker (gene) assisted selection, and sib-selection (selecting bulls on the basis of their sisters' performance).

A **breeding value** is a score assigned to a stud animal, derived from the sum of a range of individual scores for different characteristics such as milk yield or temperament. Accurate assessment of breeding values makes the selection process less subjective and more value based.

Sources: Breeds of Livestock, Oklahoma State University and Genetics Australia. Access information on both via Biozone's Biolinks.

1. Identify the two methods by which most of the genetic progress in dairy cattle has been achieved:

(a) _____ (b) _____

2. Explain what is meant by the term **genetic gain** as it applies to livestock breeding: _____

_____

3. Describe the contribution that new reproductive technologies are making to the selective breeding in cattle:

_____

4. Suggest why mixed breeds, such as Hereford-Friesian crosses, are popular for mixed beef/milk production:

_____

The Founder Effect

Occasionally, a small number of individuals from a large population may migrate away, or become isolated from, their original population. If this colonising or 'founder' population is made up of only a few individuals, it will probably have a non-representative sample of alleles from the parent population's gene pool. As a consequence of this **founder effect**, the colonising population may evolve differently from that of the parent population, particularly since the environmental conditions for the isolated population may be different. In some cases, it may be possible for certain alleles to be missing altogether from the individuals in the isolated population. Future generations of this population will not have this allele.

1. Compare the mainland population to the population which ended up on the island (use the spaces in the tables above):
   (a) Count the **phenotype** numbers for the two populations (i.e. the number of black, dark and pale beetles).
   (b) Count the **allele** numbers for the two populations: the number of dominant alleles (A) and recessive alleles (a). Calculate these as a percentage of the total number of alleles for each population.

2. Describe how the allele frequencies of the two populations are different: _____

   _____

3. Describe some possible ways in which various types of organism can be carried to an offshore island:

   (a) Plants: _____

   (b) Land animals: _____

   (c) Non-marine birds: _____

4. Since founder populations are often very small, describe another process that may further alter the allele frequencies:

   _____

   _____

Related activities: Gene Pool Exercise, Genetic Drift
Web links: Genetic Drift Simulation, Oceanic Island Colonisers

Population Bottlenecks

Populations may sometimes be reduced to low numbers by predation, disease, or periods of climatic change. A population crash may not be 'selective': it may affect all phenotypes equally. Large scale catastrophic events, such as fire or volcanic eruption, are examples of such non-selective events. Humans may severely (and selectively) reduce the numbers of some species through hunting and/or habitat destruction. These populations may recover, having squeezed through a 'bottleneck' of low numbers. The diagram below illustrates how population numbers may be reduced as a result of a catastrophic event. Following such an event, the small number of individuals contributing to the gene pool may not have a representative sample of the genes in the pre-catastrophe population, i.e. the allele frequencies in the remnant population may be severely altered. Genetic drift may cause further changes to allele frequencies. The small population may return to previous levels but with a reduced genetic diversity.

Population numbers

Low — High

Large population with plenty of genetic diversity.

Population crashes to a very low number and loses most of its genetic diversity.

Population grows to a large size again, but has lost much of its genetic diversity.

Time

The original gene pool is made up of the offspring of many lineages (family groups and sub-populations).

Lineage A   Lineage B   Lineage C

Extinction   Extinction   Extinction   Extinction   Extinction   Extinction

Only two descendents of lineage B survive the extinction event.

Genetic bottleneck

Extinction event such as a volcanic eruption.

All present day descendents of the original gene pool trace their ancestry back to individual B and therefore retain only a small sample of genes present in the original gene pool.

Modern Examples of Population Bottlenecks

Cheetahs: The world population of cheetahs currently stands at fewer than 20 000. Recent genetic analysis has found that the entire population enhibits very little genetic diversity. It appears that cheetahs may have narrowly escaped extinction at the end of the last ice age, about 10-20 000 years ago. If all modern cheetahs arose from a very limited genetic stock, this would explain their present lack of genetic diversity. The lack of genetic variation has resulted in a number of problems that threaten cheetah survival, including sperm abnormalities, decreased fecundity, high cub mortality, and sensitivity to disease.

Illinois prairie chicken: When Europeans first arrived in North America, there were millions of prairie chickens. As a result of hunting and habitat loss, the Illinois population of prairie chickens fell from about 100 million in 1900 to fewer than 50 in the 1990s. A comparison of the DNA from birds collected in the mid-twentieth century and DNA from the surviving population indicated that most of the genetic diversity has been lost.

1. Endangered species are often subjected to population bottlenecks. Explain how population bottlenecks affect the ability of a population of an endangered species to recover from its plight:

_____

_____

_____

2. Explain why the lack of genetic diversity in cheetahs has increased their sensitivity to disease:

_____

_____

3. Describe the effect of a population bottleneck on the potential of a species to adapt to changes (i.e. its ability to evolve):

_____

_____

Population Genetics and Speciation

Genetic Drift

Not all individuals, for various reasons, will be able to contribute their genes to the next generation. **Genetic drift** (also known as the Sewell-Wright Effect) refers to the *random changes in allele frequency* that occur in all populations, but are much more pronounced in small populations. In a small population, the effect of a few individuals not contributing their alleles to the next generation can have a great effect on allele frequencies. Alleles may even become **lost** from the gene pool altogether (frequency becomes 0%) or **fixed** as the only allele for the gene present (frequency becomes 100%).

The genetic makeup (allele frequencies) of the population changes randomly over a period of time

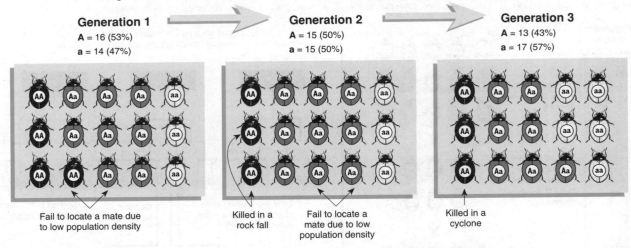

Generation 1
A = 16 (53%)
a = 14 (47%)

Fail to locate a mate due to low population density

Generation 2
A = 15 (50%)
a = 15 (50%)

Killed in a rock fall

Fail to locate a mate due to low population density

Generation 3
A = 13 (43%)
a = 17 (57%)

Killed in a cyclone

This diagram shows the gene pool of a hypothetical small population over three generations. For various reasons, not all individuals contribute alleles to the next generation. With the random loss of the alleles carried by these individuals, the allele frequency changes from one generation to the next. The change in frequency is directionless as there is no selecting force. The allele combinations for each successive generation are determined by how many alleles of each type are passed on from the preceding one.

Computer Simulation of Genetic Drift

Below are displayed the change in allele frequencies in a computer simulation showing random genetic drift. The breeding population progressively gets smaller from left to right. Each simulation was run for 140 generations.

Breeding population = 2000

Breeding population = 200

Breeding population = 20

Allele lost from the gene pool

Large breeding population
Fluctuations are minimal in large breeding populations because the large numbers buffer the population against random loss of alleles. On average, losses for each allele type will be similar in frequency and little change occurs.

Small breeding population
Fluctuations are more severe in smaller breeding populations because random changes in a few alleles cause a greater percentage change in allele frequencies.

Very small breeding population
Fluctuations in very small breeding populations are so extreme that the allele can become fixed (frequency of 100%) or lost from the gene pool altogether (frequency of 0%).

1. Explain what is meant by **genetic drift**: _____

_____

2. Describe how genetic drift affects the amount of genetic variation within very small populations: _____

_____

_____

_____

3. Identify a small breeding population of animals or plants in Britain or Europe in which genetic drift could be occurring:

_____

Related activities: Gene Pool Exercise, The Founder Effect
Web links: Genetic Drift Simulation

The Species Concept

The concept of a species is not as simple as it may first appear. Interbreeding between closely related species, such as the dog family below and 'ring species' on the next page, suggest that the boundaries of a species gene pool can be somewhat unclear. One of the best recognised definitions for a species has been proposed by the renowned evolutionary biologist Ernst Mayr: "*A species is a group of actually or potentially interbreeding natural populations that is reproductively isolated from other such groups*". Each species is provided with a unique classification name to assist with future identification.

Geographical distribution of selected *Canis* species

The global distribution of most of the species belonging to the genus *Canis* (dogs and wolves) is shown on the map to the right. The **grey wolf** (timber wolf) inhabits the cold, damp forests of North America, northern Europe and Siberia. The range of the three species of **jackal** overlap in the dry, hot, open savannah of Eastern Africa. The now-rare **red wolf** is found only in Texas, while the **coyote** is found inhabiting the open grasslands of the prairies. The **dingo** is found widely distributed throughout the Australian continent inhabiting a variety of habitats. As a result of the spread of human culture, distribution of the domesticated **dog** is global. The dog has been able to interbreed with all other members of the genus listed here, to form fertile hybrids.

Interbreeding between *Canis* species

Members of the genus to which all dogs and wolves belong present problems with the species concept. The domesticated dog is able to breed with numerous other members of the same genus to produce fertile hybrids. The coyote and red wolf in North America have ranges that overlap. They are also able to produce fertile hybrids, although these are rare. By contrast, the ranges of the three distinct species of jackal overlap in the Serengeti of Eastern Africa. These animals are highly territorial, but simply ignore members of the other jackal species and no interbreeding takes place.

For an excellent discussion of species definition among dogs see the article "The Problematic Red Wolf" in *Scientific American*, July 1995, pp. 26-31. This discusses whether or not the red wolf is a species or a long established hybrid of the grey wolf and coyote.

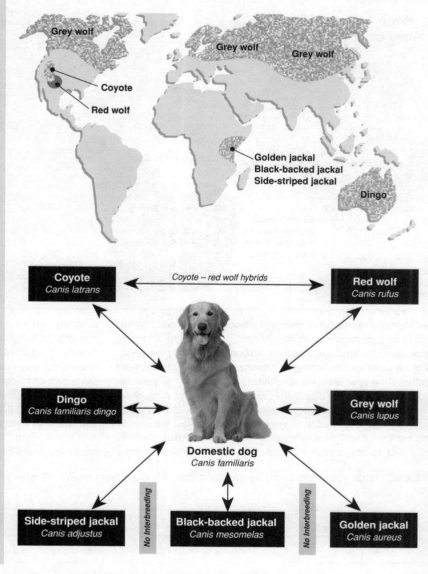

1. Describe the type of barrier that prevents the three species of jackal from interbreeding:

   _____

   _____

2. Describe the factor that has prevented the dingo from interbreeding with other *Canis* species (apart from the dog):

   _____

3. Describe a possible contributing factor to the occurrence of interbreeding between the coyote and red wolf:

   _____

   _____

4. The grey wolf is a widely distributed species. Explain why the North American population is considered to be part of the same species as the northern European and Siberian populations:

   _____

   _____

Related activities: Reproductive Isolation, Stages in Species Development, Geographical Distribution, Classification System

A 2

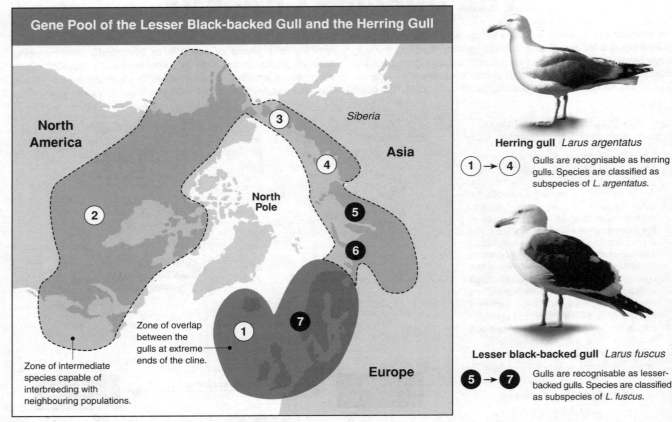

Gene Pool of the Lesser Black-backed Gull and the Herring Gull

North America

Siberia

Asia

North Pole

Zone of overlap between the gulls at extreme ends of the cline.

Zone of intermediate species capable of interbreeding with neighbouring populations.

Europe

Herring gull *Larus argentatus*

1 → 4 Gulls are recognisable as herring gulls. Species are classified as subspecies of *L. argentatus*.

Lesser black-backed gull *Larus fuscus*

5 → 7 Gulls are recognisable as lesser-backed gulls. Species are classified as subspecies of *L. fuscus*.

Species may show a gradual change in phenotype over a geographical area. Such a continuous gradual change is called a **cline**, and often occurs along the length of a country or continent. All the populations are of the same species as long as interbreeding populations link them together. **Ring species** are a special type of cline that has a circular or looped geographical distribution, resulting in the two ends of the cline overlapping. Adjacent populations can interbreed but not where the arms of the loop overlap. In the example above, four subspecies of the herring gull, and three of the lesser black-backed gull are currently recognised, forming a chain that circles the North Pole. The evidence suggests that all subspecies were derived from a single ancestral population that originated in Siberia. Members of this ancestral population migrated in opposite directions, and at the same time evolved so that, at various stages, new subspecies could be identified. Each subspecies can breed with those on either side of it. For instance, subspecies 2 can breed with subspecies 1 and 3, subspecies 4 can breed with subspecies 3 and 5, and so on. However, at the extremes of the cline, which overlaps in northern Europe, the two populations rarely interbreed; subspecies 1 and 7 behave as distinct species even though they are connected by a series of intermediate interbreeding populations.

5. Explain what you understand by the term species, identifying examples where the definition is problematic:

_____

_____

_____

6. The **ring species** (above) do not fit comfortably with the standard definition of a species. Describe the aspects of the population of gulls that:

(a) Supports the idea that they are a single species: _____

_____

_____

_____

_____

(b) Does not agree with the standard definition of a species: _____

_____

_____

_____

_____

Reproductive Isolation

The concept of a biological species is based in **reproductive isolation**, with each species isolated by factors (barriers) that prevent interbreeding (and therefore gene flow) with other species. Any factor that impedes two species from producing viable, fertile hybrids contributes to reproductive isolation. Single barriers may not completely stop gene flow, so most species have more than one type of barrier. Geographical barriers are not always classified as reproductive isolating mechanisms (RIMs) because they are not part of the species' biology. Such barriers often precede the development of other reproductive isolating mechanisms, which can operate before fertilisation (prezygotic RIMs) or after fertilisation (postzygotic RIMs).

Prezygotic Isolating Mechanisms

Spatial (geographical)

Includes physical barriers such as: mountains, rivers, altitude, oceans, isthmuses, deserts, ice sheets. There are many examples of speciation occurring as a result of isolation by oceans or by geological changes in lake basins (e.g. the proliferation of cichlid fish species in Lake Victoria). The many species of iguana from the Galapagos Islands are now quite distinct from the Central and South American species from which they arose.

Temporal (including seasonal)

Timing of mating activity for an organism may prevent contact with closely related species: nocturnal, diurnal, spring, summer, autumn, spring tide etc. Plants flower at different times of the year or even at different times of the day. Closely related animals may have quite different breeding seasons.

Ecological (habitat)

Closely related species may occupy different habitats even within the same general area. In the USA, geographically isolated species of antelope squirrels occupy different ranges either side of the Grand Canyon. The white tailed antelope squirrel inhabits the desert to the north of the canyon, while the smaller Harris's antelope squirrel has a much more limited range to the south of the canyon.

Gamete mortality

Sperm and egg fail to unite. Even if mating takes place, most gametes will fail to unite. The sperm of one species may not be able to survive in the reproductive tract of another species. Gamete recognition may be based on the presence of species specific molecules on the egg or the egg may not release the correct chemical attractants for sperm of another species.

Behavioural (ethological)

Animals attract mates with calls, rituals, dances, body language, etc. Complex displays, such as the flashes of fireflies, are quite specific. In animals, behavioural responses are a major isolating factor, preserving the integrity of mating within species. Birds exhibit a remarkable range of courtship displays that are often quite species-specific.

Structural (morphological)

Shape of the copulatory (mating) apparatus, appearance, coloration, insect attractants. Insects have a lock-and-key arrangement for their copulatory organs. Pheromone chemical attractants, which may travel many kilometres with the aid of the wind, are quite specific, attracting only members of the same species.

Land iguana: Galapagos Is.

Galapagos Is

South America

800 km

Pacific Ocean

Breeding season for species B

J F M A M J J A S O N D

Breeding season for species A

J F M A M J J A S O N D

Grand Canyon

Harris's antelope squirrel

Amphibian ovary (*Rana*)

Mammalian sperm

Peacock display of tail

Blue footed boobies courtship

Beetles mating

Damselflies mating

Population Genetics and Speciation

Related activities: The Species Concept, Stages in Species Development

A 2

Postzygotic Isolating Mechanisms

Hybrid sterility

Even if two species mate and produce hybrid offspring that are vigorous, the species are still reproductively isolated if the hybrids are sterile (genes cannot flow from one species' gene pool to the other). Such cases are common among the horse family (such as the zebra and donkey shown on the right). One cause of this sterility is the failure of meiosis to produce normal gametes in the hybrid. This can occur if the chromosomes of the two parents are different in number or structure (see the "zebronkey" karyotype on the right). The **mule**, a cross between a donkey stallion and a horse mare, is also an example of **hybrid vigour** (they are robust) as well as **hybrid sterility**. Female mules sometimes produce viable eggs but males are infertile.

Zebra stallion (2N = 44) **X** Donkey jenny (2N = 62)

Karyotype of 'Zebronkey' offspring (2N = 53)

Chromosomes contributed by zebra stallion

Chromosomes contributed by donkey jenny

Hybrid inviability

Mating between individuals of two different species may sometimes produce a zygote. In such cases, the genetic incompatibility between the two species may stop development of the fertilised egg at some embryonic stage. Fertilised eggs often fail to divide because of unmatched chromosome numbers from each gamete (a kind of aneuploidy between species). Very occasionally, the hybrid zygote will complete embryonic development but will not survive for long.

Zygote is formed

Gastrula fails to develop

Hybrid breakdown

First generation (F_1) are fertile, but the second generation (F_2) are infertile or inviable. Conflict between the genes of two species sometimes manifests itself in the second generation.

F_1

F_2

Species A Viable		Hybrid AB Reduced viability	
X		**X**	Hybrid AB Non-viable or sterile
Species B Viable		Hybrid AB Reduced viability	

1. In general terms, explain the role of reproductive isolating mechanisms in maintaining the integrity of a species:

_____

_____

2. In the following examples, classify the reproductive isolating mechanism as either **prezygotic** or **postzygotic** and describe the mechanisms by which the isolation is achieved (e.g. temporal isolation, hybrid sterility etc.):

   (a) Some different cotton species can produce fertile hybrids, but breakdown of the hybrid occurs in the next generation when the offspring of the hybrid die in their seeds or grow into defective plants:

   Prezygotic / postzygotic (delete one)   Mechanism of isolation: _____

   (b) Many plants have unique arrangements of their floral parts that stops transfer of pollen between plants:

   Prezygotic / postzygotic (delete one)   Mechanism of isolation: _____

   (c) Three species of orchid living in the same rainforest do not hybridise because they flower on different days:

   Prezygotic / postzygotic (delete one)   Mechanism of isolation: _____

   (d) Several species of the frog genus *Rana*, live in the same regions and habitats, where they may occasionally hybridise. The hybrids generally do not complete development, and those that do are weak and do not survive long:

   Prezygotic / postzygotic (delete one)   Mechanism of isolation: _____

3. Postzygotic isolating mechanisms are said to reinforce prezygotic ones. Explain why this is the case:

_____

_____

Allopatric Speciation

Allopatric speciation is a process thought to have been responsible for a great many instances of species formation. It has certainly been important in countries which have had a number of cycles of geographical fragmentation. Such cycles can occur as the result of glacial and interglacial periods, where ice expands and then retreats over a land mass. Such events are also accompanied by sea level changes which can isolate populations within relatively small geographical regions.

Stage 1: Moving into new environments

There are times when the range of a species expands for a variety of different reasons. A single population in a relatively homogeneous environment will move into new regions of their environment when they are subjected to intense competition (whether it is interspecific or intraspecific). The most severe form of competition is between members of the same species since they are competing for identical resources in the habitat. In the diagram on the right there is a 'parent population' of a single species with a common gene pool with regular 'gene flow' (theoretically any individual has access to all members of the opposite sex for mating purposes).

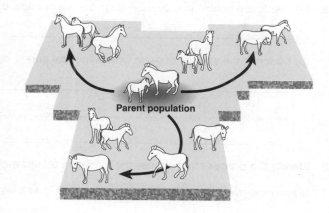

Parent population

Stage 2: Geographical isolation

Isolation of parts of the population may occur due to the formation of **physical barriers**. These barriers may cut off those parts of the population that are at the extremes of the species range and gene flow is prevented or rare. The rise and fall of the sea level has been particularly important in functioning as an isolating mechanism. Climatic change can leave 'islands' of habitat separated by large inhospitable zones that the species cannot traverse.

Example: In mountainous regions, alpine species are free to range widely over extensive habitat during cool climatic periods. During warmer periods, however, they may become isolated because their habitat is reduced to 'islands' of high ground surrounded by inhospitable lowland habitat.

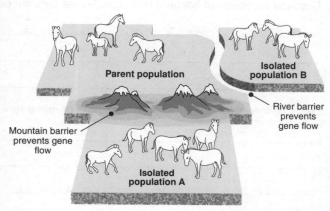

Parent population

Isolated population B

River barrier prevents gene flow

Mountain barrier prevents gene flow

Isolated population A

Stage 3: Different selection pressures

The isolated populations (A and B) may be subjected to quite different selection pressures. These will favour individuals with traits that suit each particular environment. For example, population A will be subjected to selection pressures that relate to drier conditions. This will favour those individuals with phenotypes (and therefore genotypes) that are better suited to dry conditions. They may for instance have a better ability to conserve water. This would result in improved health, allowing better disease resistance and greater reproductive performance (i.e. more of their offspring survive). Finally, as allele frequencies for certain genes change, the population takes on the status of a **subspecies**. Reproductive isolation is not yet established but the subspecies are significantly different genetically from other related populations.

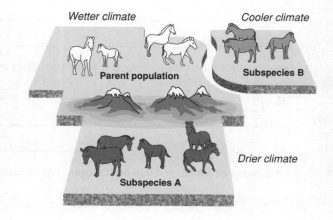

Wetter climate

Cooler climate

Parent population

Subspecies B

Drier climate

Subspecies A

Stage 4: Reproductive isolation

The separated populations (isolated subspecies) will often undergo changes in their genetic makeup as well as their behaviour patterns. These ensure that the gene pool of each population remains isolated and 'undiluted' by genes from other populations, even if the two populations should be able to remix (due to the removal of the geographical barrier). Gene flow does not occur. The arrows (in the diagram to the right) indicate the zone of overlap between two species after the new Species B has moved back into the range inhabited by the parent population. Closely-related species whose distribution overlaps are said to be **sympatric species**. Those that remain geographically isolated are called **allopatric species**.

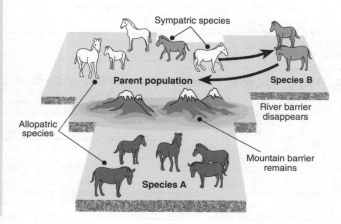

Sympatric species

Parent population

Species B

River barrier disappears

Allopatric species

Mountain barrier remains

Species A

Population Genetics and Speciation

Related activities: Reproductive Isolation, Stages in Species Development
Web links: Mechanisms of Speciation, Allopatric Speciation

RA 2

1. Describe why some animals, given the opportunity, move into new environments: _____

_____

_____

2. (a) Plants are unable to move. Explain how plants might disperse to new environments: _____

_____

(b) Describe the amount of **gene flow** within the parent population prior to and during this range expansion:

_____

_____

_____

3. Identify the **process** that causes the formation of new **mountain ranges**: _____

4. Identify the event that can cause large changes in **sea level** (up to 200 metres): _____

5. Describe six **physical barriers** that could isolate different parts of the same population: _____

_____

_____

_____

6. Describe the effect that physical barriers have on **gene flow**: _____

_____

_____

_____

7. (a) Describe four different types of **selection pressure** that could have an effect on a gene pool: _____

_____

_____

(b) Describe briefly how these selection pressures affect the isolated gene pool in terms of **allele frequencies**:

_____

_____

_____

8. Describe two **prezygotic** and two **postzygotic** reproductive isolating mechanisms (see previous activity for help):

(a) Prezygotic: _____

_____

(b) Postzygotic: _____

_____

9. Distinguish between **allopatry** and **sympatry** in populations: _____

_____

_____

_____

_____

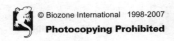

Sympatric Speciation

New species may be formed even where there is no separation of the gene pools by physical barriers. This is called **sympatric speciation**, and it is rarer than allopatric speciation, although it is not uncommon in plants which form **polyploids**. There are two situations where sympatric speciation is thought to occur. These are described below:

Speciation Through Niche Differentiation

Niche isolation

In a heterogeneous environment (one that is not the same everywhere), a population exists within a diverse collection of **microhabitats**. Some organisms prefer to occupy one particular type of 'microhabitat' most of the time, only rarely coming in contact with fellow organisms that prefer other microhabitats. Some organisms become so dependent on the resources offered by their particular microhabitat that they never meet up with their counterparts in different microhabitats.

Reproductive isolation

Finally, the individual groups have remained genetically isolated for so long because of their microhabitat preferences, that they have become reproductively isolated. They have become new species that have developed subtle differences in behaviour, structure, and physiology. Gene flow (via sexual reproduction) is limited to organisms that share a similar microhabitat preference (as shown in the diagram on the right).

Example: Some beetles prefer to find plants identical to the species they grew up on, when it is time for them to lay eggs. Individual beetles of the same species have different preferences.

An insect forced to lay its eggs on an unfamiliar plant species may give rise to a new population of flies isolated from the original population.

Original host plant species　　**New host plant species**

Original host plant species　　New host plant species

Gene flow　　No gene flow

Instant Speciation by Polyploidy

Polyploidy may result in the formation of a new species without isolation from the parent species. This event, occurring during meiosis, produces sudden reproductive isolation for the new group. Because the sex-determining mechanism is disturbed, animals are rarely able to achieve new species status this way (they are effectively sterile, e.g. tetraploid XXXX). Many plants, on the other hand, are able to reproduce vegetatively, or carry out self pollination. This ability to reproduce on their own enables such polyploid plants to produce a breeding population.

Speciation by allopolyploidy

This type of polyploidy usually arises from the doubling of chromosomes in a hybrid between two different species. The doubling often makes the hybrid fertile.

Examples: Modern wheat. Swedes are polyploid species formed from a hybrid between a type of cabbage and a type of turnip.

Origin of **polyploid event**

Parent population

New polyploid plant species spreads outwards through the existing parent population

1. Explain what is meant by **sympatric speciation** (do not confuse this with sympatric species):

_____

_____

2. Explain briefly how polyploidy may cause the formation of a new species: _____

_____

_____

3. Identify an example of a species that has been formed by polyploidy: _____

4. Explain briefly how niche differentiation may cause the formation of a new species: _____

_____

_____

Related activities: Allopatric Speciation, The Domestication of Wheat
Web links: Sympatric Speciation

A 2

Stages in Species Development

The diagram below represents a possible sequence of genetic events involved in the origin of two new species from an ancestral population. As time progresses (from top to bottom of the diagram) the amount of genetic variation increases and each group becomes increasingly isolated from the other. The mechanisms that operate to keep the two gene pools isolated from one another may begin with **geographical barriers**. This may be followed by **prezygotic** mechanisms which protect the gene pool from unwanted dilution by genes from other pools. A longer period of isolation may lead to **postzygotic** mechanisms (see the page on reproductive isolating mechanisms). As the two gene pools become increasingly isolated and different from each other, they are progressively labelled: population, race, and subspecies. Finally they attain the status of separate species.

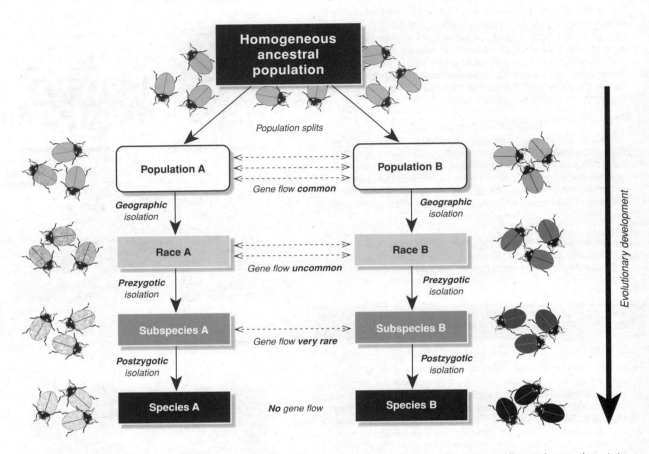

1. Explain what happens to the extent of gene flow between diverging populations as they gradually attain species status:

_____

_____

2. Early human populations about 500 000 ya were scattered across Africa, Europe, and Asia. This was a time of many regional variants, collectively called archaic *Homo sapiens*. The fossil skulls from different regions showed mixtures of characteristics, some modern and some 'primitive'. These regional populations are generally given subspecies status. Suggest reasons why gene flow between these populations may have been rare, but still occasionally occurred:

_____

_____

_____

_____

3. In the southern hemisphere, the native grey duck and the introduced mallard duck (from the Northern hemisphere) are undergoing 'species breakdown'. These two closely related species can interbreed to form hybrids.

   (a) Describe the factor preventing the two species interbreeding before the introduction of the mallards:

   _____

   (b) Describe the factor that may be deterring some of the ducks from interbreeding with the other species:

   _____

Related activities: The Species Concept, Reproductive Isolation

Patterns of Evolution

AQA-A Complete: *1-3*   **AQA-B** Complete: *1-3*   **SNAB** Complete: *1-3*   **Edexcel** Complete: *1-3*   **OCR** Complete: *1-3*

Other learning objectives are extension as appropriate

Learning Objectives

☐ 1. Compile your own glossary from the **KEY WORDS** displayed in **bold type** in the learning objectives below.

Patterns of Evolution *(pages 198-209)*

☐ 2. Distinguish patterns of species formation:
 (a) **Sequential (phyletic) speciation**
 (b) **Divergent evolution**
 (c) **Adaptive radiation** (dichotomous) speciation.
 Describe at least one example to illustrate each pattern. Recognise adaptive radiation as a form of **divergent evolution** in which a large number of species evolve to occupy vacant niches.

☐ 3. Describe how evolutionary change over time has resulted in a great diversity of forms among organisms (both living and extinct).

☐ 4. Distinguish between the two models for the pace of evolutionary change and discuss the evidence for each:
 (a) **Punctuated equilibrium**
 (b) **Gradualism**

☐ 5. Explain **convergent evolution** and describe examples of convergence in different taxa. Explain how analogous structures (analogies) may arise as a result of convergence. Recognise that analogy represents the end result of organisms evolving independently in response to the selection pressures imposed by similar environments.

☐ 6. Understand that some biologists also recognise **parallel evolution** (as distinct from convergence) to indicate evolution along similar lines in closely related groups (as opposed to more distantly related groups).

☐ 7. Using examples, distinguish between **homologous** structures and **analogous structures** arising as a result of convergent evolution. Explain the evidence for evolution provided by homologous anatomical structures, including the vertebrate pentadactyl limb.

☐ 8. Describe examples of **coevolution**, including in flowering plants and their pollinators, parasites and their hosts, and predators and their prey (including herbivory). Discuss the evidence for coevolution in species with close ecological relationships.

☐ 9. In a general way, describe how **comparative anatomy**, **embryology**, and physiology have contributed to an understanding of evolutionary relationships.

☐ 10. Discuss the significance of **vestigial organs** as indicators of evolutionary trends in some groups.

See the 'Textbook Reference Grid' on pages 8-9 for textbook page references relating to material in this topic.

Supplementary Texts

See pages 5-6 for additional details of these texts:
■ Adds, J., *et al.*, 2004. **Genetics, Evolution and Biodiversity**, (NelsonThornes), chpt. 8.

■ Clegg, C.J., 1999. **Genetics and Evolution**, (John Murray), pp. 60-78.

■ Helms, D.R. *et al.*, 1998. **Biology in the Laboratory** (W.H. Freeman), #20, #21.

■ Jones, N., *et al.*, 2001. **The Essentials of Genetics**, pp. 190-232.

See page 7 for details of publishers of periodicals:
■ **The Rise of Mammals** National Geographic, 203(4), pp. April 2003, p. 2-37. *An account of the adaptive radiation of mammals and the significance of the placenta in mammalian evolution.*

■ **Mass Extinctions** New Scientist, 11 Dec. 1999 (Inside Science). *The nature and causes of the five mass extinctions of the past. A discussion of the current sixth extinction is included.*

■ **The Sixth Extinction** National Geographic, 195(2) Feb. 1999, pp. 42-59. *High rates of extinction have occurred five times in the past. The sixth extinction is nigh, driven by human impact.*

■ **The Rise of Life on Earth: From Fins to Feet** National Geographic, 195(5) May 1999, pp. 114-127. *The evolution of diversity in the Devonian. Includes material on the evolution of tetrapods.*

■ **A Fin is a Limb is a Wing - How Evolution Fashioned its Masterworks** National Geographic, 210(5) Nov. 2006, pp. 110-135. *The evolution of complex organs and structures in animals.*

■ **Impact from the Deep** Scientific American, Oct. 2006, pp. 42-49. *Strangling heat and gases from the earth and sea, not asteroids, most likely caused several ancient mass extinctions.*

■ **A Waste of Space** New Scientist, 25 April 1998, pp. 38-39. *Vestigial organs: how they arise in an evolutionary sense and what role they may play.*

■ **Big, Bad and Furry** New Scientist, 3 Feb. 2007, pp. 32-35. *Mesozoic mammals were large and unusual at a time when dinosaurs ruled the world, and not tiny, primitive and shrew-like.*

■ **In the Blink of an Eye** New Scientist, 9 July 2005, pp. 28-31. *Rapid contemporary evolution may be widespread and humans may be driving it.*

Case study: the evolution of flight in birds

■ **Dinosaurs take Wing** National Geographic, 194(1) July 1998, pp. 74-99. *The evolution of birds from small theropod dinosaurs. This excellent account explores the homology between the typical dinosaur limb and the wing of a modern bird.*

■ **The Origin of Birds and their Flight** Scientific American, Feb. 1998, pp. 28-37. *More on avian evolution: a thorough and well illustrated account.*

■ **Which Came First: the Feather or the Bird?** Scientific American, March 2003, pp. 60-69. *An account of the evolution of feathers: how did they evolve and structure and function are related.*

See pages 10-11 for details of how to access **Bio Links** from our web site: **www.biozone.co.uk**. From Bio Links, access sites under the topics:
EVOLUTION: • BIO 414 evolution • Evolution • Evolution on the web for biology students • Harvard University biology links: evolution • The Talk.Origins archive ... *and others*

Patterns of Evolution

The diversification of an ancestral group into two or more species in different habitats is called **divergent evolution**. This process is illustrated in the diagram below, where two species have diverged from a **common ancestor**. Note that another species budded off, only to become extinct. Divergence is common in evolution. When divergent evolution involves the formation of a large number of species to occupy different niches, this is called an **adaptive** radiation. The example below (right) describes the radiation of the mammals that occurred after the extinction of the dinosaurs; an event that made niches available for exploitation. Note that the evolution of species may not necessarily involve branching: a species may accumulate genetic changes that, over time, result in the emergence of what can be recognised as a different species. This is known as **sequential evolution** (below, left).

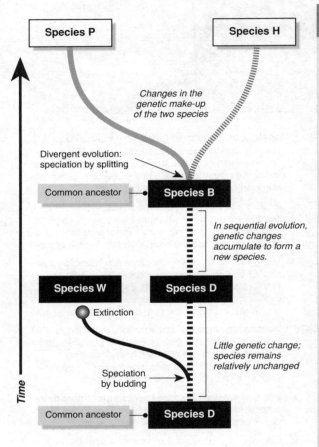

Mammalian Adaptive Radiation

Megazostrodon

***Megazostrodon*: one of the first mammals**

Megazostrodon (above) is known from fossil remains in South Africa. This shrew-like animal first appeared in the Early Jurassic period (about 195 million years ago) and probably had an insectivorous diet.

The earliest true mammals evolved about 195 million years ago, long before they underwent their major adaptive radiation some 65-50 million years ago. These ancestors to the modern forms were very small (12 cm), many were nocturnal and fed on insects and other invertebrate prey. It was climatic change as well as the extinction of the dinosaurs (and their related forms) that suddenly left many niches vacant for exploitation by such an adaptable 'generalist'. All modern mammal orders developed very quickly and early.

1. In the hypothetical example of divergent evolution illustrated above, left:

(a) Classify the type of evolution that produced species B from species D: _____

(b) Classify the type of evolution that produced species P and H from species B: _____

(c) Name all species that evolved from: **Common ancestor D**: _____ **Common ancestor B**: _____

(d) Suggest why species B, P, and H all possess a physical trait not found in species D or W: _____

_____

2. (a) Explain the distinction between **divergence** and **adaptive radiation**: _____

_____

_____

(b) Discuss the differences between **sequential evolution** and **divergent evolution**: _____

_____

_____

_____

_____

_____

_____

Related activities: The Species Concept, The Modern Theory of Evolution, Adaptive Radiation in Mammals

Geographical Distribution

The camel family, Camelidae, consists of six modern-day species that have survived on three continents: Asia, Africa and South America. They are characterised by having only two functional toes, supported by expanded pads for walking on sand or snow. The slender snout bears a cleft upper lip. The recent distribution of the camel family is fragmented. Geophysical forces such as plate tectonics and the ice age cycles have controlled the extent of their distribution. South America, for example, was separated from North America until the end of the Pliocene, about 2 million years ago. Three general principles about the dispersal and distribution of land animals are:

■ When very closely related animals (as shown by their anatomy) were present at the same time in widely separated parts of the world, it is highly probable that there was no barrier to their movement in one or both directions between the localities in the past.
■ The most effective barrier to the movement of land animals (particularly mammals) was a sea between continents (as was caused by changing sea levels during the ice ages).
■ The scattered distribution of modern species may be explained by the movement out of the area they originally occupied, or extinction in those regions between modern species.

Origin and Dispersal of the Camel Family

Recent distribution
Tertiary distribution

Ancestor of camel family originated in North America during the **tertiary period** about 40 million years ago.

Arabian camel from North Africa and the Middle East

Four llama species, including the domesticated llama and alpaca, as well as the wild guanaco and vicuña, exist in the mountainous regions of South America.

Formation of a land bridge across the Bering Strait allows passage into Asia by about 1 million years ago.

Bactrian camels in the Gobi Desert region of central Asia.

Arabian camels were introduced into Australia from the Middle east in the 1850s. An estimated 100 000 roam wild throughout Australia's sandy deserts.

Arabian camel
Camelus dromedarius

Bactrian camel
Camelus bactrianus

Vicuña
Vicugna vicugna

Llama
Lama glama

Guanaco
Lama guanicoe

1. The early camel ancestors were able to move into the tropical regions of Central and South America. Explain why this did not happen in southern Asia and southern Africa:

   _____

2. Arabian camels are found wild in the Australian outback. Explain how they got there and why they were absent during prehistoric times:

   _____

3. The camel family originated in North America. Explain why there are no camels in North America now:

   _____

4. Explain how early camels managed to get to Asia from North America: _____

   _____

5. Describe the present distribution of the camel family and explain why it is scattered (discontinuous):

   _____

   _____

Related activities: The Founder Effect, Allopatric Speciation

RA 2

Patterns of Evolution

The Rate of Evolutionary Change

The pace of evolution has been much debated, with two models being proposed: **gradualism** and **punctuated equilibrium**. Some scientists believe that both mechanisms may operate at different times and in different circumstances. Interpretations of the fossil record will vary depending on the time scales involved. During its formative millennia, a species may have accumulated its changes gradually (e.g. over 50 000 years). If that species survives for 5 million years, the evolution of its defining characteristics would have been compressed into just 1% of its (species) lifetime. In the fossil record, the species would appear quite suddenly.

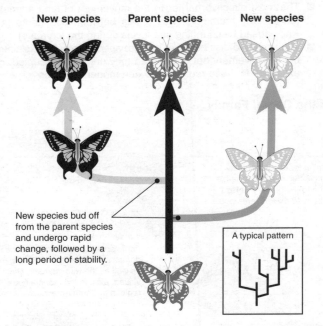

New species bud off from the parent species and undergo rapid change, followed by a long period of stability.

A typical pattern

Each species undergoes gradual changes in its genetic makeup and phenotype.

New species diverges from the parent species.

A typical pattern

Punctuated Equilibrium	**Gradualism**
There is abundant evidence in the fossil record that, instead of gradual change, species stayed much the same for long periods of time (called stasis). These periods were punctuated by short bursts of evolution which produce new species quite rapidly. According to the punctuated equilibrium theory, most of a species' existence is spent in stasis and little time is spent in active evolutionary change. The stimulus for evolution occurs when some crucial factor in the environment changes.	Gradualism assumes that populations slowly diverge from one another by accumulating adaptive characteristics in response to different selective pressures. If species evolve by gradualism, there should be transitional forms seen in the fossil record, as is seen with the evolution of the horse. Trilobites, an extinct marine arthropod, are another group of animals that have exhibited gradualism. In a study in 1987 a researcher found that they changed gradually over a three million year period.

1. Discuss the nature of the environments that would support each of the following paces of evolutionary change:

   (a) Punctuated equilibrium: _____

   _____

   _____

   (b) Gradualism: _____

   _____

   _____

2. In the fossil record of early human evolution, species tend to appear suddenly, linger for often very extended periods before disappearing suddenly. There are few examples of smooth inter-gradations from one species to the next. Explain which of the above models best describes the rate of human evolution:

   _____

3. Some species apparently show little evolutionary change over long periods of time (hundreds of millions of years).

   (a) Name two examples of such species: _____

   (b) State the term given to this lack of evolutionary change: _____

   (c) Explain why such species have changed little over evolutionary time: _____

   _____

Homologous Structures

The evolutionary relationships between groups of organisms is determined mainly by structural similarities called **homologous structures** (homologies), which suggest that they all descended from a common ancestor with that feature. The bones of the forelimb of air-breathing vertebrates are composed of similar bones arranged in a comparable pattern. This is indicative of a common ancestry. The early land vertebrates were amphibians and possessed a limb structure called the **pentadactyl limb**: a limb with five fingers or toes (below left). All vertebrates that descended from these early amphibians, including reptiles, birds and mammals, have limbs that have evolved from this same basic pentadactyl pattern. They also illustrate the phenomenon known as **adaptive radiation**, since the basic limb plan has been adapted to meet the requirements of different niches.

Generalised Pentadactyl Limb

The forelimbs and hind limbs have the same arrangement of bones but they have different names. In many cases bones in different parts of the limb have been highly modified to give it a specialised locomotory function.

Specialisations of Pentadactyl Limbs

Bird's wing

Mole's forelimb

Bat's wing

Dog's front leg

Seal's flipper

Human arm

1. Briefly describe the purpose of the major anatomical change that has taken place in each of the limb examples above:

   (a) Bird wing: *Highly modified for flight. Forelimb is shaped for aerodynamic lift and feather attachment*

   (b) Human arm: _____

   (c) Seal flipper: _____

   (d) Dog foot: _____

   (e) Mole forelimb: _____

   (f) Bat wing: _____

2. Describe how homology in the pentadactyl limb is evidence for adaptive radiation: _____

   _____

   _____

   _____

3. Homology in the behaviour of animals (for example, sharing similar courtship or nesting rituals) is sometimes used to indicate the degree of relatedness between groups. Suggest how behaviour could be used in this way:

   _____

   _____

   _____

Related activities: Adaptations and Fitness, Patterns of Evolution
Web links: All in the Family

A 2

202

Vestigial Organs

Some classes of characters are more valuable than others as reliable indicators of common ancestry. Often, the less any part of an animal is used for specialised purposes, the more important it becomes for classification. This is because common ancestry is easier to detect if a particular feature is unaffected by specific adaptations arising later during the evolution of the species. Vestigial organs are an example of this because, if they have no clear function and they are no longer subject to natural selection, they will remain unchanged through a lineage. It is sometimes argued that some vestigial organs are not truly vestigial, i.e. they may perform some small function. While this may be true in some cases, the features can still be considered vestigial if their new role is a minor one, unrelated to their original function.

Ancestors of Modern Whales

1.8 m long

2.5 m long

20-25 m long

Pakicetus (early Eocene) a carnivorous, four limbed, early Eocene whale ancestor, probably rather like a large otter. It was still partly terrestrial and not fully adapted for aquatic life.

Protocetus (mid Eocene). Much more whale-like than *Pakicetus*. The hind limbs were greatly reduced and although they still protruded from the body (arrowed), they were useless for swimming.

Basilosaurus (late Eocene). A very large ancestor of modern whales. The hind limbs contained all the leg bones, but were vestigial and located entirely within the main body, leaving a tissue flap on the surface (arrowed).

Vestigial organs are common in nature. The vestigial hind limbs of modern whales (right) provide anatomical evidence for their evolution from a carnivorous, four footed, terrestrial ancestor. The oldest known whale, *Pakicetus*, from the early Eocene (~54 mya) still had four limbs. By the late Eocene (~40 mya), whales were fully marine and had lost almost all traces of their former terrestrial life. For fossil evidence, see *Whale Origins* at: www.neoucom.edu/Depts/Anat/whaleorigins.htm

Femur

Pelvis

Vestigial hindlimb

Forelimb

Modern right whale up to 20 m

RM-DoC

Vestigial organs in birds and reptiles

In all snakes (far left), one lobe of the lung is vestigial (there is not sufficient room in the narrow body cavity for it). In some snakes there are also vestiges of the pelvic girdle and hind limbs of their walking ancestors. Like all ratites, kiwis (left) are flightless. However, more than in other ratites, the wings of kiwis are reduced to tiny vestiges. Kiwis evolved in the absence of predators to a totally ground dwelling existence.

1. In terms of natural selection explain how structures, that were once useful to an organism, could become vestigial:

   _____

   _____

   _____

   _____

2. Suggest why a vestigial structure, once it has been reduced to a certain size, may not disappear altogether:

   _____

   _____

3. Whale evolution shows the presence of **transitional forms** (fossils that are intermediate between modern forms and very early ancestors). Suggest how vestigial structures indicate the common ancestry of these forms:

   _____

   _____

   _____

A 2   **Related activities**: Natural Selection

© Biozone International 1998-2007
Photocopying Prohibited

Convergent Evolution

Not all similarities between species are inherited from a common ancestor. Species from different evolutionary branches may come to resemble each other if they have similar ecological roles and natural selection has shaped similar adaptations. This is called **convergent evolution** or **convergence**. Similarity of form due to convergence is called **analogy**.

Convergence in Swimming Form

Although similarities in body form and function can arise because of common ancestry, it may also be a result of **convergent evolution**. Selection pressures in a particular environment may bring about similar adaptations in unrelated species. These selection pressures require the solving of problems in particular ways, leading to the similarity of body form or function. The development of succulent forms in unrelated plant groups (*Euphorbia* and the cactus family) is an example of convergence in plants. In the example (right), the selection pressures of the aquatic environment have produced a similar **streamlined** body shape in unrelated vertebrate groups. Icthyosaurs, penguins, and dolphins each evolved from terrestrial species that took up an aquatic lifestyle. Their general body form has evolved to become similar to that of the shark, which has always been aquatic. Note that flipper shape in mammals, birds, and reptiles is a result of convergence, but its origin from the pentadactyl limb is an example of **homology**.

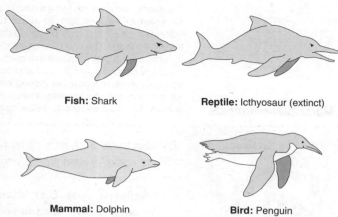

Fish: Shark   **Reptile:** Icthyosaur (extinct)

Mammal: Dolphin   **Bird:** Penguin

Analogous Structures

Analogous structures are those that have the same function and often the same basic external appearance, but **quite different origins**. The example on the right illustrates how a complex eye structure has developed independently in two unrelated groups. The appearance of the **eye** is similar, but there is no genetic relatedness between the two groups (mammals and cephalopod molluscs). The **wings** of birds and insects are also an example of analogy. The wings perform the same function, but the two groups share no common ancestor. *Longisquama*, a lizard-like creature that lived about 220 million years ago, also had 'wings' that probably allowed gliding between trees. These 'wings' were not a modification of the forearm (as in birds), but highly modified long scales or feathers extending from its back.

Mammalian eye   **Octopus eye**

Iris
Lens
Retina — Cornea

Iris
Lens
Retina — Cornea

1. In the example above illustrating convergence in swimming form, describe two ways in which the body form has evolved in response to the particular selection pressures of the aquatic environment:

   (a) _____

   _____

   (b) _____

   _____

2. Describe two of the selection pressures that have influenced the body form of the swimming animals above:

   (a) _____

   (b) _____

3. When early taxonomists encountered new species in the Pacific region and the Americas, they were keen to assign them to existing taxonomic families based on their apparent similarity to European species. In recent times, many of the new species have been found to be quite unrelated to the European families they were assigned to. Explain why the traditional approach did not reveal the true evolutionary relationships of the new species:

   _____

   _____

   _____

   _____

Related activities: Homologous Structures

RA 2

Patterns of Evolution

4. For each of the paired examples (b)-(f), briefly describe the adaptations of body shape, diet, and locomotion that appear similar in both forms, and the likely selection pressures that are acting on these mammals to produce similar body forms:

Convergence Between Marsupials and Placentals

Marsupials and **placental** mammals were separated from each other very early in mammalian evolution (about 120 mya). Marsupials were initially widely distributed throughout the ancient supercontinent of Gondwana, and there are some modern species still living in the American continent. Gondwana split up about 100 million years ago. As the placentals developed, they displaced the marsupials in most habitats around the world. The island continent of Australia, because of its early isolation by the sea, escaped this competition and placentals did not reach the continent until the arrival of humans 35 000 to 50 000 years ago. The Australian marsupials evolved into a wide variety of forms (below left) that bear a remarkable resemblance to ecologically equivalent species of North American placentals (below right).

Australia

North America

Marsupial mammals

Placental mammals

Marsupial		Placental
Wombat	(a) Adaptations: Both have rodent-like teeth, eat roots and above ground plants, and excavate burrows. Selection pressures: Diet requires chisel-like teeth for gnawing. The need to seek safety from predators on open grassland.	Wood chuck
Flying phalanger	(b) Adaptations: Selection pressures:	Flying squirrel
Marsupial mole	(c) Adaptations: Selection pressures:	Mole
Marsupial mouse	(d) Adaptations: Selection pressures:	Mouse
Tasmanian wolf (tiger)	(e) Adaptations: Selection pressures:	Wolf
Long-eared bandicoot	(f) Adaptations: Selection pressures:	Jack rabbit

Adaptive Radiation in Mammals

Adaptive radiation is diversification (both structural and ecological) among the descendants of a single ancestral group to occupy different niches. Immediately following the sudden extinction of the dinosaurs, the mammals underwent an adaptive radiation. Most of the modern mammalian groups became established very early. The diagram below shows the divergence of the mammals into major orders; many occupying niches left vacant by the dinosaurs. The vertical extent of each grey shape

shows the time span for which that particular mammal order has existed (note that the scale for the geological time scale in the diagram is not linear). Those that reach the top of the chart have survived to the present day. The width of a grey shape indicates how many species were in existence at any given time (narrow means there were few, wide means there were many). The dotted lines indicate possible links between the various mammal orders for which there is no direct fossil evidence.

1. In general terms, discuss the **adaptive radiation** that occurred in mammals: _____

_____

_____

2. Name the term that you would use to describe the animal groups at point **C** (above): _____

3. Explain what occurred at point **B** (above): _____

_____

4. Describe two things that the animal orders labelled **D** (above) have in common:

   (a) _____

   (b) _____

5. Identify the two orders that appear to have been most successful in terms of the number of species produced:

_____

6. Explain what has happened to the mammal orders labelled **A** in the diagram above: _____

_____

7. Identify the **epoch** during which there was the most adaptive radiation: _____

Related activities: Patterns of Evolution

Patterns of Evolution

RDA 3

8. Describe two key features that distinguish mammals from other vertebrates:

(a) _____ (b) _____

9. Describe the principal reproductive features distinguishing each of the major mammalian lines (sub-classes):

(a) Monotremes: _____

(b) Marsupials: _____

(c) Placentals: _____

10. There are 18 orders of placental mammals (or 17 in schemes that include the pinnipeds within the Carnivora). Their names and a brief description of the type of mammal belonging to each group is provided below. Identify and label each of the diagrams with the correct name of their Order:

Orders of Placental Mammals

Order	Description
Insectivora	Insect-eating mammals
Macroscelidae	Elephant shrews (formerly classified with insectivores)
Chiroptera	Bats
Cetacea	Whales and dolphins
Pholidota	Pangolins
Rodentia	Rodents
Probiscidea	Elephants
Sirenia	Sea-cows (manatees)
Artiodactyla	Even-toed hoofed mammals
Dermoptera	Colugos
Primates	Primates
Xenarthra	Anteaters, sloths, and armadillos
Lagomorpha	Pikas, hares, and rabbits
Carnivora	Flesh-eating mammals (canids, raccoons, bears, cats)
Pinnipedia	Seals, sealions, walruses. (Often now included as a sub-order of Carnivora).
Tubulidentata	Aardvark
Hyracoidea	Hyraxes
Perissodactyla	Odd-toed hoofed mammals

1 _____ 2 _____ 3 _____

4 _____ 5 _____ 6 _____

7 _____ 8 _____ 9 _____ 10 _____ 11 _____ 12 _____

13 _____ 14 _____ 15 _____ 16 _____ 17 _____ 18 _____

11. For each of three named **orders** of placental mammal, describe one **adaptive feature** that allows it to exploit a different niche from other placentals, and describe a **biological advantage** conferred by the adaptation:

(a) Order: _____ Adaptive feature: _____

Biological advantage: _____

(b) Order: _____ Adaptive feature: _____

Biological advantage: _____

(c) Order: _____ Adaptive feature: _____

Biological advantage: _____

Coevolution

The term **coevolution** is used to describe cases where two (or more) species reciprocally affect each other's evolution. Each party in a coevolutionary relationship exerts selective pressures on the other and, over time, the species develop a relationship that may involve mutual dependency. Coevolution is a likely consequence when different species have close ecological interactions with one another. These ecological relationships include predator-prey and parasite-host relationships and mutualistic relationships such as those between plants and their pollinators (see *Pollination Syndromes*). There are many examples of coevolution amongst parasites or pathogens and their hosts, and between predators and their prey, as shown on the following page.

Swollen-thorn *Acacia* lack the cyanogenic glycosides found in related *Acacia* spp. and the thorns are large and hollow, providing living space for the aggressive, stinging *Pseudomyrmex* ants which patrol the plant and protect it from browsing herbivores. The *Acacia* also provides the ants with protein rich food.

Hummingbirds (above) are important pollinators in the tropics. Their needle-like bills and long tongues can take nectar from flowers with deep tubes. Their ability to hover enables them to feed quickly from dangling flowers. As they feed, their heads are dusted with pollen, which is efficiently transferred between flowers.

Butterflies find flowers by vision and smell them after landing to judge their nectar source. Like bees, they can remember characteristics of desirable flowers and so exhibit constancy, which benefits both pollinator and plant. Butterfly flowers are very fragrant and are blue, purple, deep pink, red, or orange.

Bees are excellent pollinators; they are strong enough to enter intricate flowers and have medium length tongues which can collect nectar from many flower types. They have good colour vision, which extends into the UV, but they are red-blind, so bee pollinated flowers are typically blue, purplish, or white and they may have nectar guides that are visible as spots.

Beetles represent a very ancient group of insects with thousands of modern species. Their high diversity has been attributed to extensive coevolution with flowering plants. Beetles consume the ovules as well as pollen and nectar and there is evidence that ovule herbivory by beetles might have driven the evolution of protective carpels in angiosperms.

NZ's short tailed bat pollinates *Dactylanthus* flowers on the forest floor

Bats are nocturnal and colour-blind but have an excellent sense of smell and are capable of long flights. Flowers that have coevolved with bat pollinators are open at night and have light or drab colours that do not attract other pollinators. Bat pollinated flowers also produce strong fragrances that mimic the smell of bats and have a wide bell shape for easy access.

1. Using examples, explain what you understand by the term coevolution: _____

_____

_____

_____

_____

2. Describe some of the strategies that have evolved in plants to attract pollinators: _____

_____

_____

_____

_____

Patterns of Evolution

Related activities: Pollination Syndromes
Web links: The Evolutionary Arms Race, The Coevolutionary Arms Race

RA 3

Predators, Parasites, and Coevolution

Missouri Botanical Gardens

Trypanosoma brucei

CDC

Predators have obviously evolved to exploit their prey, with effective offensive weapons and hunting ability being paramount. Prey have evolved numerous strategies to protect themselves from predators, including large size and strength, protective coverings, defensive weapons, and toxicity. Lions have evolved the ability to hunt cooperatively to increase their chance of securing a kill from swift herding species such as zebra and gazelles.

Female *Helicornius* butterflies will avoid laying their eggs on plants already occupied by eggs, because their larvae are highly cannibalistic. Passionfruit plants (*Passiflora*) have exploited this by creating fake, yellow eggs on leaves and buds. *Passiflora* has many chemical defences against herbivory, but these have been breached by *Heliconius*. It has thus counter-evolved new defences against this herbivory by this genus.

Trypanosomes provide a good example of **host-parasite coevolution**. Trypanosomes must evolve strategies to evade their host's defences, but their virulence is constrained by needing to keep their host alive. Molecular studies show that *Trypanosoma brucei* coevolved in Africa with the first hominids around 5 mya, but *T. cruzi* contact with human hosts occurred in South America only after settlements were made by nomadic cultures.

3. Explain how coevolution could lead to an increase in biodiversity: _____

_____

_____

_____

_____

_____

4. Discuss some of the possible consequences of species competition: _____

_____

_____

_____

_____

_____

_____

5. The analogy of an "arms race" is often used to explain the coevolution of exploitative relationships such as those of a parasite and its host. Form a small group to discuss this idea and then suggest how the analogy is flawed:

_____

_____

_____

_____

_____

_____

_____

_____

_____

Pollination Syndromes

The mutualistic relationship between plants and their pollinators represents a classic case of coevolution. Flower structure has evolved in many different ways in response to the many types of animal pollinators. Flowers and pollinators have coordinated traits known as **pollination syndromes**. This makes it relatively easy to deduce pollinators type from the appearance of flowers (and vice versa). Plants and animals involved in such pollination associations often become highly specialised in ways that improve pollination efficiency: innovation by one party leads to some response from the other.

Controlling Pollinator Access

Flowers control pollinator access by flower shape and position.

Dandelion

Rigid inflorescences offer a stable landing platform to small or heavy insects, such as bumblebees.

Fuschia

Only animals that can hover can collect rewards from and pollinate flowers that hang upside down.

Attracting Pollinators

Flowers advertise the presence of nectar and pollen, with colour, scent, shape, and arrangement.

Rose

Daisy

Nectar guides help the pollinator to locate nectar and pollen. In this flower, the inner petals reflect UV.

While many flowers, like roses, are fragrant, flowers pollinated by flies (right) can give off dung or rotten meat smells.

Common Pollination Syndromes: Insects

Beetles

Ancient insect group
Good sense of smell
Hard, smooth bodies

Beetle-pollinated flowers

Ancient plant groups
Strong, fruity odours
Large, often flat, with easy access

Nectar-feeding flies

Sense nectar with feet
Tubular mouthparts

Nectar-feeding fly-pollinated flowers

Simple flowers with easy access
red or light colour, little odour

Moths

Many active at night
Good sense of smell
Feed with long, narrow tongues
Some need landing platforms

Moth-pollinated flowers

Flowers may be open at night
Fragrant; with heavy, musky scent
Nectar in narrow, deep tubes
landing platforms often provided

Carrion flies

Attracted by heat, odours, or
or colour of carrion or dung.
Food in the form of nectar or
pollen not required.

Carrion fly-pollinated flowers

Coloured to resemble dung or carrion
Produce heat or foul odours
No nectar or pollen reward offered

Common Pollination Syndromes: Vertebrates

Birds

Most require a perching site
Good colour vision, including red
Poor sense of smell
Feed during daylight
High energy requirements

Bird-pollinated flowers

Large and damage resistant
Often red or other bright colours
Not particularly fragrant
Open during the day
Copious nectar produced

Bats

Active at night
High food requirements
Colour blind
Good sense of smell
Cannot fly in foliage
High blossom intelligence

Bat-pollinated flowers

Open at night
Plentiful nectar and pollen offered
Light or dingy colours
Strong, often bat-like odours
Open shape, easy access
Pendulous or on the trunks of trees

Non-flying mammals

Relatively large size
High energy requirements
Colour vision may be lacking
Good sense of smell

Non-bat mammal-pollinated flowers

Robust, damage resistant
Copious, sugar-rich nectar
Dull coloured
Odorous, but not necessarily fragrant

1. (a) Describe a common pollination syndrome of an insect: _____

_____

(b) Describe a common pollination syndrome of a vertebrate: _____

_____

2. Suggest how knowledge of pollination syndromes might be used to develop testable predictions about plant and animal pollination relationships:

_____

_____

_____

_____

Patterns of Evolution

Related activities: Coevolution

A 2

Biodiversity and Classification

AQA-A	AQA-B	SNAB	Edexcel	OCR
Complete:	Complete:	Complete:	Complete:	Complete:
1-5, 7	1-6, 7-8	1-11	1-7	1-8
Extension: 7,	Extension: 7,		Extension:	Extension:
8-11	10-11		8-11	10-11

Learning Objectives

☐ 1. Compile your own glossary from the **KEY WORDS** displayed in **bold type** in the learning objectives below.

Biodiversity *(pages 189-190)*

☐ 2. Explain the importance of classification in recognising, appreciating, and conserving the **biodiversity** on Earth. Understand the concept of a **species** in terms of their reproductive isolation and potential for breeding.

Classification Systems

The five kingdoms *(pages 211, 213-230)*

☐ 3. Describe the principles and importance of scientific classification. Recognise **taxonomy** as the study of the theory and practice of classification.

☐ 4. Describe the **distinguishing features** of each kingdom in the **five kingdom classification system**:
- **Prokaryotae** (Monera): bacteria and cyanobacteria.
- **Protoctista**: includes the algae and protozoans.
- **Fungi**: includes yeasts, moulds, and mushrooms.
- **Plantae**: includes mosses, liverworts, tracheophytes.
- **Animalia**: all invertebrate phyla and the chordates.

Note that the **six kingdom classification system separates out the Prokaryotae** into two separate kingdoms, i.e. **Archaebacteria**: the archaebacteria and **Eubacteria**: the "true" bacteria.

☐ 5. Recognise at least seven major **taxonomic categories**: **kingdom, phylum, class, order, family, genus,** and **species**. Distinguish taxonomic categories from **taxa**, which are groups of organisms: "genus" is a taxonomic category, whereas the genus *Drosophila* is a taxon.

☐ 6. Understand the basis for assigning organisms to different taxonomic categories. Recall what is meant by a **distinguishing feature**. Appreciate that species are classified on the basis of **shared derived characters** rather than primitive (ancestral) characters. *For example, within the vertebrates, the presence of a backbone is a derived, therefore a distinguishing, feature. Within the mammals, the backbone is an ancestral feature and is not distinguishing, whereas mammary glands (a distinguishing feature) are derived.*

☐ 7. Explain how **binomial nomenclature** is used to classify organisms. Appreciate the problems associated with using **common names** to describe organisms.

☐ 8. Explain the relationship between classification and phylogeny. Appreciate that newer classification schemes attempt to better reflect the **phylogeny** of organisms.

New classification schemes *(pages 211-212)*

☐ 9. Recognise recent reclassifications of organisms, e.g. into six kingdoms or into three **domains** (**Archaea, Eubacteria,** and **Eukarya**). Explain the basis and rationale for these classifications.

☐ 10. Appreciate that **cladistics** provides a method of classification based on relatedness, and emphasises the presence of **shared derived characters**. Discuss the benefits and disadvantages of cladistic schemes.

Classification keys *(pages 231-233)*

☐ 11. Explain what a **classification key** is and what it is used for. Describe the essential features of a classification key. Use a simple taxonomic key to recognise and classify some common organisms.

See the 'Textbook Reference Grid' on pages 8-9 for textbook page references relating to material in this topic.

See page 7 for details of publishers of periodicals.

STUDENT'S REFERENCE

■ **How Many Mammals in Britain?** Biol. Sci. Rev., 12(4) March 2000, pp. 18-22. *The nature of species and abundance of mammals in Britain.*

■ **The Species Enigma** New Scientist, 13 June 1998 (Inside Science). *An account of the nature of species, ring species, and the status of hybrids.*

■ **Biological Taxonomy** Biol. Sci. Rev., 8(3) Jan. 1996, pp. 34-37. *Taxonomy, including a clear, concise treatment of cladistics and phenetics.*

■ **Taxonomy: The Naming Game Revisited** Biol. Sci. Rev., 9(5) May 1997, pp. 31-35. *New tools for taxonomy and how they are used (includes the exemplar of the reclassification of the kingdoms).*

TEACHER'S REFERENCE

■ **What's in a Name?** Scientific American, Nov. 2004, pp. 20-21. *A proposed classification system called phyloclode, based solely on phylogeny.*

■ **Family Feuds** New Scientist, 24 January 1998, pp. 36-40. *Molecular and morphological analysis used for determining species inter-relatedness.*

■ **The Loves of the Plants** Scientific American, Feb. 1996, pp. 98-103. *The classification of plants and the development of keys to plant identification.*

■ **The Problematic Red Wolf** Sci. American, July 1995, pp. 26-31. *Is the red wolf a species or a long-established hybrid? Correctly naming and recognising species can affect conservation efforts.*

■ **Crumbling Foundations** SSR, Dec. 2005, pp. 65-74. *The current diminution of taxonomic expertise is nothing short of a disaster.*

See pages 10-11 for details of how to access **Bio Links** from our web site: **www.biozone.co.uk**. From Bio Links, access sites under the topics:
BIODIVERSITY > Taxonomy and Classification:
• Birds and DNA • Taxonomy: Classifying life • The phylogeny of life... *and others*
MICROBIOLOGY > General Microbiology: • British Mycological Society • Major groups of prokaryotes • The microbial world ... *and others*
PLANT BIOLOGY > Classification and Diversity: • Flowering plant diversity ... *and others*

The New Tree of Life

With the advent of more efficient genetic (DNA) sequencing technology, the genomes of many bacteria began to be sequenced. In 1996, the results of a scientific collaboration examining DNA evidence confirmed the proposal that life comprises three major evolutionary lineages (domains) and not two as was the convention.

The recognised lineages were the **Eubacteria**, the **Eukarya** and the **Archaea** (formerly the Archaebacteria). The new classification reflects the fact that there are very large differences between the archaea and the eubacteria. All three domains probably had a distant common ancestor.

A Five (or Six) Kingdom World (right)

The diagram (right) represents the **five kingdom system** of classification commonly represented in many biology texts. It recognises two basic cell types: prokaryote and eukaryote. The domain Prokaryota includes all bacteria and cyanobacteria. Domain Eukaryota includes protoctists, fungi, plants, and animals. More recently, based on 16S ribosomal RNA sequence comparisons, Carl Woese divided the prokaryotes into two kingdoms, the Eubacteria and Archaebacteria. Such **six-kingdom systems** are also commonly recognised in texts.

A New View of the World (below)

In 1996, scientists deciphered the full DNA sequence of an unusual bacterium called *Methanococcus jannaschii*. An **extremophile**, this methane-producing archaebacterium lives at 85°C; a temperature lethal for most bacteria as well as eukaryotes. The DNA sequence confirmed that life consists of three major evolutionary lineages, not the two that have been routinely described. Only 44% of this archaebacterium's genes resemble those in bacteria or eukaryotes, or both.

Kingdom Animalia

Kingdom Fungi

Kingdom Plantae

Kingdom Protoctista

Includes: algae, ciliates, amoebae, and flagellates

Kingdom Prokaryotae

Includes the **Eubacteria** and **Archaebacteria**. In the six kingdom system, these two groups are formally divided into separate kingdoms.

Domain Eukaryota

Domain Prokaryota

Domain Eubacteria

Lack a distinct nucleus and cell organelles. Generally prefer less extreme environments than Archaea. Includes well-known pathogens, many harmless and beneficial species, and the cyanobacteria (photosynthetic bacteria containing the pigments chlorophyll *a* and phycocyanin).

Domain Archaea

Closely resemble eubacteria in many ways but cell wall composition and aspects of metabolism are very different. Live in extreme environments similar to those on primeval Earth. They may utilise sulfur, methane, or halogens (chlorine, fluorine), and many tolerate extremes of temperature, salinity, or pH.

Domain Eukarya

Complex cell structure with organelles and nucleus. This group contains four of the kingdoms classified under the more traditional system. Note that Kingdom Protoctista is separated into distinct groups: e.g. amoebae, ciliates, flagellates.

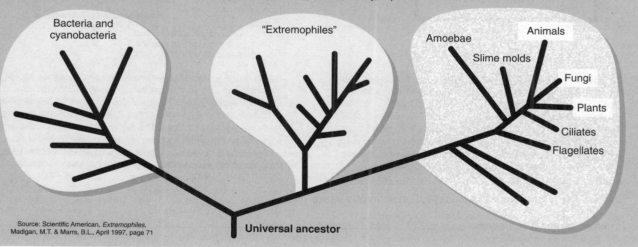

Bacteria and cyanobacteria

"Extremophiles"

Amoebae

Slime molds

Animals

Fungi

Plants

Ciliates

Flagellates

Source: Scientific American, *Extremophiles*, Madigan, M.T. & Marrs, B.L., April 1997, page 71

Universal ancestor

1. Explain why some scientists have recommended that the conventional classification of life be revised so that the Archaea, Eubacteria and Eukarya are three separate domains:

_____

_____

2. Describe one feature of the three domain system that is very different from the five kingdom classification:

_____

_____

3. Describe one way in which the three domain system and the six kingdom classification are alike: _____

_____

Related activities: New Classification Schemes, Features of Taxonomic Groups, Features of the Five Kingdoms

New Classification Schemes

Taxonomy is the study of classification. Ever since Darwin, the aim of classification has been to organise species, and to reflect their evolutionary history (**phylogeny**). Each successive group in the taxonomic hierarchy should represent finer and finer branching from a common ancestor. In order to reconstruct evolutionary history, phylogenetic trees must be based on features that are due to shared ancestry (homologies). Traditional taxonomy has relied mainly on **morphological characters** to do this. Modern technology has assisted taxonomy by providing **biochemical evidence** (from proteins and DNA) for the relatedness of species. The most familiar approach to classifying organisms is to use **classical evolutionary taxonomy**. It considers branching sequences and overall likeness. A more recent approach has been to use **cladistics**: a technique which emphasises phylogeny

or relatedness, usually based on biochemical evidence (and largely ignoring morphology or appearance). Each branch on the tree marks the point where a new species has arisen by evolution. Traditional and cladistic schemes do not necessarily conflict, but there have been reclassifications of some taxa (notably the primates, but also the reptiles, dinosaurs, and birds). Traditional taxonomists criticise cladistic schemes because they do not recognise the amount of visible change in morphology that occurs in species after their divergence from a common ancestor. Popular classifications will probably continue to reflect similarities and differences in appearance, rather than a strict evolutionary history. In this respect, they are a compromise between phylogeny and the need for a convenient filing system for species diversity.

A Classical Taxonomic View

Hominidae

Pongidae

The 'great apes'

Humans | Chimpanzees | Gorillas | Orangutans

On the basis of overall anatomical similarity (e.g. bones and limb length, teeth, musculature), apes are grouped into a family (Pongidae) that is separate from humans and their immediate ancestors (Hominidae). The family Pongidae (the great apes) is not monophyletic (of one phylogeny), because it stems from an ancestor that also gave rise to a species in another family (i.e. humans). This traditional classification scheme is now at odds with schemes derived after considering genetic evidence.

A Cladistic View

Hominidae

Homininae

Ponginae

Humans | Chimpanzees | Gorillas | Orangutans

1.4%

1.8%

A small genetic difference implies a recent common ancestor

3.6%

A greater genetic difference implies that two groups are more distantly related

Based on the evidence of genetic differences (% values above), chimpanzees and gorillas are more closely related to humans than to orangutans, and chimpanzees are more closely related to humans than they are to gorillas. Under this scheme there is no true family of great apes. The family Hominidae includes two subfamilies: Ponginae and Homininae (humans, chimpanzees, and gorillas). This classification is monophyletic: the Hominidae includes all the species that arise from a common ancestor.

1. Briefly explain the benefits of classification schemes based on:

   (a) Morphological characters: _____

   _____

   _____

   (b) Relatedness in time (from biochemical evidence): _____

   _____

   _____

2. Describe the contribution of biochemical evidence to taxonomy: _____

   _____

3. Based on the diagram above, state the family to which the chimpanzees belong under:

   (a) A traditional scheme: _____ (b) A cladistic scheme: _____

Related activities: Features of Taxonomic Groups, Classification System

Features of Taxonomic Groups

In order to distinguish organisms, it is desirable to classify and name them (a science known as **taxonomy**). An effective classification system requires features that are distinctive to a particular group of organisms. The distinguishing features of some major taxonomic groups are provided in the following pages by means of diagrams and brief summaries. Revised classification systems, recognising three domains (rather than five kingdoms) are now recognised as better representations of the true diversity of life. However, for the purposes of describing the groups with which we are most familiar, the five kingdom system (used here) is still appropriate. Note that most animals show **bilateral symmetry** (body divisible into two halves that are mirror images). **Radial symmetry** (body divisible into equal halves through various planes) is a characteristic of cnidarians and ctenophores. Definitions of specific terms relating to features of structure or function can be found in any general biology text.

Kingdom: PROKARYOTAE (Bacteria)

- Also known as monerans or prokaryotes.
- Two major bacterial lineages are recognised: the primitive **Archaebacteria** and the more advanced **Eubacteria**.
- All have a prokaryotic cell structure: they lack the nuclei and chromosomes of eukaryotic cells, and have smaller (70S) ribosomes.
- Have a tendency to spread genetic elements across species barriers by sexual conjugation, viral transduction and other processes.
- Can reproduce rapidly by binary fission in the absence of sex.

- Have evolved a wider variety of metabolism types than eukaryotes.
- Bacteria grow and divide or aggregate into filaments or colonies of various shapes.
- They are taxonomically identified by their appearance (form) and through biochemical differences.

Species diversity: 10 000 + Bacteria are rather difficult to classify to the species level because of their relatively rampant genetic exchange, and because their reproduction is usually asexual.

Eubacteria

- Also known as 'true bacteria', they probably evolved from the more ancient Archaebacteria.
- Distinguished from Archaebacteria by differences in cell wall composition, nucleotide structure, and ribosome shape.
- Very diverse group comprises most bacteria.
- The **gram stain** provides the basis for distinguishing two broad groups of bacteria. It relies on the presence of peptidoglycan (unique to bacteria) in the cell wall. The stain is easily washed from the thin peptidoglycan layer of gram negative walls but is retained by the thick peptidoglycan layer of gram positive cells, staining them a dark violet colour.

Gram-Positive Bacteria

The walls of gram positive bacteria consist of many layers of peptidoglycan forming a thick, single-layered structure that holds the gram stain.

Bacillus alvei: a gram positive, flagellated bacterium. Note how the cells appear dark.

Gram-Negative Bacteria

The cell walls of gram negative bacteria contain only a small proportion of peptidoglycan, so the dark violet stain is not retained by the organisms.

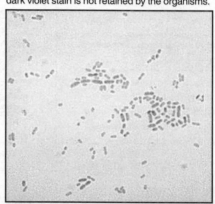

Photos: CDC

Alcaligenes odorans: a gram negative bacterium. Note how the cells appear pale.

Kingdom: FUNGI

- Heterotrophic.
- Rigid cell wall made of chitin.
- Vary from single celled to large multicellular organisms.
- Mostly saprotrophic (i.e. feeding on dead or decaying material).
- Terrestrial and immobile.

Examples:
Mushrooms/toadstools, yeasts, truffles, morels, moulds, and lichens.

Species diversity: 80 000 +

Reproduction by means of spores

Gills

- **Lichens** are symbiotic associations of a fungus (provides protection) and an alga (provides the food).

Puffballs

Filaments called hyphae form the main body of the fungus

Mushrooms

Lichens

Kingdom: PROTOCTISTA

- A diverse group of organisms that do not fit easily into other taxonomic groups.
- Unicellular or simple multicellular.
- Widespread in moist or aquatic environments.

Examples of algae: green, brown, and red algae, dinoflagellates, diatoms.

Examples of protozoa: amoebas, foraminiferans, radiolarians, ciliates.

Species diversity: 55 000 +

Algae 'plant-like' protoctists

- Autotrophic (photosynthesis)
- Characterised by the type of chlorophyll present

Cell walls of cellulose, sometimes with silica

Diatom

Protozoa 'animal-like' protoctists

- Heterotrophic nutrition and feed via ingestion
- Most are microscopic (5 μm-250 μm)

Lack cell walls

Move via projections called pseudopodia

Amoeba

Kingdom: PLANTAE

- Multicellular organisms (the majority are photosynthetic and contain chlorophyll).
- Cell walls made of cellulose; Food is stored as starch.
- Subdivided into two major divisions based on tissue structure: *Bryophytes* (non-vascular) and *Tracheophytes* (vascular) plants.

Non-Vascular Plants:

- Non vascular, lacking transport tissues (no xylem or phloem).
- They are small and restricted to moist, terrestrial environments.
- Do not possess 'true' roots, stems or leaves.

Phylum Bryophyta: Mosses, liverworts, and hornworts.

Species diversity: 18 600 +

Phylum: Bryophyta

Sexual reproductive structures

Flattened thallus (leaf like structure)

Sporophyte: reproduce by spores

Rhizoids anchor the plant into the ground

Liverworts

Mosses

Vascular Plants:

- Vascular: possess transport tissues.
- Possess true roots, stems, and leaves, as well as stomata.
- Reproduce via spores, not seeds.
- Clearly defined *alternation of sporophyte and gametophyte generations*.

Seedless Plants:

Spore producing plants, includes:

Phylum Filicinophyta: Ferns
Phylum Sphenophyta: Horsetails
Phylum Lycophyta: Club mosses

Species diversity: 13 000 +

Phylum: Lycophyta
Leaves

Club moss

Phylum: Sphenophyta
Leaves

Reproduce via spores on the underside of leaf

Horsetail

Phylum: Filicinophyta
Large dividing leaves called fronds

Rhizome

Adventitious roots

Fern

Seed Plants:

Also called Spermatophyta. Produce seeds housing an embryo. Includes:

Gymnosperms

- Lack enclosed chambers in which seeds develop.
- Produce seeds in cones which are exposed to the environment.

Phylum Cycadophyta: Cycads
Phylum Ginkgophyta: Ginkgoes
Phylum Coniferophyta: Conifers

Species diversity: 730 +

Phylum: Cycadophyta
Palm-like leaves

Cone

Cycad

Phylum: Ginkophyta
Flat leaves

Ginkgo

Phylum: Coniferophyta
Needle-like leaves

Male cones

Woody stems

Female cones

Conifer

Angiosperms

Phylum: Angiospermophyta

- Seeds in specialised reproductive structures called flowers.
- Female reproductive ovary develops into a fruit.
- Pollination usually via wind or animals.

Species diversity: 260 000 +

The phylum Angiospermophyta may be subdivided into two classes:

Class *Monocotyledoneae* (Monocots)

Class *Dicotyledoneae* (Dicots)

Angiosperms: **Monocotyledons**

Flower parts occur in multiples of 3

Leaves have parallel veins

- Only have one cotyledon (food storage organ)
- Normally herbaceous (non-woody) with no secondary growth

Lily

Examples: cereals, lilies, daffodils, palms, grasses.

Angiosperms: **Dicotyledons**

Leaves have branching veins

- Flower parts occur in multiples of 4 or 5
- Possible to have secondary growth (either herbaceous or woody)

Have two cotyledons inside the seed (acorn)

Oak

Examples: many annual plants, trees and shrubs.

Kingdom: ANIMALIA

- Over 800 000 species described in 33 existing phyla.
- Multicellular, heterotrophic organisms.
- Animal cells lack cell walls.

- Further subdivided into various major phyla on the basis of body symmetry, type of body cavity, and external and internal structures.

Phylum: Rotifera

- A diverse group of small organisms with sessile, colonial, and planktonic forms.
- Most freshwater, a few marine.
- Typically reproduce via cyclic parthenogenesis.
- Characterised by a wheel of cilia on the head used for feeding and locomotion, a large muscular pharynx (mastax) with jaw like trophi, and a foot with sticky toes.

Species diversity: 1500 +

Cilia
Head
Mastax
Foot
Toes

Bdelloid: non planktonic, creeping rotifer

Spines for protection against predators

Lorica

Ovary

Eggs

Planktonic forms swim using their crown of cilia

Phylum: Porifera

- Lack organs.
- All are aquatic (mostly marine).
- Asexual reproduction by budding.
- Lack a nervous system.

Examples: sponges.

Species diversity: 8000 +

Body wall perforated by pores through which water enters

Water leaves by a larger opening - the osculum

Sponge

- Capable of regeneration (the replacement of lost parts)
- Possess spicules (needle-like internal structures) for support and protection

Tube sponge

Sessile (attach to ocean floor)

Phylum: Cnidaria

- Two basic body forms:

  Medusa: umbrella shaped and free swimming by pulsating bell.

  Polyp: cylindrical, some are sedentary, others can glide, or somersault or use tentacles as legs.

- Some species have a life cycle that alternates between a polyp stage and a medusa stage.
- All are aquatic (most are marine).

Examples: Jellyfish, sea anemones, hydras, and corals.

Species diversity: 11 000 +

Some have air-filled floats

Nematocysts (stinging cells)

Jellyfish (Portuguese man-o-war)

Single opening acts as mouth and anus

Sea anemone

Polyps stick to seabed

Polyps may aggregate in colonies

Brain coral

Contraction of the bell propels the free swimming medusa

Colonial polyps

Phylum: Platyhelminthes

- Unsegmented body.
- Flattened body shape.
- Mouth, but no anus.
- Many are parasitic.

Examples: Tapeworms, planarians, flukes.

Species diversity: 20 000 +

Liver fluke

Hooks

Detail of head (scolex)

Tapeworm

Planarian

Phylum: Nematoda

- Tiny, unsegmented roundworms.
- Many are plant/animal parasites

Examples: Hookworms, stomach worms, lung worms, filarial worms

Species diversity: 80 000 - 1 million

Muscular pharynx
Ovary
Anus
Mouth
Intestine

A general nematode body plan

A roundworm parasite

Phylum: Annelida

- Cylindrical, segmented body with chaetae (bristles).
- Move using hydrostatic skeleton and/or parapodia (appendages).

Examples: Earthworms, leeches, polychaetes (including tubeworms).

Species diversity: 15 000 +

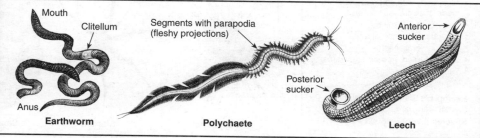

Mouth
Clitellum
Anus
Earthworm

Segments with parapodia (fleshy projections)

Polychaete

Anterior sucker

Posterior sucker

Leech

Kingdom: ANIMALIA *(continued)*

Phylum: Mollusca

- Soft bodied and unsegmented.
- Body comprises head, muscular foot, and visceral mass (organs).
- Most have radula (rasping tongue).
- Aquatic and terrestrial species.
- Aquatic species possess gills.

Examples: Snails, mussels, squid.
Species diversity: 110 000 +

Class: Bivalvia

Radula lost in bivalves
Mantle secretes shell
Two shells hinged together
Scallop

Class: Gastropoda

Mantle secretes shell
Muscular foot for locomotion
Head
Land snail

Class: Cephalopoda

Well developed eyes
Tentacles with eyes
Head
Foot divided into tentacles
Squid

Phylum: Arthropoda

- Exoskeleton made of chitin.
- Grow in stages after moulting.
- Jointed appendages.
- Segmented bodies.
- Heart found on dorsal side of body.
- Open circulation system.
- Most have compound eyes.

Species diversity: 1 million +
Make up 75% of all living animals.

Arthropods are subdivided into the following classes:

Class: Crustacea (crustaceans)
- Mainly marine.
- Exoskeleton impregnated with mineral salts.
- Gills often present.
- Includes: Lobsters, crabs, barnacles, prawns, shrimps, isopods, amphipods
- **Species diversity:** 35 000 +

Class: Arachnida (chelicerates)
- Almost all are terrestrial.
- 2 body parts: cephalothorax and abdomen (except horseshoe crabs).
- Includes: spiders, scorpions, ticks, mites, horseshoe crabs.
- **Species diversity:** 57 000 +

Class: Insecta (insects)
- Mostly terrestrial.
- Most are capable of flight.
- 3 body parts: head, thorax, abdomen.
- Include: Locusts, dragonflies, cockroaches, butterflies, bees, ants, beetles, bugs, flies, and more
- **Species diversity:** 800 000 +

Myriapods (=many legs)
Class Diplopoda (millipedes)
- Terrestrial.
- Have a rounded body.
- Eat dead or living plants.
- **Species diversity:** 2000 +

Class Chilopoda (centipedes)
- Terrestrial.
- Have a flattened body.
- Poison claws for catching prey.
- Feed on insects, worms, and snails.
- **Species diversity:** 7000 +

Class: Crustacea

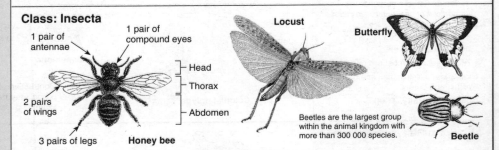

2 pairs of antennae
3 pairs of mouthparts
Cheliped (first leg)
Shrimp
Walking legs
Cephalothorax (fusion of head and thorax)
Abdomen
Swimmerets
Crab
Amphipod

Class: Arachnida

Abdomen
Simple eyes
4 pairs of walking legs
No antennae
2 pairs of feeding appendages
Cephalothorax
Spider
Scorpion
Tick
Abdomen
Telson (tail)
Carapace
Horseshoe crab
(exception to group: has 3 body parts)

Class: Insecta

1 pair of antennae
1 pair of compound eyes
Head
Thorax
Abdomen
2 pairs of wings
3 pairs of legs
Honey bee
Locust
Butterfly
Beetles are the largest group within the animal kingdom with more than 300 000 species.
Beetle

Class: Diplopoda

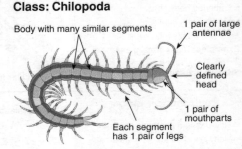

Body with many similar segments
Clearly defined head
1 pair of antennae
Each segment has 2 pairs of legs
1 pair of mouthparts

Class: Chilopoda

Body with many similar segments
1 pair of large antennae
Clearly defined head
1 pair of mouthparts
Each segment has 1 pair of legs

Phylum: Echinodermata

- Rigid body wall, internal skeleton made of calcareous plates.
- Many possess spines.
- Ventral mouth, dorsal anus.
- External fertilisation.
- Unsegmented, marine organisms.
- Tube feet for locomotion.
- Water vascular system.

Examples: Starfish, brittlestars, feather stars, sea urchins, sea lilies.
Species diversity: 6000 +

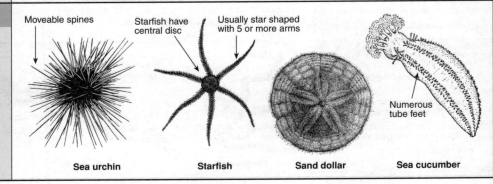

Moveable spines
Starfish have central disc
Usually star shaped with 5 or more arms
Numerous tube feet
Sea urchin
Starfish
Sand dollar
Sea cucumber

Kingdom: ANIMALIA (continued)

Phylum: Chordata

- Dorsal notochord (flexible, supporting rod) present at some stage in the life history.
- Post-anal tail present at some stage in their development.
- Dorsal, tubular nerve cord.
- Pharyngeal slits present.
- Circulation system closed in most.
- Heart positioned on ventral side.

Species diversity: 48 000 +

- A very diverse group with several sub-phyla:
  - Urochordata (sea squirts, salps)
  - Cephalochordata (lancelet)
  - Craniata (vertebrates)

Sub-Phylum Craniata (vertebrates)
- Internal skeleton of cartilage or bone.
- Well developed nervous system.
- Vertebral column replaces notochord.
- Two pairs of appendages (fins or limbs) attached to girdles.

Further subdivided into:

Class: Chondrichthyes (cartilaginous fish)
- Skeleton of cartilage (not bone).
- No swim bladder.
- All aquatic (mostly marine).
- Include: Sharks, rays, and skates.

Species diversity: 850 +

Class: Osteichthyes (bony fish)
- Swim bladder present.
- All aquatic (marine and fresh water).

Species diversity: 21 000 +

Class: Amphibia (amphibians)
- Lungs in adult, juveniles may have gills (retained in some adults).
- Gas exchange also through skin.
- Aquatic and terrestrial (limited to damp environments).
- Include: Frogs, toads, salamanders, and newts.

Species diversity: 3900 +

Class Reptilia (reptiles)
- Ectotherms with no larval stages.
- Teeth are all the same type.
- Eggs with soft leathery shell.
- Mostly terrestrial.
- Include: Snakes, lizards, crocodiles, turtles, and tortoises.

Species diversity: 7000 +

Class: Aves (birds)
- Terrestrial endotherms.
- Eggs with hard, calcareous shell.
- Strong, light skeleton.
- High metabolic rate.
- Gas exchange assisted by air sacs.

Species diversity: 8600 +

Class: Mammalia (mammals)
- Endotherms with hair or fur.
- Mammary glands produce milk.
- Glandular skin with hair or fur.
- External ear present.
- Teeth are of different types.
- Diaphragm between thorax/abdomen.

Species diversity: 4500 +
Subdivided into three subclasses:
Monotremes, marsupials, placentals.

Class: Chondrichthyes (cartilaginous fish)

Lateral line sense organ

Asymmetrical tail fin provides lift

Skin with toothlike scales

Ectotherms with endoskeleton made of cartilage

Pelvic fin

Pectoral fin

Hammerhead shark

No operculum (bony flap) over gills

Stingray

Class: Osteichthyes (bony fish)

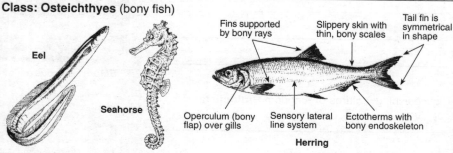

Eel

Seahorse

Fins supported by bony rays

Slippery skin with thin, bony scales

Tail fin is symmetrical in shape

Operculum (bony flap) over gills

Sensory lateral line system

Ectotherms with bony endoskeleton

Herring

Class: Amphibia

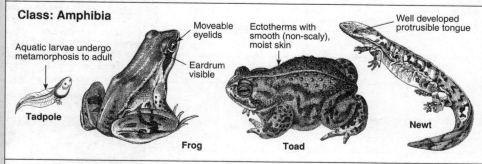

Moveable eyelids

Ectotherms with smooth (non-scaly), moist skin

Well developed protrusible tongue

Aquatic larvae undergo metamorphosis to adult

Eardrum visible

Tadpole

Frog

Toad

Newt

Class: Reptilia

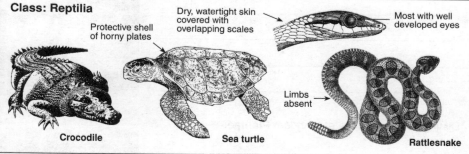

Dry, watertight skin covered with overlapping scales

Most with well developed eyes

Protective shell of horny plates

Limbs absent

Crocodile

Sea turtle

Rattlesnake

Class: Aves

Some birds are flightless

Feathers

Forelimbs modified as wings

Horny beak with no teeth

Horny scales on feet only

Penguin

Kiwi

Seagull

Class: Mammalia

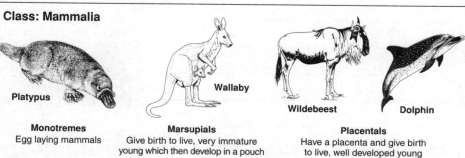

Wallaby

Platypus

Wildebeest

Dolphin

Monotremes
Egg laying mammals

Marsupials
Give birth to live, very immature young which then develop in a pouch

Placentals
Have a placenta and give birth to live, well developed young

Features of the Five Kingdoms

The classification of organisms into taxonomic groups is based on how biologists believe they are related in an evolutionary sense. Organisms in a taxonomic group share features which set them apart from other groups. By identifying these features, it is possible to develop an understanding of the evolutionary history of the group. The focus of this activity is to summarise the **distinguishing features** of each of the five kingdoms in the five kingdom classification system.

1. Distinguishing features of Kingdom **Prokaryotae**:

_____

_____

_____

_____

_____

2. Distinguishing features of Kingdom **Protoctista**:

_____

_____

_____

_____

_____

3. Distinguishing features of Kingdom **Fungi**:

_____

_____

_____

_____

_____

_____

4. Distinguishing features of Kingdom **Plantae**:

_____

_____

_____

_____

_____

5. Distinguishing features of Kingdom **Animalia**:

_____

_____

_____

_____

_____

_____

_____

Staphylococcus dividing

Helicobacter pylori

Trypanosoma parasite — Red blood cell

Amoeba

Mushrooms

Yeast cells in solution

Maple seeds

Pea plants

Cicada moulting

Gibbon

Related activities: The New Tree of Life, Features of Taxonomic Groups

Classification System

The classification of organisms is designed to reflect how they are related to each other. The fundamental unit of classification of living things is the **species**. Its members are so alike genetically that they can interbreed. This genetic similarity also means that they are almost identical in their physical and other characteristics. Species are classified further into larger, more comprehensive categories (higher taxa). It must be emphasised that all such higher classifications are human inventions to suit a particular purpose.

1. The table below shows part of the classification for humans using the seven major levels of classification. For this question, use the example of the classification of the European hedgehog, on the next page, as a guide.

    (a) Complete the list of the classification levels on the left hand side of the table below:

	Classification level	Human classification
1.	_____	_____
2.	_____	_____
3.	_____	_____
4.	_____	_____
5.	Family	Hominidae
6.	_____	_____
7.	_____	_____

    (b) The name of the Family that humans belong to has already been entered into the space provided. Complete the classification for humans (*Homo sapiens*) on the table above.

2. Describe the two-part scientific naming system (called the **binomial system**) that is used to name organisms:

    _____

    _____

3. Give two reasons why the classification of organisms is important:

    (a) _____

    _____

    (b) _____

    _____

4. Traditionally, the classification of organisms has been based largely on similarities in physical appearance. More recently, new methods involving biochemical comparisons have been used to provide new insights into how species are related. Describe an example of a biochemical method for comparing how species are related:

    _____

    _____

    _____

5. As an example of physical features being used to classify organisms, mammals have been divided into three major sub-classes: monotremes, marsupials, and placentals. Describe the main physical feature distinguishing each of these taxa:

    (a) Monotreme: _____

    _____

    (b) Marsupial: _____

    _____

    (c) Placental: _____

    _____

Related activities: New Classification Schemes, Classification Keys, Features of Taxonomic Groups

RA 2

Classification of the European Hedgehog

Below is the classification for the **European hedgehog**. Only one of each group is subdivided in this chart showing the levels that can be used in classifying an organism. Not all possible subdivisions have been shown here. For example, it is possible to indicate such categories as **super-class** and **sub-family**. The only natural category is the **species**, often separated into geographical **races**, or **sub-species**, which generally differ in appearance.

Kingdom: **Animalia**
Animals: one of five kingdoms

Phylum: **Chordata**
Animals with a notochord (supporting rod of cells along the upper surface).
tunicates, salps, lancelets, and vertebrates

23 other phyla

Sub-phylum: **Vertebrata**
Animals with backbones.
fish, amphibians, reptiles, birds, mammals

Class: **Mammalia**
Animals that suckle their young on milk from mammary glands.
placentals, marsupials, monotremes

Sub-class: **Eutheria**
Mammals whose young develop for some time in the female's reproductive tract gaining nourishment from a placenta.
placentals

Order: **Insectivora**
Insect eating mammals.
An order of over 300 species of primitive, small mammals that feed mainly on insects and other small invertebrates.

17 other orders

Sub-order: **Erinaceomorpha**
The hedgehog-type insectivores. One of the three suborders of insectivores. The other suborders include the tenrec-like insectivores (*tenrecs and golden moles*) and the shrew-like insectivores (*shrews, moles, desmans, and solenodons*).

Family: **Erinaceidae**
The only family within this suborder. Comprises two subfamilies: the true or spiny hedgehogs and the moonrats (gymnures). Representatives in the family include the desert hedgehog, long-eared hedgehog, and the greater and lesser moonrats.

Genus: ***Erinaceus***
One of eight genera in this family. The genus *Erinaceus* includes four Eurasian species and another three in Africa.

7 other genera

Species: ***europaeus***
The European hedgehog. Among the largest of the spiny hedgehogs. Characterised by a dense covering of spines on the back, the presence of a big toe (hallux) and 36 teeth.

6 other species

The order *Insectivora* was first introduced to group together shrews, moles, and hedgehogs. It was later extended to include tenrecs, golden moles, desmans, tree shrews, and elephant shrews, and the taxonomy of the group became very confused. Recent reclassification of the elephant shrews and tree shrews into their own separate orders has made the Insectivora a more cohesive group taxonomically.

European hedgehog
Erinaceus europaeus

Features of Microbial Groups

A microorganism (or microbe) is literally a microscopic organism. The term is usually reserved for the organisms studied in microbiology: bacteria, fungi, microscopic protoctistans, and viruses. Most of these taxa also have macroscopic representatives. This is especially the case within the fungi. The distinction between a macrofungus and a microfungus is an artificial but convenient one. Unlike microfungi, which are made conspicuous by the diseases or decay they cause, macrofungi are the ones most likely to be observed with the naked eye. Examples of microfungi, which include yeasts and pathogenic species, are illustrated in this activity. Macrofungi, which include mushrooms, toadstools, and lichens, are illustrated in *Features of Macrofungi and Plants*.

Biodiversity and Classification

1. Distinguishing features of Kingdom **Prokaryotae**:

_____
_____
_____
_____
_____
_____
_____
_____
_____

2. Distinguishing features of Kingdom **Protoctista**:

_____
_____
_____
_____
_____
_____
_____
_____
_____
_____

3. Distinguishing features of Kingdom **Fungi** (microfungi):

_____
_____
_____
_____
_____
_____
_____
_____
_____
_____
_____

Spirillum bacteria

Staphylococcus

Anabaena cyanobacterium

Foraminiferan

Spirogyra algae

Diatoms: *Pleurosigma*

Curvularia sp. conidiophore

Yeast cells in solution

Microsporum distortum (a pathogenic fungus)

Related activities: The New Tree of Life, Features of Taxonomic Groups, Features of Macrofungi and Plants

ERA 1

Features of Macrofungi and Plants

Although plants and fungi are some of the most familiar organisms in our environment, their classification has not always been straightforward. We know now that the plant kingdom is monophyletic, meaning that it is derived from a common ancestor. The variety we see in plant taxa today is a result of their enormous diversification from the first plants. Although the fungi were once grouped together with the plants, they are unique organisms that differ from other eukaryotes in their mode of nutrition, structural organisation, growth, and reproduction. The focus of this activity is to summarise the features of the fungal kingdom, the major divisions of the plant kingdom, and the two classes of flowering plants (angiosperms).

Lichen

Bracket fungus

Liverwort

Moss

Fern frond

Ground fern

Pine tree cone

Cycad

Coconut palms

Wheat plants

Deciduous tree

Flowering plant

1. **Macrofungi** features:

2. **Moss and liverwort** features:

3. **Fern** features:

4. **Gymnosperm** features:

5. **Monocot angiosperm** features:

6. **Dicot angiosperm** features:

Features of Animal Taxa

The animal kingdom is classified into about 35 major **phyla**. Representatives of the more familiar taxa are illustrated below: **cnidarians** (includes jellyfish, sea anemones, and corals), **annelids** (segmented worms), **arthropods** (insects, crustaceans, spiders, scorpions, centipedes and millipedes), **molluscs** (snails, bivalve shellfish, squid and octopus), **echinoderms** (starfish and sea urchins), **vertebrates** from the phylum **chordates** (fish, amphibians, reptiles, birds, and mammals). The **arthropods** and the **vertebrates** have been represented in more detail, giving the **classes** for each of these **phyla**. This activity asks you to describe the **distinguishing features** of each of the taxa represented below.

Sea anemones

Jellyfish

Tubeworms

Earthworm

Long-horned beetle

Butterfly

Crab

Woodlouse

Scorpion

Spider

Centipede

Millipede

1. **Cnidarian** features: _____

2. **Annelid** features: _____

3. **Insect** features: _____

4. **Crustacean** features: _____

5. **Arachnid** features: _____

6. **Myriapod** (class Chilopoda and Diplopoda) features: _____

Related activities: The New Tree of Life, Features of Taxonomic Groups

ER 1

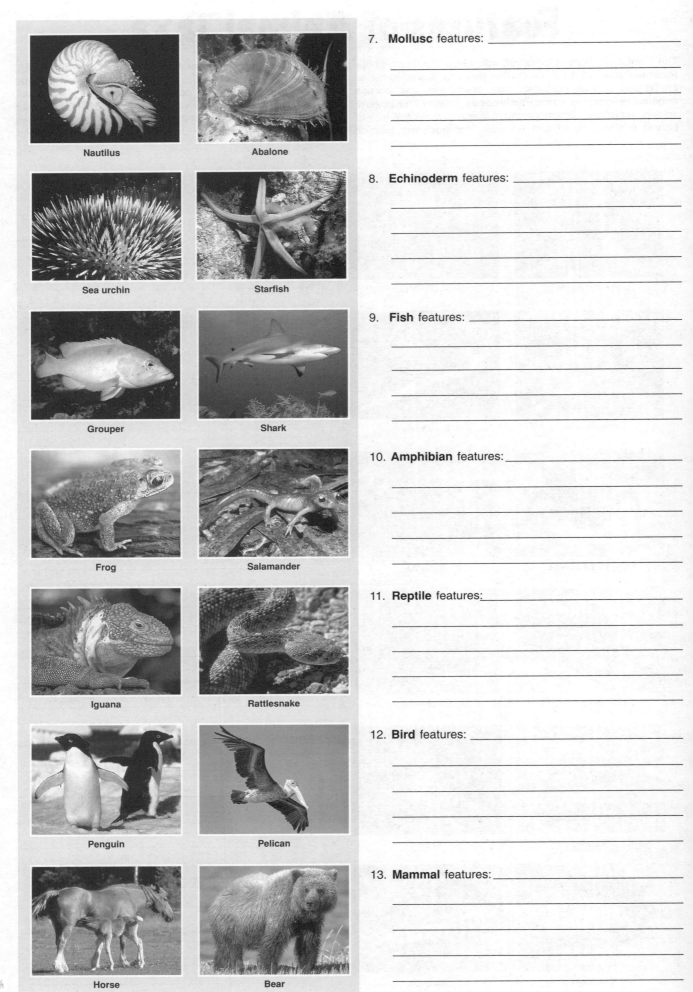

Nautilus

Abalone

Sea urchin

Starfish

Grouper

Shark

Frog

Salamander

Iguana

Rattlesnake

Penguin

Pelican

Horse

Bear

7. **Mollusc** features: _____

_____

_____

_____

_____

8. **Echinoderm** features: _____

_____

_____

_____

_____

9. **Fish** features: _____

_____

_____

_____

_____

10. **Amphibian** features: _____

_____

_____

_____

_____

11. **Reptile** features: _____

_____

_____

_____

_____

12. **Bird** features: _____

_____

_____

_____

_____

13. **Mammal** features: _____

_____

_____

_____

_____

The Classification of Life

For this activity, cut away the two pages of diagrams that follow from your book. The five kingdoms that all living things are grouped into, are listed on this page and the following page.

1. Cut out all of the images of different living organisms (cut around each shape closely, taking care to include their names).

2. Sort them into their classification groups by placing them into the spaces provided on this and the following page.

3. To fix the images in place, first use a temporary method, so that you can easily reposition them if you need to. Make a permanent fixture when you are completely satisfied with your placements on the page.

Kingdom Prokaryotae (Monera)

Kingdom Protoctista

Kingdom Fungi

Kingdom Plantae

Phylum Bryophyta

Phylum Filicinophyta

Phylum Angiospermophyta

Class Monocotyledoneae   Class Dicotyledoneae

Phylum Cycadophyta

Phylum Coniferophyta

Related activities: Features of Taxonomic Groups, Features of Microbial Groups, Features of Macrofungi and Plants, Features of Animal Taxa, Classification Keys

PR 1

Kingdom Animalia

Phylum Porifera	Phylum Cnidaria	Phylum Platyhelminthes	Phylum Annelida

Phylum Mollusca

Class Gastropoda   Class Bivalvia   Class Cephalopoda

Phylum Echinodermata

Phylum Arthropoda
Superclass Crustacea

Classes Chilopoda/Diplopoda   Class Arachnida   Class Insecta

Class Chondrichthyes   **Phylum Chordata**   Class Osteichthyes   Class Amphibia

Class Reptilia
Order Squamata   Order Crocodilia   Order Chelonia   Class Aves

Class Mammalia
Subclass Prototheria   Subclass Metatheria   Subclass Eutheria

Cut out the organisms on this page and paste them into the spaces provided at the start of this activity. Organisms are not to scale.

Marine jellyfish

Liverwort

Earthworm

Blowfly

Seahorse

Centipede

Parrot

Sea star

Tube sponge

Tapeworm

Peccary

Echidna

Staphylococcus bacteria

Freshwater eel

Stingray

Scorpion

Frog

Garden snail

Euglena

Tortoise

Cycad

Kangaroo

Amoeba

African violets

This page has been deliberately left blank

Cut out the images
on the other side of this page

Cut out the organisms on this page and paste them into the spaces provided at the start of this activity. Organisms are not to scale.

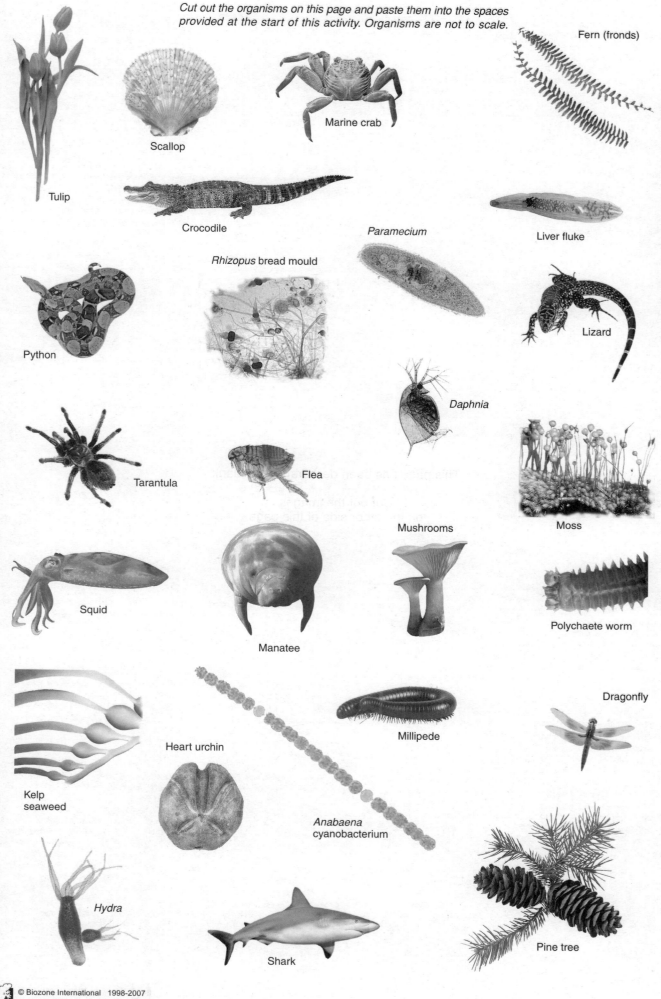

Fern (fronds)

Scallop

Marine crab

Tulip

Crocodile

Paramecium

Liver fluke

Rhizopus bread mould

Python

Lizard

Daphnia

Tarantula

Flea

Moss

Mushrooms

Squid

Manatee

Polychaete worm

Dragonfly

Kelp seaweed

Heart urchin

Millipede

Anabaena cyanobacterium

Hydra

Shark

Pine tree

This page has been deliberately left blank

Cut out the images
on the other side of this page

Classification Keys

Classification systems provide biologists with a way in which to identify species. They also indicate how closely related, in an evolutionary sense, each species is to others. An organism's classification should include a clear, unambiguous **description**, an accurate **diagram**, and its unique name, denoted by the **genus** and **species**. Classification keys are used to identify an organism and assign it to the correct species (assuming that the organism has already been formally classified and is included in the key). Typically, keys are **dichotomous** and involve a series of linked steps. At each step, a choice is made between two features; each alternative leads to another question until an identification is made. If the organism cannot be identified, it may be a new species or the key may need revision. Two examples of **dichotomous keys** are provided here. The first (below) describes features for identifying the larvae of various genera within the order Trichoptera (caddisflies). From this key you should be able to assign a generic name to each of the caddisfly larvae pictured. The key on the next page identifies aquatic insect orders.

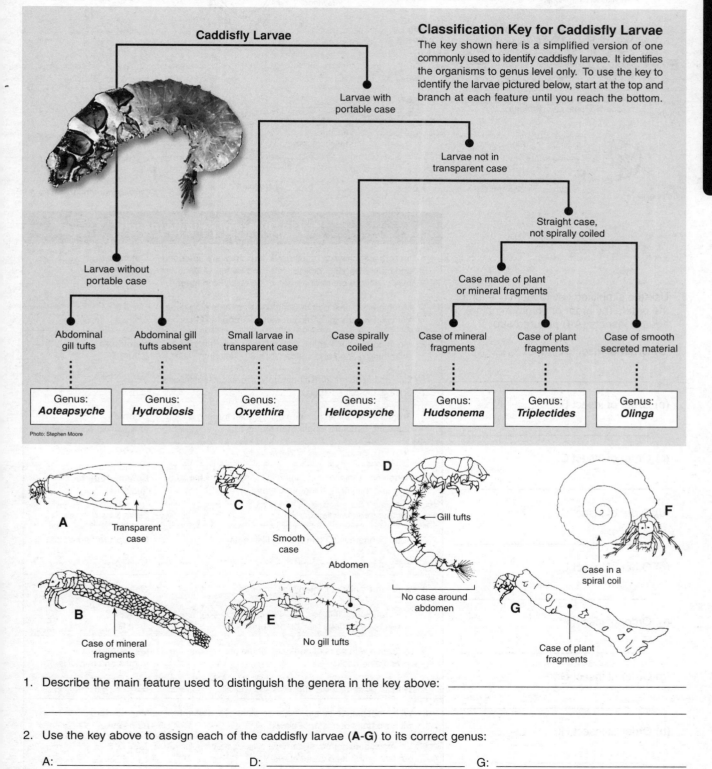

Classification Key for Caddisfly Larvae

The key shown here is a simplified version of one commonly used to identify caddisfly larvae. It identifies the organisms to genus level only. To use the key to identify the larvae pictured below, start at the top and branch at each feature until you reach the bottom.

Photo: Stephen Moore

1. Describe the main feature used to distinguish the genera in the key above: _____

_____

2. Use the key above to assign each of the caddisfly larvae (**A-G**) to its correct genus:

A: _____   D: _____   G: _____

B: _____   E: _____

C: _____   F: _____

Related activities: Keying Put Plant Species
Web links: Tree ID, What is the Key to Classification?

A 2

Key to Orders of Aquatic Insects

1	Insects with chewing mouthparts; forewings are hardened and meet along the midline of the body when at rest (they may cover the entire abdomen or be reduced in length).	**Coleoptera** (beetles)
	Mouthparts piercing or sucking and form a pointed cone	*Go to 2*
	With chewing mouthparts, but without hardened forewings	*Go to 3*
2	Mouthparts form a short, pointed beak; legs fringed for swimming or long and spaced for suspension on water.	**Hemiptera** (bugs)
	Mouthparts do not form a beak; legs (if present) not fringed or long, or spaced apart.	*Go to 3*
3	Prominent upper lip (labium) extendable, forming a food capturing structure longer than the head.	**Odonata** (dragonflies & damselflies)
	Without a prominent, extendable labium	*Go to 4*
4	Abdomen terminating in three tail filaments which may be long and thin, or with fringes of hairs.	**Ephemeroptera** (mayflies)
	Without three tail filaments	*Go to 5*
5	Abdomen terminating in two tail filaments	**Plecoptera** (stoneflies)
	Without long tail filaments	*Go to 6*
6	With three pairs of jointed legs on thorax	*Go to 7*
	Without jointed, thoracic legs (although non-segmented prolegs or false legs may be present).	**Diptera** (true flies)
7	Abdomen with pairs of non-segmented prolegs bearing rows of fine hooks.	**Lepidoptera** (moths and butterflies)
	Without pairs of abdominal prolegs	*Go to 8*
8	With eight pairs of finger-like abdominal gills; abdomen with two pairs of posterior claws.	**Megaloptera** (dobsonflies)
	Either, without paired, abdominal gills, or, if such gills are present, without posterior claws.	*Go to 9*
9	Abdomen with a pair of posterior prolegs bearing claws with subsidiary hooks; sometimes a portable case.	**Trichoptera** (caddisflies)

3. Use the simplified key to identify each of the orders (by order or common name) of aquatic insects (**A-I**) pictured above:

(a) Order of insect A:

_____

(b) Order of insect B:

_____

(c) Order of insect C:

_____

(d) Order of insect D:

_____

(e) Order of insect E:

_____

(f) Order of insect F:

_____

(g) Order of insect G:

_____

(h) Order of insect H:

_____

(i) Order of insect I:

_____

Keying Out Plant Species

Dichotomous keys are a useful tool in biology and can enable identification to the species level provided the characteristics chosen are appropriate for separating species. Keys are extensively used by botanists as they are quick and easy to use in the field, although they sometime rely on the presence of particular plant parts such as fruits or flowers. Some also require some specialist knowledge of plant biology. The following simple activity requires you to identify five species of the genus *Acer* from illustrations of the leaves. It provides valuable practice in using characteristic features to identify plants to species level.

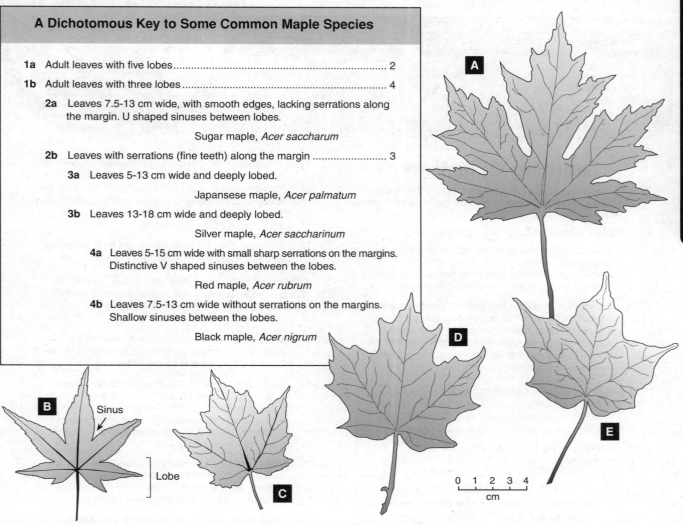

A Dichotomous Key to Some Common Maple Species

1a Adult leaves with five lobes... 2

1b Adult leaves with three lobes 4

   2a Leaves 7.5-13 cm wide, with smooth edges, lacking serrations along the margin. U shaped sinuses between lobes.

              Sugar maple, *Acer saccharum*

   2b Leaves with serrations (fine teeth) along the margin 3

      3a Leaves 5-13 cm wide and deeply lobed.

              Japansese maple, *Acer palmatum*

      3b Leaves 13-18 cm wide and deeply lobed.

              Silver maple, *Acer saccharinum*

      4a Leaves 5-15 cm wide with small sharp serrations on the margins. Distinctive V shaped sinuses between the lobes.

              Red maple, *Acer rubrum*

      4b Leaves 7.5-13 cm wide without serrations on the margins. Shallow sinuses between the lobes.

              Black maple, *Acer nigrum*

1. Use the dichotomous key to the common species of *Acer* to identify the species illustrated by the leaves (drawn to scale). Begin at the top of the key and make a choice as to which of the illustrations best fits the description:

(a) Species A: _____

(b) Species B: _____

(c) Species C: _____

(d) Species D: _____

(e) Species E: _____

2. Identify a feature that could be used to identify maple species when leaves are absent: _____

3. Suggest why it is usually necessary to consider a number of different features in order to classify plants to species level:

_____

_____

4. When identifying a plant, suggest what you should be sure of before using a key to classify it to species level:

_____

_____

Related activities: Classification Keys

A 2

Homeostasis

Learning Objectives

☐ 1. Compile your own glossary from the **KEY WORDS** displayed in **bold type** in the learning objectives below.

Principles of Homeostasis *(pages 236-239, 247)*

☐ 2. Understand the role of **homeostasis** in providing independence from the fluctuating external environment. Identify factors that require regulation in order for an organism to maintain a steady state.

☐ 3. Explain how **negative feedback** stabilises systems against excessive change. Describe the role of **receptors**, **effectors**, and negative feedback in homeostasis. Appreciate the interdependence of the hormonal and nervous regulatory systems with which mammals achieve homeostasis.

Nervous and hormonal regulation *(pages 239-246)*

☐ 4. Outline the general structure of a **nervous system**. Relate this structure to the way in which animals receive **stimuli** and generate a **response**.

☐ 5. Understand the principles of nerve impulse transmission. Explain the importance of negative feedback in nervous systems. Contrast the speed of nervous and endocrine responses (see #7).

☐ 6. Understand the terms **endocrine gland**, **hormone**, and **target tissue**. Describe the structure of the **endocrine system** and identify its role in homeostasis, including the role of the **hypothalamus** and **pituitary**.

☐ 7. Understand how hormones exert their effects and why they are able to have wide-ranging effects on the body. Explain the role of negative feedback in the control of hormone release.

☐ 8. EXTENSION: Describe types of **cell signalling** and identify the involvement of signalling molecules in a **signal transduction** pathway. Describe an example of a **steroid hormone** and a **peptide hormone**, and distinguish between the mode of action of each type.

Case Studies in Homeostasis

☐ 9. Demonstrate your understanding of homeostasis by discussing the control of body temperature or blood glucose. Consider the role of negative feedback and the requirement to coordinate control mechanisms.

Control of body temperature *(pages 251-255)*

☐ 10. Explain how body temperature is regulated in a mammal. Include reference to the role of:
    (a) **Hypothalamus** (heat gain and heat loss centres).
    (b) **Autonomic nervous system** and **involuntary activity** such as shivering.
    (c) **Thermoreceptors** in the skin (hot/cold receptors).
    (d) **Voluntary activity** such as changes in behaviour to seek shade or take exercise.

☐ 11. Contrast the mechanisms for **thermoregulation** in an ectotherm (e.g. a reptile) and in an endothermic homeotherm (e.g. a placental mammal). Describe the mechanisms involved in heat production, conservation, and loss, and identify the degree of tolerance to fluctuations in body temperature.

☐ 12. Explain the role of a **countercurrent heat exchange** system in regulating heat balance in a vertebrate.

Control of blood glucose *(pages 247-248)*

☐ 13. Identify the normal range over which blood glucose levels fluctuate and describe the factors that lead to variation in blood glucose levels.

☐ 14. Describe the cellular structure of an **islet of Langerhans** from the pancreas, and outline the role of the pancreas as an **endocrine organ**.

☐ 15. Explain how the regulation **blood glucose** level is achieved in humans, including reference to:
    (a) Negative feedback mechanisms.
    (b) The hormones **insulin** and **glucagon**.
    (c) The role of the liver in glucose-glycogen conversions.
    (d) The hormone adrenaline.

☐ 16. Explain how the actions of insulin and glucagon are mediated. Include reference to the role of membrane receptors and the effect of hormones on the enzyme controlled reactions that alter blood glucose level.

☐ 17. Describe the causes and symptoms of diabetes mellitus (Types 1 and 2). Describe the control of diabetes through the use of insulin injection and regulation of carbohydrate intake.

☐ 18. Explain the advantages of treating diabetes with human insulin produced using **genetic engineering**. Identify the source of insulin for diabetics in the past and describe problems associated with it.

The homeostatic role of the liver *(pages 249-250)*

☐ 19. Appreciate the homeostatic role of the liver, including:
    (a) Its role in **carbohydrate metabolism** and the production of glucose from amino acids.
    (b) Its role in protein metabolism: **deamination**, **transamination**, and the formation of **urea**.
    (c) Its role in **fat metabolism**, including the use of fats in respiration, the synthesis of triglycerides and cholesterol, and the transport of lipids.

☐ 20. Describe the gross structure and histology of the liver. Identify the following: liver lobule, hepatocytes, portal triad, Kupffer cells, sinusoids, and central vein.

Excretion and Water Balance in Animals

☐ 21. Clearly explain what is meant by the terms **excretion** and **osmoregulation**. Appreciate why organisms must regulate water balance and dispose of the toxic waste products of metabolism.

Osmoregulation *(pages 257-258)*

☐ 22. Describe the different homeostatic problems associated with living in salt and fresh water. Distinguish clearly between **osmoregulators** and **osmoconformers**.

☐ 23. Describe the mechanisms by which animals in fresh and salt water regulate their water and ion balance.

Water budget in a desert mammal *(page 265)*

☐ 24. Describe the control of water budget in a desert rodent with reference to:
- Physiological mechanisms that minimise water loss.
- How water requirements are met without drinking.
- Behavioural adaptations that reduce need for water.

Excretion *(pages 256, 259-264)*

☐ 25. Identify the major **nitrogenous waste products** excreted by animals and identify their origin. Appreciate how the excretory product is related to the life history and the environment of the animal.

☐ 26. Identify the non-nitrogenous excretory products in mammals. Name the organs involved in their disposal.

☐ 27. The mammalian **kidney** provides a good example of a homeostatic organ. Outline two main functions of the **kidney** in the regulation of body fluids in mammals.

☐ 28. Describe the deamination of amino acids and the subsequent production of urea. Identify this metabolic cycle and its location (see #19b). Identify the portion of the amino acid that is used as a respiratory substrate.

☐ 29. Describe the gross structure of the mammalian kidney to include the cortex, medulla, and renal pelvis. Interpret features of the histology of the kidney from sections viewed with a light microscope.

☐ 30. Using a labelled diagram, describe the structure and arrangement of a nephron and its associated blood vessels in relation to its function in producing urine. Include reference to **glomerulus**, proximal and distal **convoluted tubules**, and **collecting duct**.

☐ 31. Explain concisely how the kidney nephron produces urine. Include reference to:
- (a) The process of **ultrafiltration** in the **glomerulus**
- (b) The ultrastructure of the glomerulus and **renal capsule** in relation to ultrafiltration.
- (c) The **selective reabsorption** of water and solutes in the **proximal convoluted tubule**.
- (d) The ultrastructure of the proximal convoluted tubule in relation to reabsorption.
- (e) The role of the **loop of Henle** and the **countercurrent multiplier system** in creating and maintaining the ionic (salt) gradient in the kidney.
- (f) The role of the ionic gradient in the kidney in producing a concentrated urine and in fluid balance.

☐ 32. Explain how the water and solute content of the blood is controlled. Include reference to the role of **osmoreceptors** in the hypothalamus, release of antidiuretic hormone (**ADH**) from the **posterior pituitary**, the action of the ADH on the kidney, and the role of **negative feedback** in regulating ADH output.

☐ 33. Recognise the role of **aldosterone** in promoting sodium reabsorption in the kidney.

☐ 34. EXTENSION: Explain under what circumstances **renal dialysis** is used and what is involved in the process.

See the 'Textbook Reference Grid' on pages 8-9 for textbook page references relating to material in this topic.

Supplementary Texts

See pages 5-7 for additional details of these texts:

■ Clegg, C.J., 1998. **Mammals: Structure and Function** (John Murray), pp. 42-57, 70-71.

■ Jones, M. and G. Jones, 2002. **Mammalian Physiology & Behaviour**, (CUP), pp. 17-28 (liver).

■ Rowett, H., 1999. **Basic Anatomy & Physiology**, (John Murray), pp. 62, 70, 91-92, 99, 117-123.

See page 7 for details of publishers of periodicals:

STUDENT'S REFERENCE

■ **Homeostasis** Biol. Sci. Rev., 12(5) May 2000, pp. 2-5. *Homeostasis: what it is, how it is achieved through feedback mechanisms, the involvement of the autonomic nervous system, and adaptations for homeostasis in different environments.*

■ **Urea: A Product of Excess Dietary Protein** Biol. Sci. Rev., 17(4) April 2005, pp. 6-8. *An account of urea formation; nitrogen balance, urea and the nitrogen cycle, and human use of urea.*

■ **Glucose Centre Stage** Biol. Sci. Rev., 19(2), Nov. 2006, pp. 14-17. *An excellent account of the homeostatic control of blood glucose.*

■ **Diabetes** Biol. Sci. Rev., 15(2), Nov. 2002, pp. 30-35. *The homeostatic imbalance that results in diabetes. The role of the pancreas in the hormonal regulation of blood glucose is discussed.*

■ **Metabolic Powerhouse** New Scientist, 11 November 2000 (Inside Science). *The myriad roles of the liver in metabolic processes, including discussion of amino acid and glucose metabolism.*

■ **The Liver in Health and Disease** Biol. Sci. Rev., 14(2) Nov. 2001, pp. 14-20. *The various roles of the liver: production of bile, and metabolism of protein, lipids, carbohydrates, and drugs.*

■ **A Fair Exchange** Biol. Sci. Rev., 13(1), Sept. 2000, pp. 2-5. *Formation and reabsorption of tissue fluid (includes disorders of fluid balance).*

■ **Temperature Regulation** Biol. Sci. Rev., 17(2) Nov. 2004, pp. 2-7. *Thermoregulation in animals and the principles of endo-and ectothermy. An excellent account, which includes a discussion of how animals cope with extremes of temperature.*

■ **Hair Growth in Mammals** Biol. Sci. Rev., 17(4) April 2005, pp. 37-40. *How cycles of hair growth are timed so that the properties of the pelage meet the needs of the changing seasons.*

■ **The Kidney** Biol. Sci. Rev., 16(2) Nov. 2003, pp. 2-6. *The structure of the kidneys, and their role in regulating extracellular fluid volume, blood pressure, acid-base balance, and metabolic wastes. The operation of the kidney nephron as a countercurrent multiplier is also discussed.*

■ **Nitrogen Excretion in Animals** Biol. Sci. Rev., 8(5) May 1996, pp. 27-31. *Excretory products in animals, including a discussion of the urea cycle.*

■ **Uric Acid - Life Saver and Liability** Biol. Sci. Rev., 18(2) Sept. 2005, pp. 7-10. *Nitrogen excretion and the situations in which uric acid is produced.*

■ **Acute Liver Failure** Biol. Sci. Rev., 18(3) Feb. 2006, pp. 30-33. *Liver structure and function and the consequences to homeostasis of liver failure.*

■ **Countercurrent Exchange Mechanisms** Biol. Sci. Rev., 9(1) Sept. 1996, pp. 2-6. *The role of countercurrent multipliers in biological systems: including operation in the kidney nephron.*

■ **Basement Membranes** Biol. Sci. Rev., 13(4) March 2001, pp. 36-39. *The structure, function, and diversity of basement membranes, with an account of their structural role in the glomerulus.*

TEACHER'S REFERENCE

■ **Chemical Toxins and Body Defences** Biologist, 49(1) February 2002. *The various and fascinating ways in which mammals are able to eliminate toxic substances from their bodies.*

See pages 10-11 for details of how to access **Bio Links** from our web site: **www.biozone.co.uk**. From Bio Links, access sites under the topics:

GENERAL BIOLOGY ONLINE RESOURCES
• Biology I interactive animations • Instructional multimedia, University of Alberta ... and others
> **Online Textbooks and Lecture Notes**: • Human biology help • Learn.co.uk • S-Cool! A level biology revision guide ... and others > **Glossaries**: Animal anatomy glossary... and others

ANIMAL BIOLOGY: • Anatomy and physiology • Comparative vertebrate anatomy lecture notes • Human physiology lecture notes > **Excretion**: • Comparative physiology of vertebrate kidneys • Excretory system • The kidney • Urinary system > **Homeostasis**: • Ask the experts: What is homeostasis? • Homeostasis • Homeostasis: general principles • Physiological homeostasis

Presentation MEDIA to support this topic:

HEALTH & DISEASE:
• Non-Infectious Disease

Principles of Homeostasis

Homeostasis is the condition where the body's internal environment remains relatively constant, despite external fluctuations. Homeostasis of the internal environment is an essential feature of complex animals and it is the job of the body's **organ systems** to maintain it, even as they make necessary exchanges with the environment. Homeostatic control systems have three functional components: a receptor to detect change, a control centre, and an effector to direct an appropriate response. In negative feedback systems, movement away from an ideal state triggers a mechanism to counteract further change in that direction. Using feedback systems, the body counteracts disturbances and restores the steady state.

Organ systems maintain a constant internal environment that provides for the needs of all the body's cells, making it possible for animals to move through different and often highly variable external environments. This representation of a mammal shows how organ systems permit exchanges with the environment. The exchange surfaces of organ systems are usually internal, but may be connected to the environment via openings on the body surface.

The finger-like villi of the small intestine greatly expand the surface area for nutrient absorption.

Lung tissue provides an expansive, moist surface for gas exchange.

Kidney tubules exchange chemicals with the blood through capillaries.

Negative Feedback and Control Systems

1 A stress or disturbance takes the internal environment away from optimum

2 Stress is detected by receptors and corrective mechanisms are activated

3 The corrective mechanisms act to restore conditions back to the set value

Negative feedback acts to eliminate any deviation from preferred conditions. It is part of almost all the control systems in living things. The diagram (above left) shows how a stress or disturbance is counteracted by corrective mechanisms that act to restore conditions back to an optimum value. The diagram (above right) illustrates this principle for a biological system.

1. Identify the three main components of a regulatory control system in the human body: _____

_____

2. Explain how animals use feedback mechanisms to maintain a steady state despite their constant exchanges with a variable environment:

_____

_____

_____

Related activities: Maintaining Homeostasis

Maintaining Homeostasis

The various organ systems of the body act to maintain homeostasis through a combination of hormonal and nervous mechanisms. In everyday life, the body must regulate respiratory gases, protect itself against agents of disease (pathogens), maintain fluid and salt balance, regulate energy and nutrient supply, and maintain a constant body temperature. All these must be coordinated and appropriate responses made to incoming stimuli. In addition, the body must be able to repair itself when injured and be capable of reproducing (leaving offspring).

Regulating Respiratory Gases

Oxygen demand changes with activity level and environment (e.g. altitude).

CO₂ production changes with activity level and environment.

Capacity for O₂ transport depends on blood haemoglobin.

Muscular activity increases oxygen demand and carbon dioxide production.

Oxygen must be delivered to all cells and carbon dioxide (a waste product of cellular respiration) must be removed. Breathing (inhalation and exhalation) brings in oxygen and expels CO₂. The rate of breathing is varied according to the oxygen requirement. Both gases are transported around the body in the blood; the oxygen mostly bound to haemoglobin.

Coping with Pathogens

Lymph tissue

Attack by pathogens inhaled or eaten with food and drink.

Infections of the reproductive system (STIs) from yeasts, viruses, and bacteria.

Attack on skin and mucous membranes from fungal pathogens.

All of us are under constant attack from pathogens (disease causing organisms). The body has a number of mechanisms that help to prevent the entry of pathogens and limit the damage they cause if they do enter the body. The skin, the digestive system and the immune system are all involved in limiting damage.

Maintaining Nutrient Supply

Digestion in the gut provides the building materials for the body to grow and repair tissue.

Food and drink provides energy and nutrients, but supply is pulsed at mealtimes with little in between.

Water must be reabsorbed from the digested material.

The solid waste products of digestion (faeces) must be eliminated.

Food and drink must be taken in to maintain the body's energy supplies. Steady levels of energy (as glucose) is available to cells through hormonal regulation of blood sugar levels. Insulin, released by the endocrine cells of the pancreas, causes cells to take up glucose after a meal. Glucagon causes the release of glucose from the liver.

Repairing Injuries

Wounds result in bleeding. Clotting begins soon after and phagocytes prevent the entry of pathogens.

Muscle and tendon injuries through excessive activity.

Hernias can be caused by strain as in heavy lifting.

Bone fractures caused by falls and blows.

Damage to body tissues triggers the inflammatory response. There is pain, swelling, redness, and heat. Phagocytes and other white blood cells move to the injury site. The inflammatory response is started (and ended) by chemical signals (e.g. from histamine and prostaglandins) released when tissue is damaged.

Related activities: Principles of Homeostasis, Nervous and Hormonal Regulatory Systems, Control of Kidney Function

RA 2

Homeostasis

Maintaining Fluid and Ion Balance

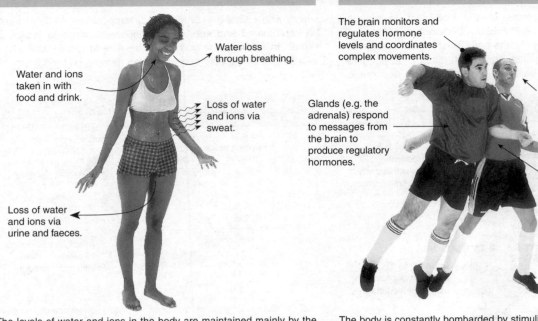

Water loss through breathing.

Water and ions taken in with food and drink.

Loss of water and ions via sweat.

Loss of water and ions via urine and faeces.

Coordinating Responses

The brain monitors and regulates hormone levels and coordinates complex movements.

Environmental stimuli bombard the senses through ears, nose, eyes, skin, and mouth.

Glands (e.g. the adrenals) respond to messages from the brain to produce regulatory hormones.

Simple reflexes, such as pain withdrawal, allow rapid responses to stimuli.

The levels of water and ions in the body are maintained mainly by the kidneys, although the skin is also important. Osmoreceptors monitor the fluid and ion levels of the blood and bring about the release of regulatory hormones; the kidneys regulate reabsorption of water and sodium from blood in response to levels of the hormones ADH and aldosterone.

The body is constantly bombarded by stimuli from the environment. The brain sorts these stimuli into those that require a response and those that do not. Responses are coordinated via nervous or hormonal controls. Simple nervous responses (reflexes) act quickly. Hormonal responses take longer to produce a response and the response is more prolonged.

1. Describe two mechanisms that operate to restore homeostasis after infection by a pathogen:

    (a) _____

    _____

    _____

    (b) _____

    _____

    _____

2. Describe two mechanisms by which responses to stimuli are brought about and coordinated:

    (a) _____

    _____

    (b) _____

    _____

3. Explain two ways in which water and ion balance are maintained. Name the organ(s) and any hormones involved:

    (a) _____

    _____

    (b) _____

    _____

4. Explain two ways in which the body regulates its respiratory gases during exercise:

    (a) _____

    _____

    (b) _____

    _____

Nervous Regulatory Systems

An essential feature of living organisms is their ability to coordinate their activities. In multicellular animals, such as mammals, detecting and responding to environmental change, and regulating the internal environment (homeostasis) is brought about by two coordinating systems: the nervous and endocrine systems. Although structurally these two systems are quite different, they frequently interact to coordinate behaviour and physiology. The nervous system contains cells called neurones (or nerve cells). Neurones are specialised to transmit information in the form of electrochemical impulses (action potentials). The nervous system is a signalling network with branches carrying information directly to and from specific target tissues. Impulses can be transmitted over considerable distances and the response is very precise and rapid. Whilst it is extraordinarily complex, comprising millions of neural connections, its basic plan (below) is quite simple. Further detail on nervous system structure and function is provided in the topic *Responses and Coordination*.

Coordination by the Nervous System

The vertebrate nervous system consists of the central nervous system (brain and spinal cord), and the nerves and receptors outside it (peripheral nervous system). Sensory input to receptors comes via stimuli. Information about the effect of a response is provided by feedback mechanisms so that the system can be readjusted. The basic organisation of the nervous system can be simplified into a few key components: the sensory receptors, a central nervous system processing point, and the effectors which bring about the response (below):

External stimuli → **Internal stimuli** →

Receptors (sense organs)
e.g. eyes, ears, taste buds, stretch and pressure receptors

Sensory input is received by the sensory structures (via stimuli) and converted into an electrical response.

Impulses are transmitted by sensory neurones to the central nervous system

Brain

Central nervous system (CNS)
processing of sensory input and coordination of a response (brain and spinal cord)

Muscles and glands bring about appropriate action

Motor output: impulses are transmitted by motor neurones to effectors

Effectors (muscles and glands)

RESPONSE

Feedback information

Motor cortex coordinates appropriate response

Eye perceives the stimulus and sends messages via sensory neurones to the brain

Motor neurones carry the message to effectors (the muscles of hand and arm)

In the example above, the approach of the frisbee is perceived by the eye. The motor cortex of the brain integrates the sensory message. Coordination of hand and body orientation is brought about through motor neurones to the muscles.

Homeostasis

Comparison of nervous and hormonal control

	Nervous control	Hormonal control
Communication	Impulses across synapses	Hormones in the blood
Speed	Very rapid (within a few milliseconds)	Relatively slow (over minutes, hours, or longer)
Duration	Short term and reversible	Longer lasting effects
Target pathway	Specific (through nerves) to specific cells	Hormones broadcast to target cells everywhere
Action	Causes glands to secrete or muscles to contract	Causes changes in metabolic activity

1. Identify the three basic components of a nervous system and explain how they function to maintain homeostasis:

_____

_____

_____

2. Describe two differences between nervous control and endocrine (hormonal) control of body systems:

(a) _____

(b) _____

Related activities: The Mammalian Nervous System, Hormonal Regulatory Systems

A 1

Hormonal Regulatory Systems

The endocrine system regulates the body's processes by releasing chemical messengers (hormones) into the bloodstream. Hormones are potent chemical regulators: they are produced in minute quantities yet can have a large effect on metabolism. The endocrine system comprises endocrine cells (organised into endocrine glands), and the hormones they produce. Unlike exocrine glands (e.g. sweat and salivary glands), endocrine glands are ductless glands, secreting hormones directly into the bloodstream rather than through a duct or tube. Some organs (e.g. the pancreas) have both endocrine and exocrine regions, but these are structurally and functionally distinct. The basis of hormonal control and the role of negative feedback mechanisms in regulating hormone levels are described below.

The Mechanism of Hormone Action

Endocrine cells produce hormones (chemical messengers) and secrete them into the bloodstream where they are distributed throughout the body. Although hormones are broadcast throughout the body, they affect only specific target cells. These target cells have receptors on the plasma membrane which recognise and bind the hormone (see inset, below right). The binding of hormone and receptor triggers the response in the target cell. Cells are unresponsive to a hormone if they do not have the appropriate receptors.

Target cells

Hormone travels in the bloodstream throughout the body

Endocrine cell secretes hormone into bloodstream

Cytoplasm of cell

Plasma membrane

Hormone molecule

Hormone receptor

Receptors on the target cell receive the hormone

Antagonistic Hormones

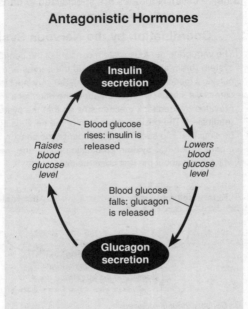

Insulin secretion

Blood glucose rises: insulin is released

Raises blood glucose level

Lowers blood glucose level

Blood glucose falls: glucagon is released

Glucagon secretion

The effects of one hormone are often counter-acted by a hormone with an opposing action. Feedback mechanisms adjust the balance of the two hormones to maintain a physiological function. Example: insulin decreases blood glucose and glucagon raises it.

1. (a) Explain what is meant by **antagonistic hormones** and describe an example of how two such hormones operate:

   _____

   _____

   Example: _____

   (b) Explain the role of feedback mechanisms in adjusting hormone levels (explain using an example if this is helpful):

   _____

   _____

2. Explain how a hormone can bring about a response in target cells even though all cells may receive the hormone:

   _____

   _____

3. Explain why hormonal control differs from nervous system control with respect to the following:

   (a) The speed of hormonal responses is slower: _____

   _____

   (b) Hormonal responses are generally longer lasting: _____

   _____

Related activities: Cell Signalling, Signal Transduction, The Endocrine System, Control of Blood Glucose

Cell Signalling

Cells use **signals** (chemical messengers) to gather information about, and respond to, changes in their cellular environment and for communication between cells. The signalling and response process is called the **signal transduction pathway**, and often involves a number of enzymes and molecules in a **signal cascade** which causes a large response in the target cell. Cell signalling pathways are categorised primarily on the distance over which the signal molecule travels to reach its target cell, and generally fall into three categories. The **endocrine** pathway involves the transport of hormones over large distances through the circulatory system. During **paracrine** signalling, the signal travels an intermediate distance to act upon neighbouring cells. **Autocrine** signalling involves a cell producing and reacting to its own signal. These three pathways are diagrammed below.

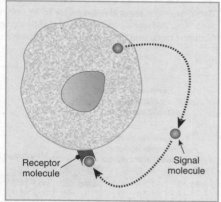

Endocrine signalling: Hormone signals are released by ductless endocrine glands and carried long distances through the body by the circulatory system to the target cells. Examples include sex hormones, growth factors and neurohormones such as dopamine.

Paracrine signalling: Signals released from a cell act upon target cells within the immediate vicinity. The chemical messenger can be transferred through the extracellular fluid (e.g. at synapses) or directly between cells, which is important during embryonic development.

Autocrine signalling: Cells produce and react to their own signals. In vertebrates, when a foreign antibody enters the body, some T-lymphocytes produce a growth factor to stimulate their own production. The increased number of T-lymphocytes helps to fight the infection.

Signalling Receptors and Signalling Molecules

Insulin like growth factor 1 (IGF-1)

Progesterone

Examples of cell signalling molecules

The binding sites of cell receptors are very specific; they only bind certain **ligands** (signal molecules). This stops them from reacting to every signal bombarding the cell. Receptors fall into two main categories:

▶ **Cytoplasmic receptors**
Cytoplasmic receptors, located within the cell cytoplasm, bind ligands which are able to cross the plasma membrane unaided.

▶ **Transmembrane receptors**
These span the cell membrane and bind ligands which cannot cross the plasma membrane on their own. They have an extra-cellular domain outside the cell, and an intracellular domain within the cell cytosol. Ion channels, protein kinases and G-protein linked receptors are examples of transmembrane receptors (see diagram on right).

Structure of a transmembrane receptor

1. Briefly describe the three types of cell signalling:

   (a) _____

   (b) _____

   (c) _____

2. Identify the components that all three cell signalling types have in common: _____

   _____

   _____

Related activities: Hormonal Regulatory Systems, Signal Transduction

A 2

Homeostasis

Signal Transduction

Once a hormone signal is released, it is carried in the blood to target cells that respond specifically to that hormone. Water soluble hormones are carried free in the blood, whilst steroid and thyroid hormones are carried bound to plasma proteins. Target cells contain receptors that bind the hormone so that it is able to exert its effect. The binding occurs in one of two ways. One involves an interaction of the hormone with transmembrane receptors followed by activation of a second messenger. The other involves an interaction of the hormone with intracellular cytoplasmic receptors, without the involvement of a second messenger. Once the target cells respond, the response is recognised by the hormone-producing cell through a feedback signal and the hormone is degraded.

Hormone Action By Second Messenger

Cyclic AMP is a **second messenger** linking the hormone to the cellular response. Cellular concentration of CAMP increases markedly once a hormone binds and the cascade of enzyme-driven reactions is initiated.

Gene Activation by Steroids

Steroid hormones alter cell function through direct activation of genes. Once inside the target cell, they bind to intracellular receptor sites, creating hormone-receptor complexes that activate specific genes.

1. Describe the two mechanisms by which a hormone can bring about a cellular response:

(a) _____

_____

(b) _____

_____

2. State in what way these two mechanisms are alike: _____

_____

3. Explain how a very small amount of hormone is able to exert a disproportionately large effect on a target cell:

_____

_____

4. Explain how the binding of a hormone to a target cell can be likened to an enzyme-substrate reaction: _____

_____

_____

A 3

Related activities: Hormonal Regulatory Systems, Cell Signalling
Web links: First and Second Messenger System

The Endocrine System

Homeostasis is achieved through the activity of the nervous and endocrine systems, which interact in the regulation of the body's activities. The nervous system is capable of rapid responses to stimuli. Slower responses, and long term adjustments of the body (growth, reproduction, and adaptation to stress), are achieved through endocrine control. The endocrine system comprises **endocrine glands** and their **hormones**. Endocrine glands are ductless glands and are distributed throughout the body. Under appropriate stimulation, they secrete chemical messengers, called **hormones**, which are carried in the blood to **target** cells, where they have a specific metabolic effect. After exerting their effect, hormones are broken down and excreted from the body. Although a hormone circulates in the blood, only the targets will respond. Hormones may be amino acids, peptides, proteins (often modified), fatty acids, or steroids. Some basic features of the human endocrine system are explained below and on the next page.

Hypothalamus
Coordinates nervous and endocrine systems. Secretes releasing hormones, which regulate the hormones of the anterior pituitary. Produces oxytocin and ADH, which are released from the posterior pituitary.

Parathyroid glands
On the surface of the thyroid, they secrete PTH (parathyroid hormone), which regulates blood calcium levels and promotes the release of calcium from bone. High levels of calcium in the blood inhibit PTH secretion.

Pancreas
Specialised α and β endocrine cells in the pancreas produce glucagon and insulin. Together, these control blood sugar levels.

Ovaries (in females)
At puberty the ovaries increase their production of oestrogen and progesterone. These hormones control and maintain female characteristics (breast development and pelvic widening), stimulate the menstrual cycle, maintain pregnancy, and prepare the mammary glands for lactation.

Pituitary gland
The pituitary is located below the hypothalamus. It secretes at least nine hormones that regulate the activities of other endocrine glands.

Thyroid gland
Secretes thyroxine, an iodine containing hormone needed for normal growth and development. Thyroxine stimulates metabolism and growth via protein synthesis.

Adrenal glands
The **adrenal medulla** produces adrenalin and noradrenalin; responsible for the fight or flight response. The **adrenal cortex** produces various steroid hormones, including aldosterone (sodium regulation) and cortisol (response to stress).

Testes (in males)
At puberty (the onset of sexual maturity) the testes of males produce testosterone in greater amounts. Testosterone controls and maintains "maleness" (muscular development and deeper voice), and promotes sperm production.

Homeostasis

1. Distinguish between a neurotransmitter and a hormone: _____

_____

_____

2. Explain why it is an advantage for hormones to be carried in the blood: _____

_____

_____

3. The dotted arrows in the diagram above indicate the development of secondary sexual characteristics. Identify the male and female hormones involved in the development of these characteristics and state their role:

   (a) Male: _____

   _____

   (b) Female: _____

   _____

Related activities: Hormonal Regulation Systems, The Pituitary, Control of Blood Glucose
Web links: Drag and Drop Hormone Match

RA 3

The Role of the Hypothalamus

The hypothalamus contains **neurosecretory cells**, specialised secretory neurones, which are both nerve cells and endocrine cells. They produce hormones (usually peptides) in the cell body, which are packaged into droplets and transported along the axon. At the axon terminal, the **neurohormone** is released into the blood in response to nerve impulses.

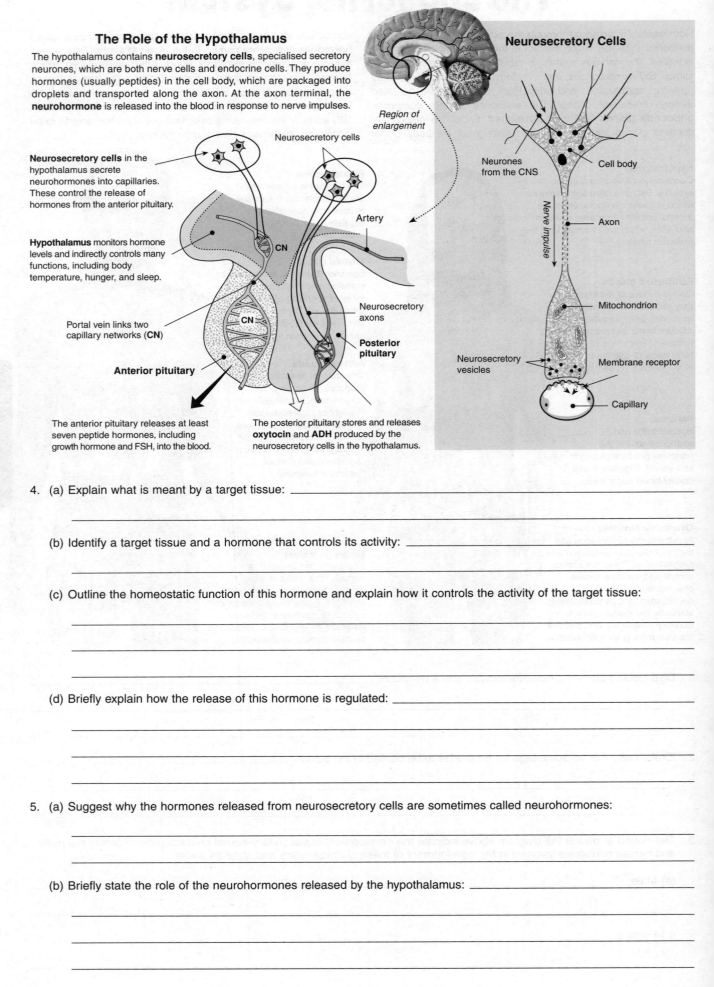

Region of enlargement

Neurosecretory Cells

Neurones from the CNS

Cell body

Nerve impulse

Axon

Mitochondrion

Neurosecretory vesicles

Membrane receptor

Capillary

Neurosecretory cells

Neurosecretory cells in the hypothalamus secrete neurohormones into capillaries. These control the release of hormones from the anterior pituitary.

Hypothalamus monitors hormone levels and indirectly controls many functions, including body temperature, hunger, and sleep.

Portal vein links two capillary networks (**CN**)

Artery

CN

CN

Anterior pituitary

Neurosecretory axons

Posterior pituitary

The anterior pituitary releases at least seven peptide hormones, including growth hormone and FSH, into the blood.

The posterior pituitary stores and releases **oxytocin** and **ADH** produced by the neurosecretory cells in the hypothalamus.

4. (a) Explain what is meant by a target tissue: _____

_____

(b) Identify a target tissue and a hormone that controls its activity: _____

_____

(c) Outline the homeostatic function of this hormone and explain how it controls the activity of the target tissue:

_____

_____

_____

(d) Briefly explain how the release of this hormone is regulated: _____

_____

_____

_____

5. (a) Suggest why the hormones released from neurosecretory cells are sometimes called neurohormones:

_____

_____

(b) Briefly state the role of the neurohormones released by the hypothalamus: _____

_____

_____

_____

The Pituitary

The **hypothalamus** is located at the base of the brain, just above the pituitary gland. Information comes to the hypothalamus through sensory pathways from the sense organs. On the basis of this information, the hypothalamus controls and integrates many basic physiological activities (e.g. temperature regulation, food and fluid intake, and sleep), including the reflex activity of the **autonomic nervous system**. The pituitary gland comprises two regions: the **posterior pituitary** and the **anterior pituitary**.

The **posterior pituitary** is neural (nervous) in origin and is essentially an extension of the hypothalamus. Its neurosecretory cells release oxytocin and ADH directly into the bloodstream in response to nerve impulses. The **anterior pituitary** is connected to the hypothalamus by blood vessels and receives releasing and inhibiting factors from the hypothlamus via a capillary network. These releasing factors regulate the secretion of the anterior pituitary's hormones.

ANTERIOR PITUITARY

The anterior pituitary releases at least seven **peptide hormones** (below) into the blood from simple secretory cells. The release of these hormones is regulated by releasing and inhibiting factors from the hypothalamus.

POSTERIOR PITUITARY

The posterior pituitary develops as an extension of the hypothalamus. The release of its two hormones, oxytocin and antidiuretic hormone, occurs directly as a result of nervous input to the hypothalamus.

Hypothalamus

Hormones from the anterior pituitary

Hormones from the posterior pituitary

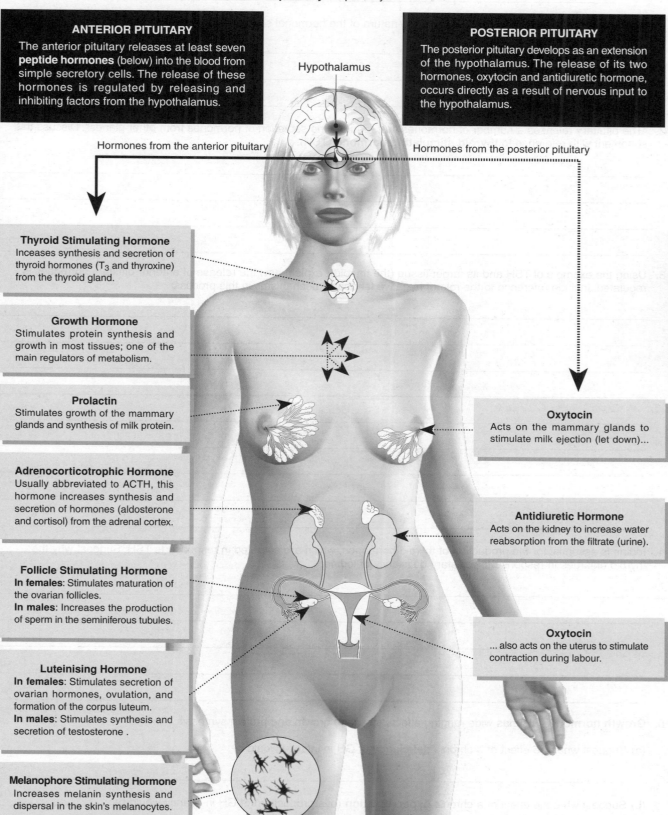

Homeostasis

Thyroid Stimulating Hormone
Inceases synthesis and secretion of thyroid hormones (T_3 and thyroxine) from the thyroid gland.

Growth Hormone
Stimulates protein synthesis and growth in most tissues; one of the main regulators of metabolism.

Prolactin
Stimulates growth of the mammary glands and synthesis of milk protein.

Adrenocorticotrophic Hormone
Usually abbreviated to ACTH, this hormone increases synthesis and secretion of hormones (aldosterone and cortisol) from the adrenal cortex.

Follicle Stimulating Hormone
In females: Stimulates maturation of the ovarian follicles.
In males: Increases the production of sperm in the seminiferous tubules.

Luteinising Hormone
In females: Stimulates secretion of ovarian hormones, ovulation, and formation of the corpus luteum.
In males: Stimulates synthesis and secretion of testosterone .

Melanophore Stimulating Hormone
Increases melanin synthesis and dispersal in the skin's melanocytes.

Oxytocin
Acts on the mammary glands to stimulate milk ejection (let down)...

Antidiuretic Hormone
Acts on the kidney to increase water reabsorption from the filtrate (urine).

Oxytocin
... also acts on the uterus to stimulate contraction during labour.

Related activities: The Endocrine System

RA 2

1. (a) Explain how the anterior and posterior pituitary differ with respect to their relationship to the hypothalamus:

_____

_____

_____

_____

(b) Explain how these differences relate to the nature of the hormonal secretions for each region: _____

_____

_____

_____

2. "The pituitary releases a number of hormones that regulate the secretion of hormones from other glands". Discuss this statement with reference to **thyroid stimulating hormone** (TSH):

_____

_____

_____

_____

3. Using the example of TSH and its target tissue (the thyroid), explain how the release of anterior pituitary hormones is regulated. Include reference to the role of negative feedback mechanisms in this process:

_____

_____

_____

_____

_____

_____

_____

_____

_____

4. Iodine is essential for the production of thyroid hormones, which are secreted in response to TSH. Suggest why the thyroid enlarges in response to a dietary deficiency of iodine:

_____

_____

_____

_____

_____

5. **Growth hormone** (GH) has wide ranging effects on tissue growth and protein synthesis.

(a) Suggest what the effect of a chronic **deficiency** of GH in infancy would be: _____

_____

(b) Suggest what the effect of a chronic **hypersecretion** (overproduction) of GH in infancy would be: _____

_____

Control of Blood Glucose

The endocrine portion of the **pancreas** produces the hormones that regulate blood glucose. Two hormones, insulin and glucagon, maintain blood glucose at a steady state through **negative feedback**. Insulin promotes a decrease in blood glucose synthesizing glycogen and promoting cellular uptake of glucose. Glucagon promotes an increase in blood glucose through the breakdown of glycogen and the synthesis of glucose from amino acids. When normal blood glucose levels are restored, negative feedback stops hormone secretion. Regulating blood glucose

to within narrow limits allows energy to be available to cells as needed. Extra energy is stored, as glycogen or fat, and is mobilized to meet energy needs as required. The liver is pivotal in these carbohydrate conversions. One of the consequences of a disruption to this system is the disease **diabetes mellitus**. In type 1 diabetes, the insulin producing β cells of the pancreatic tissue are destroyed as a result of autoimmune activity and insulin is not produced. In type 2 diabetes, the pancreatic cells produce insulin, but the body's cells cease responding its message.

Effects of Insulin and Glucagon

The hormones insulin and glucagon together regulate blood glucose levels. When food is unavailable for long periods of time, glucose can be obtained firstly from the metabolism of stored glycogen in the liver and, when this becomes depleted, from stored fats. Data are based on a morning meal with no further food (fasting) for 12 hours or more.

In type 1 diabetes mellitus, the β cells of the pancreas are destroyed and insulin must be delivered to the bloodstream by injection. Type 2 diabetics produce insulin, but their cells do not respond to it.

Negative Feedback in Blood Glucose Regulation

1. (a) Identify the stimulus for the release of insulin: _____

(b) Identify the stimulus for the release of glucagon: _____

(c) Explain how glucagon brings about an increase in blood glucose level: _____

(d) Explain how insulin brings about a decrease in blood glucose level: _____

2. Outline the role of negative feedback in the control of blood glucose: _____

3. Explain why fats are metabolised after a long period without food: _____

Related activities: The Liver's Homeostatic Role, Diabetes Mellitus

A 2

Homeostasis

Diabetes Mellitus

Diabetes is a general term for a range of disorders sharing two common symptoms: production of large amounts of urine and excessive thirst. Other symptoms may be present as well, depending on the type of diabetes. **Diabetes mellitus** is the most common form of diabetes and is characterised by high blood sugar. **Type 1** has a juvenile onset while **Type 2** affects adults.

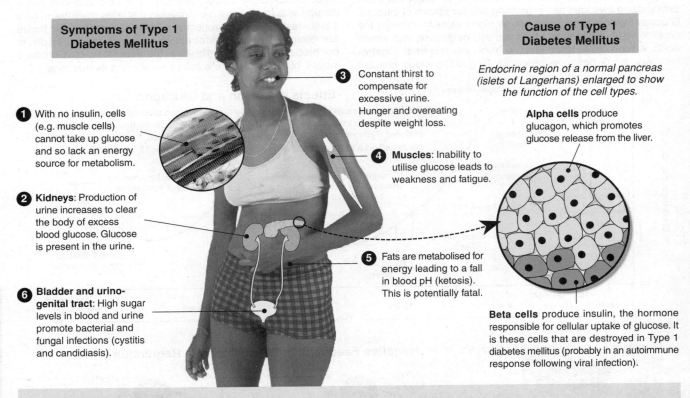

Symptoms of Type 1 Diabetes Mellitus

1 With no insulin, cells (e.g. muscle cells) cannot take up glucose and so lack an energy source for metabolism.

2 Kidneys: Production of urine increases to clear the body of excess blood glucose. Glucose is present in the urine.

6 Bladder and urino-genital tract: High sugar levels in blood and urine promote bacterial and fungal infections (cystitis and candidiasis).

3 Constant thirst to compensate for excessive urine. Hunger and overeating despite weight loss.

4 Muscles: Inability to utilise glucose leads to weakness and fatigue.

5 Fats are metabolised for energy leading to a fall in blood pH (ketosis). This is potentially fatal.

Cause of Type 1 Diabetes Mellitus

Endocrine region of a normal pancreas (islets of Langerhans) enlarged to show the function of the cell types.

Alpha cells produce glucagon, which promotes glucose release from the liver.

Beta cells produce insulin, the hormone responsible for cellular uptake of glucose. It is these cells that are destroyed in Type 1 diabetes mellitus (probably in an autoimmune response following viral infection).

Type 1 Diabetes mellitus (insulin dependent)

Incidence: About 10-15% of all diabetics.

Age at onset: Early; often in childhood (often called juvenile onset diabetes).

Symptoms: Symptoms are severe. Insulin deficiency accelerates fat breakdown and leads to a number of metabolic complications: hyperglycaemia (high blood sugar), excretion of glucose in the urine, increased urine production, excessive thirst and hunger, weight loss, and ketosis.

Cause: Absolute deficiency of insulin due to lack of insulin production (pancreatic beta cells are destroyed in an autoimmune reaction). There is a genetic component but usually a childhood viral infection triggers the development of the disease. Mumps, coxsackie, and rubella are implicated.

Treatments

Present treatments: Regular insulin injections combined with dietary management to keep blood sugar levels stable. Blood glucose is monitored regularly with testing kits to guard against sudden falls in blood glucose (hypoglycaemia).

Until recently, insulin was extracted from dead animals. Now, genetically engineered yeast or bacterial cells containing the gene for human insulin are grown in culture, providing abundant, low cost insulin, without the side effects associated with animal insulin.

New treatments: Cell therapy involves the transplant of insulin producing islet cells. To date, approximately 400 patients worldwide have received islet cell transplants from donor pancreases, with varying degrees of success. Cell therapy promises to be a practical and effective way to provide sustained relief for Type 1 diabetics.

Future treatments: In the future, gene therapy, where the gene for insulin is inserted into the diabetic's cells, may be possible.

1. Describe the **symptoms** of diabetes mellitus Type 1 and relate these to the physiological cause of the disease:

_____

_____

_____

_____

2. Summarise the **treatments** for diabetes mellitus Type 1 (list key words/phrases only):

(a) Present treatment: _____

(b) New treatments: _____

(c) Future treatments: _____

Related activities: Control of Blood Glucose

The Liver's Homeostatic Role

The liver, located just below the diaphragm and making up 3-5% of body weight, is the largest homeostatic organ. It performs a vast number of functions including production of bile, storage and processing of nutrients, and detoxification of poisons and metabolic wastes. The liver has a **unique double blood supply** and up to 20% of the total blood volume flows through it at any one time.

This rich vascularisation makes it the central organ for regulating activities associated with the blood and circulatory system. In spite of the complexity of its function, the liver tissue and the liver cells themselves are structurally relatively simple. Features of liver structure and function are outlined below. The histology of the liver in relation to its role is described on the next page.

Homeostatic Functions of the Liver

The liver is one of the largest and most complex organs in the body. It has a central role as an organ of homeostasis and performs many functions, particularly in relation to the regulation of blood composition. General functions of the liver are outlined below. Briefly summarised, the liver:

1. Secretes bile, important in emulsifying fats in digestion.
2. Metabolises amino acids, fats, and carbohydrates (below).
3. Synthesises glucose from non-carbohydrate sources when glycogen stores are exhausted (gluconeogenesis).
4. Stores iron, copper, and some vitamins (A, D, E, K, B_{12}).
5. Converts unwanted amino acids to urea (urea cycle).
6. Manufactures heparin and plasma proteins (e.g. albumin).
7. Detoxifies poisons or turns them into less harmful forms.
8. Some liver cells phagocytose worn-out blood cells.
9. Synthesises cholesterol from acetyl coenzyme A.

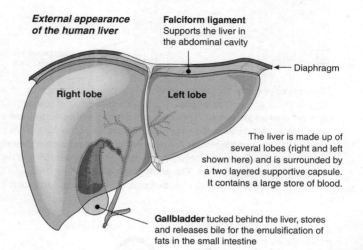

External appearance of the human liver

Falciform ligament Supports the liver in the abdominal cavity

Diaphragm

Right lobe / Left lobe

The liver is made up of several lobes (right and left shown here) and is surrounded by a two layered supportive capsule. It contains a large store of blood.

Gallbladder tucked behind the liver, stores and releases bile for the emulsification of fats in the small intestine

Summary of Liver Functions

The Internal Structure of the Liver

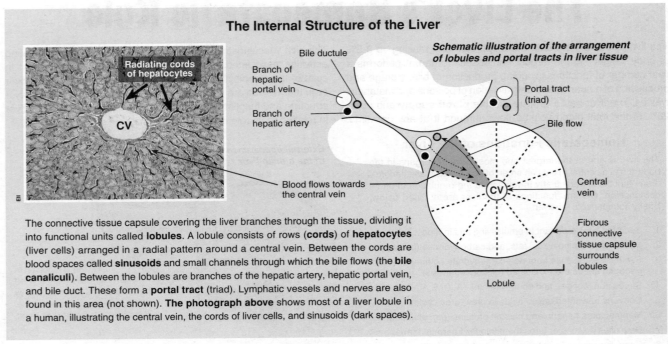

Schematic illustration of the arrangement of lobules and portal tracts in liver tissue

Bile ductule

Branch of hepatic portal vein

Branch of hepatic artery

Blood flows towards the central vein

Portal tract (triad)

Bile flow

Central vein

Fibrous connective tissue capsule surrounds lobules

Lobule

Radiating cords of hepatocytes

CV

The connective tissue capsule covering the liver branches through the tissue, dividing it into functional units called **lobules**. A lobule consists of rows (**cords**) of **hepatocytes** (liver cells) arranged in a radial pattern around a central vein. Between the cords are blood spaces called **sinusoids** and small channels through which the bile flows (the **bile canaliculi**). Between the lobules are branches of the hepatic artery, hepatic portal vein, and bile duct. These form a **portal tract** (triad). Lymphatic vessels and nerves are also found in this area (not shown). **The photograph above** shows most of a liver lobule in a human, illustrating the central vein, the cords of liver cells, and sinusoids (dark spaces).

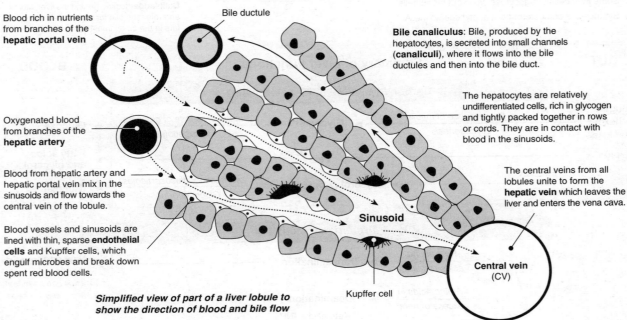

Blood rich in nutrients from branches of the **hepatic portal vein**

Bile ductule

Bile canaliculus: Bile, produced by the hepatocytes, is secreted into small channels (**canaliculi**), where it flows into the bile ductules and then into the bile duct.

Oxygenated blood from branches of the **hepatic artery**

The hepatocytes are relatively undifferentiated cells, rich in glycogen and tightly packed together in rows or cords. They are in contact with blood in the sinusoids.

Blood from hepatic artery and hepatic portal vein mix in the sinusoids and flow towards the central vein of the lobule.

The central veins from all lobules unite to form the **hepatic vein** which leaves the liver and enters the vena cava.

Blood vessels and sinusoids are lined with thin, sparse **endothelial cells** and Kupffer cells, which engulf microbes and break down spent red blood cells.

Sinusoid

Central vein (CV)

Kupffer cell

Simplified view of part of a liver lobule to show the direction of blood and bile flow

1. State the two sources of blood supply to the liver, describing the primary physiological purpose of each supply:

   (a) Supply 1: _____ Purpose: _____

   (b) Supply 2: _____ Purpose: _____

2. Briefly describe the role of the following structures in liver tissue:

   (a) Bile canaliculi: _____

   (b) Phagocytic Kupffer cells: _____

   (c) Central vein: _____

   (d) Sinusoids: _____

3. Briefly explain three important aspects of **either** protein metabolism **or** carbohydrate metabolism in the liver:

   (a) _____

   (b) _____

   (c) _____

Temperature Regulation

The process of controlling body temperature is called **thermoregulation**. For many years, animals were classified as either homeotherms or homoiotherms (= constant body temperature) or poikilotherms (= variable body temperature). Unfortunately, these terms are not particularly accurate for many animals: for example, some mammals (typical homeotherms) may have unstable body temperatures. A more recent, thermal classification of animals is based on the source of the body heat: whether it is largely from the environment (ectothermic) or from metabolic activity (endothermic). This classification can be more accurately applied to most animals but, in reality, many animals still fall somewhere between the two extremes.

How Body Temperature Varies

Aquatic invertebrates like jellyfish are true poikilotherms: their temperature is the same as the environment.

Tuna and some of the larger sharks can maintain body temperatures up to 14°C above the water temperature.

Hibernating rodents and bats let their body temperature drop to well below what is typical for most mammals.

Most birds and mammals maintain a body temperature that varies less than 2°C: they are true homeotherms.

Increasingly homeothermic →

Poikilothermic

Body temperature varies with the environmental temperature. Traditionally includes all animals other than birds and mammals, but many reptiles, some large insects and some large fish are not true poikilotherms because they may maintain body temperatures that are different from the surrounding environment.

Homeothermic

Body temperature remains almost constant despite environmental fluctuations. Traditionally includes birds and mammals, which typically maintain body temperatures close to 37-38°C. Many reptiles are partially homeothermic and achieve often quite constant body temperatures through behavioural mechanisms.

Source of Body Heat

With a few exceptions, most fish are fully ectothermic. Unlike many reptiles they do not usually thermoregulate.

Snakes use heat energy from the environment to increase their body temperature for activity.

Some large insects like bumblebees may raise their temperature for short periods through muscular activity.

Mammals (and birds) achieve high body temperatures through metabolic activity and reduction of heat losses.

Increasingly endothermic →

Ectothermic

Ectotherms depend on the environment for their heat energy. The term ectotherm is often equated with poikilotherm, although they are not the same. Poikilotherms are also ectotherms but many ectotherms may regulate body temperature (often within narrow limits) by changing their behaviour (e.g. snakes and lizards).

Endothermic

Endotherms rely largely on metabolic activity for their heat energy. Since they usually maintain a constant body temperature, most endotherms are also homeotherms. As well as birds and mammals, some fast swimming fish, like tuna, and some large insects may also use muscular activity to maintain a high body temperature.

Daily Temperature Variations in Ectotherms and Endotherms

Ectotherm: Diurnal lizard (top right)

Body temperature is regulated by behaviour so that it does not rise above 40°C. Basking increases heat uptake from the sun. Activity occurs when body temperature is high. Underground burrows are used for retreat.

Endotherm: Human (bottom right)

Body temperature fluctuates within narrow limits over a 24 hour period. Exercise and eating increase body temperature for a short time. Body temperature falls during rest and is partly controlled by an internal rhythm.

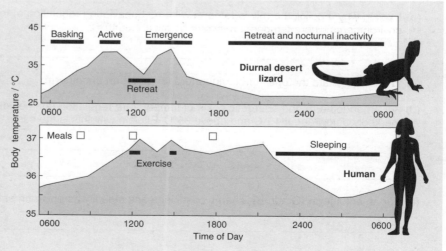

Homeostasis

Related activities: Thermoregulation in Mammals, Mechanisms of Thermoregulation

RDA 2

1. (a) Explain what is meant by an homeothermic endotherm: _____

_____

(b) Explain why the term "poikilotherm" is not a good term for classifying many terrestrial lizards and snakes:

_____

_____

2. Ectotherms will often maintain high, relatively constant body temperatures for periods in spite of environmental
fluctuations, yet they also tolerate declines in body temperature to levels lower than are tolerated by endotherms.
(a) Describe the advantages of letting body temperature fluctuate with the environment (particularly at low temperature):

_____

(b) Suggest why ectothermy is regarded as an adaptation to low or variable food supplies: _____

_____

_____

3. Some endotherms do not always maintain a high body temperature. Some, such as small rodents, allow their body
temperatures to fall during hibernation. Explain the advantage of this behaviour:

_____

_____

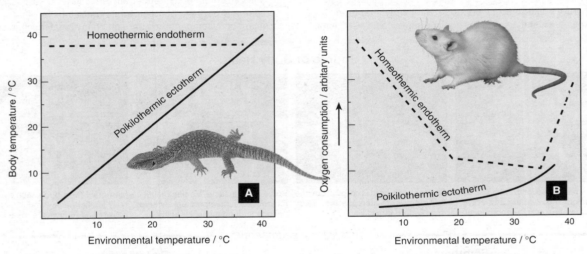

4. The two graphs above illustrate the differences in temperature regulation between a endothermic homeotherm and a
ectothermic poikilotherm (such as a fish). Graph A shows change in body temperature with environmental temperature.
Graph B shows change in oxygen consumption with environmental temperature. Use the graphs to answer the following:
(a) Explain how ectotherms and endotherms differ in their response to changes in environmental temperature (graph **A**):

_____

(b) Explain why a poikilothermic ectotherm, with no behavioural regulation of temperature, would be limited to
environments where temperatures were below about 40°C:

_____

(c) In graph **B**, state the optimum temperature range for an endotherm: _____

(d) For an endotherm, the energetic costs of temperature regulation (as measured by oxygen consumption) increase
markedly below about 15°C and above 35°C. Explain why this is the case:

_____

_____

(e) For an ectotherm (Graph B), energy costs increase steadily as environmental temperature increases. Explain why:

_____

_____

Thermoregulation in Mammals

For a body to maintain a constant temperature heat losses must equal heat gains. Heat exchanges with the environment occur via three mechanisms: **conduction** (direct heat transfer), **radiation** (indirect heat transfer), and **evaporation**. The importance of each of these depends on environment. For example, there is no evaporative loss in water, but such losses in air can be very high. Coverage of temperature regulation in humans, including the role of the skin, can be found in the next activity.

Water has a great capacity to transfer heat away from organisms; its cooling power can be more than 50 times greater than that of air. For most aquatic animals (with the exception of aquatic birds and mammals and a few fish) heat retention is impossible. Instead, they carry out their metabolic activities at the ambient temperature. For most marine organisms this does not fluctuate much.

Seals, whales and dolphins have heavily insulated surfaces of fat or blubber (up to 60% of body thickness). Blood is diverted to the outside of the blubber in warm water or if heat needs to be lost.

Large body size reduces heat loss by lowering the surface area to volume ratio.

In all animals, metabolic activity generates heat.

Heat loss from flippers and tail flukes is minimised by the use of **countercurrent heat exchangers** (see top of next page for a full explanation)

Temperature regulation mechanisms in water

- Low metabolic activity
- Heat generation from metabolic activity
- Insulation layer of blubber
- Changes in circulation patterns when swimming
- Large body size
- Heat exchange systems in limbs or high activity muscle

Temperature regulation mechanisms in air

- Behaviour or habitat choice
- Heat generation from metabolic activity
- Insulation (fat, fur, feathers)
- Changes in blood flow
- Large body size
- Sweating and panting
- Tolerance of fluctuation in body temperature

Homeostasis

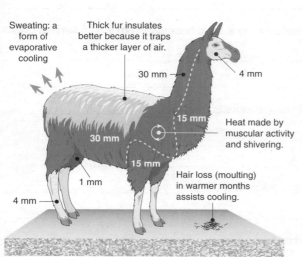

Sweating: a form of evaporative cooling

Thick fur insulates better because it traps a thicker layer of air.

30 mm

4 mm

15 mm

30 mm

Heat made by muscular activity and shivering.

15 mm

1 mm

Hair loss (moulting) in warmer months assists cooling.

4 mm

Environmental temperature ranges

°C
100
70
50
20
0
-20
-50
-70

Sea and freshwater

Air temperature on land

For most mammals, the thickness of the fur or hair varies around the body (as indicated above). Thermoregulation is assisted by adopting body positions that expose or cover areas of thin fur (the figures above are for the llama-like guanaco).

Animals adapted to temperature extremes (hot or cold) can often tolerate large fluctuations in their body temperature before they become stressed. In camels, the body temperature may fluctuate up to 7°C (34°C to 41°C) over a 24 hour period.

Dog

Brown bear

Elephant seal

Musk oxen

Panting to lose accumulated heat is important in dogs, which have sweat glands only on the pads of their feet.

Thick hair, fur or wool traps air in a layer next to the skin. This insulating air layer reduces heat loss and slows heat gain.

Thick blubber and large body size in seals and other marine mammals provide an effective insulation.

Mammals and birds in cold climates, like the musk oxen above, cluster together to retain body heat.

1. Explain how water differs from air in the way in which it transmits heat away from the body of an organism:

_____

_____

2. Describe two ways in which a mammal (a endothermic homeotherm) maintains its internal body temperature in water:

(a) _____

(b) _____

Related activities: Temperature Regulation, Mechanisms of Thermoregulation

RA 2

Countercurrent Heat Exchange Systems

Blood flow back to the body core

Blood in the vein gains heat from the warmer artery as it flows back towards the body.

Capillary bed at the end of the limb

36°C   30°C   24°C   18°C   12°C

Vein

Heat transfer

Artery

37°C   31°C   25°C   19°C   13°C

Blood flow from the body core

Blood in the artery enters the limb at or near body core temperature. It cools as it flows towards the end of the limb, losing heat to the vein that flows alongside.

Cool environmental temperature: 10°C or below

On reaching the **capillary bed**, the (now cooler) arterial blood has less heat to lose to the environment.

Countercurrent heat exchange systems occur in both aquatic and terrestrial animals as an adaptation to maintaining a stable core temperature. The diagram illustrates the general principle of countercurrent heat exchangers: heat is exchanged between incoming and outgoing blood. In the flippers and fins of whales and dolphins, and the legs of aquatic birds, they minimise heat loss. In some terrestrial animals adapted to hot climates, the countercurrent exchange mechanism works in the opposite way to prevent the head from overheating: venous blood cools the arterial blood before it supplies the brain.

3. (a) Explain how a large body size assists in maintaining body temperature in both aquatic and terrestrial mammals:

_____

(b) Describe a way in which small terrestrial mammals compensate for more rapid heat loss from a high surface area:

_____

_____

4. (a) Explain how thick hair or fur assists in the regulation of body temperature in mammals: _____

_____

_____

(b) Explain why fur/hair thickness varies over different regions of a mammal's body: _____

_____

(c) Explain how you would expect fur thickness to vary between related mammal species at high and low altitude:

_____

(d) Explain how marine mammals compensate for lack of thick hair or fur: _____

_____

5. Giving an example, explain how countercurrent heat exchange systems assist in temperature regulation in mammals:

_____

_____

_____

6. (a) Describe the role that group behaviour plays in temperature regulation in some mammals: _____

_____

(b) Name an animal, other than musk oxen, that uses this behaviour: _____

(c) Describe another behaviour, not reliant on a group, that is important in thermoregulation. For the behaviour, suggest when and where it would occur, and comment on its adaptive value:

_____

_____

_____

Mechanisms of Thermoregulation

In humans and other placental mammals, the temperature regulation centre of the body is in the **hypothalamus**. In humans, it has a 'set point' temperature of 36.7°C. The hypothalamus responds directly to changes in core temperature and to nerve impulses from receptors in the skin. It then coordinates appropriate nervous and hormonal responses to counteract the changes and restore normal body temperature. Like a thermostat, the hypothalamus detects a return to normal temperature and the corrective mechanisms are switched off (negative feedback). Toxins produced by pathogens, or substances released from some white blood cells, cause the set point to be set to a higher temperature. This results in fever and is an important defence mechanism in the case of infection.

Counteracting Heat Loss

Heat promoting centre* in the hypothalamus monitors fall in skin or core temperature below 35.8°C and coordinates responses that generate and conserve heat. These responses are mediated primarily through the **sympathetic nerves** of the autonomic nervous system.

Thyroxine (together with adrenaline) **increases metabolic rate**.

Under conditions of *extreme* cold, adrenaline and thyroxine increase the energy releasing activity of the liver. Under normal conditions, the liver is thermally neutral.

Muscular activity (including **shivering**) produces internal heat.

Erector muscles of hairs contract to raise hairs and increase insulating layer of air. Blood flow to skin decreases (**vasoconstriction**).

Factors causing heat loss

• Wind chill factor accelerates heat loss through conduction.

• Heat loss due to temperature difference between the body and the environment.

• The rate of heat loss from the body is increased by being wet, by inactivity, dehydration, inadequate clothing, or shock.

Factors causing heat gain

• Gain of heat directly from the environment through radiation and conduction.

• Excessive fat deposits make it harder to lose the heat that is generated through activity.

• Heavy exercise, especially with excessive clothing.

*NOTE: The heat promoting centre is also called the "cold centre" and the heat losing centre is also called the "hot centre". We have used the terminology descriptive of the activities promoted by the centre in each case.

Counteracting Heat Gain

Heat losing centre* in the hypothalamus monitors any rise in skin or core temperature above 37.5°C and coordinates responses that increase heat loss. These responses are mediated primarily through the **parasympathetic nerves** of the autonomic nervous system.

Sweating increases. Sweat cools by evaporation.

Muscle tone and **metabolic rate** are decreased. These mechanisms reduce the body's heat output.

Blood flow to skin (**vasodilation**) increases. This increases heat loss.

Erector muscles of hairs relax to flatten hairs and decrease insulating air layer.

Homeostasis

The skin and thermoregulation

Thermoreceptors in the dermis (probably free nerve endings) detect changes in skin temperature outside the normal range and send nerve impulses to the hypothalamus, which mediates a response. Thermoreceptors are of two types: **hot thermoreceptors** detect a rise in skin temperature above 37.5°C while the **cold thermoreceptors** detect a fall below 35.8°C. Temperature regulation by the skin involves **negative feedback** because the output is fed back to the skin receptors and becomes part of a new stimulus-response cycle.

Note that the thermoreceptors detect the temperature change, but the hair erector muscles and blood vessels are the **effectors** for mediating a response.

Cross section through the skin of the scalp.

Blood vessels in the dermis dilate (vasodilation) or constrict (vasoconstriction) to respectively promote or restrict heat loss.

Hairs raised or lowered to increase or decrease the thickness of the insulating air layer between the skin and the environment.

Sweat glands produce sweat in response to parasympathetic stimulation from the hypothalamus. Sweat cools through evaporation.

Fat in the subdermal layers insulates the organs against heat loss.

---

1. State two mechanisms by which body temperature could be reduced after intensive activity (e.g. hard exercise):

   (a) _____   (b) _____

2. Briefly state the role of the following in regulating internal body temperature:

   (a) The hypothalamus: _____

   (b) The skin: _____

   (c) Nervous input to effectors: _____

   (d) Hormones: _____

Related activities: Temperature Regulation, Thermoregulation in Mammals

A 2

Nitrogenous Wastes in Animals

Waste materials are generated by the metabolic activity of cells. If allowed to accumulate, they would reach toxic concentrations and so must be continually removed. Excretion is the process of removing waste products and other toxins from the body. Waste products include carbon dioxide and water, and the nitrogenous (nitrogen containing) wastes that result from the breakdown of amino acids and nucleic acids. The simplest breakdown product of nitrogen containing compounds is ammonia, a small molecule that cannot be retained for long in the body because of its high toxicity. Most aquatic animals excrete ammonia immediately into the water where it is washed away. Other animals convert the ammonia to a less toxic form that can remain in the body for a short time before being excreted via special excretory organs. The form of the excretory product in terrestrial animals (urea or uric acid) depends on the type of organism and its life history. Terrestrial animals that lay eggs produce uric acid rather than urea, because it is non-toxic and very insoluble. It remains as an inert solid mass in the egg until hatching.

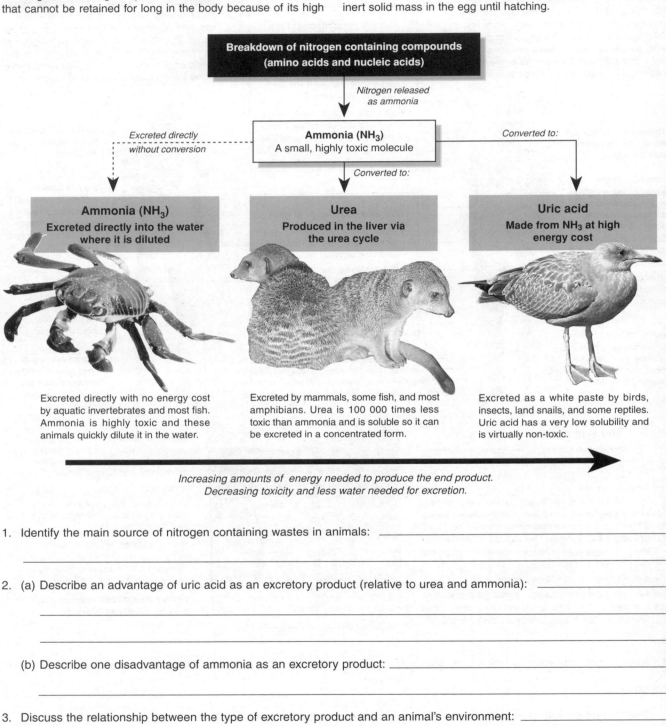

Breakdown of nitrogen containing compounds
(amino acids and nucleic acids)

Nitrogen released as ammonia

Ammonia (NH_3)
A small, highly toxic molecule

Excreted directly without conversion

Converted to:

Converted to:

Ammonia (NH_3)
Excreted directly into the water where it is diluted

Excreted directly with no energy cost by aquatic invertebrates and most fish. Ammonia is highly toxic and these animals quickly dilute it in the water.

Urea
Produced in the liver via the urea cycle

Excreted by mammals, some fish, and most amphibians. Urea is 100 000 times less toxic than ammonia and is soluble so it can be excreted in a concentrated form.

Uric acid
Made from NH_3 at high energy cost

Excreted as a white paste by birds, insects, land snails, and some reptiles. Uric acid has a very low solubility and is virtually non-toxic.

Increasing amounts of energy needed to produce the end product.
Decreasing toxicity and less water needed for excretion.

1. Identify the main source of nitrogen containing wastes in animals: _____

   _____

2. (a) Describe an advantage of uric acid as an excretory product (relative to urea and ammonia): _____

   _____

   _____

   (b) Describe one disadvantage of ammonia as an excretory product: _____

   _____

3. Discuss the relationship between the type of excretory product and an animal's environment: _____

   _____

   _____

   _____

   _____

Related activities: Water Balance in Animals, Waste Products in Humans

Water Balance in Animals

Many aspects of metabolism, e.g. enzyme activity, membrane transport, and nerve conduction are dependent on particular concentrations of ions and metabolites. To achieve this balance, the salt and water content of the internal environment must be regulated; a process called **osmoregulation**. The mechanisms for obtaining, retaining, and eliminating water and solutes (including excretion of nitrogenous wastes) in marine, freshwater and terrestrial organisms vary considerably. Differences reflect both the constraints of the environment and the evolutionary inheritance of the organism.

Mechanisms against water loss

- Behaviour and habitat choice
- Waxy cuticle of insects
- Oily secretion from glands in the skin
- Insulating fur or feathers
- Dry, scaly skin of reptiles

Large losses from airways except in arthropods.

Metabolism of stored fat (as in the camel's hump) can be used to provide water. In camels, the hump disappears as the fat is used up.

The efficient kidneys of mammals produce a concentrated urine containing **urea**. Most reptiles, birds, insects and land snails excrete nitrogenous waste as uric acid (requires very little water but more energy in its excretion).

Drinks water
Eats food

Stomach

Water is extracted from the diet, even when this is dry food (e.g. seeds).

Some insects, frogs, and arachnids (e.g. scorpions) can take up water directly from the atmosphere.

Water loss by evaporation from skin or exoskeleton.

Water lost in faeces and urine.

Animals adapted to arid regions and those that hibernate can tolerate large losses of body weight as they metabolise their fat reserves to produce water.

····> Losses
——> Gains

The primary water balance problem for terrestrial animals is water loss. Water is required for all metabolism, including the metabolism of food and the disposal of waste products. Mammals must **drink water** regularly, although some are able to survive for long periods without drinking by generating water from the metabolism of fats. Mammals have efficient **kidneys** and produce a concentrated urine high in urea. In most land arthropods, water is conserved by limiting losses to the environment. The **chitinous exoskeleton** itself does not reduce water loss much but the waxy cuticle of insects retards water loss very effectively. The respiratory structures of arthropods are chitinous, internal tubes and there is little loss from these. All animals show behavioural adaptations to limiting water loss by seeking out damper environments. This is particularly important for desert animals and for arthropods with little resistance to dehydration.

Homeostasis

Tolerance of water loss and burrowing behaviour in desert frogs.

Behavioural adaptations, efficient kidneys, and thick fur in kangaroos.

Humidity seeking behavioural adaptations in woodlice.

Chitinous exoskeleton with waxy, waterproof cuticle in insects.

1. (a) Briefly describe four ways in which animals obtain water: _____

   _____

   (b) Identify three ways in which water is lost: _____

2. Discuss structural and behavioural adaptations for reducing water loss in a named arthropod **or** a named mammal:

   _____

   _____

   _____

   _____

   _____

   _____

   _____

Marine Environments

Kidneys with relatively few nephrons often lacking glomeruli (filtering units). Urine is produced by secretion of ions (not filtration).

Gain of water and salts from food and by drinking seawater.

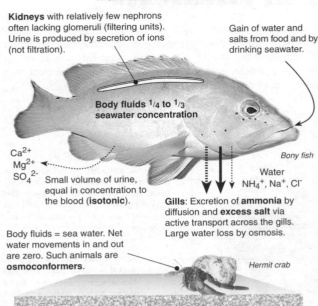

Body fluids $1/4$ to $1/3$ seawater concentration

Bony fish

Ca^{2+}
Mg^{2+}
SO_4^{2-}

Small volume of urine, equal in concentration to the blood (**isotonic**).

Water
NH_4^+, Na^+, Cl^-

Gills: Excretion of **ammonia** by diffusion and **excess salt** via active transport across the gills. Large water loss by osmosis.

Body fluids = sea water. Net water movements in and out are zero. Such animals are **osmoconformers**.

Hermit crab

Freshwater Environments

....▸ Losses	⟶ Gains
⟹ Active transport of ions	
⇛ Major losses of nitrogen	

Body fluids more concentrated than the freshwater

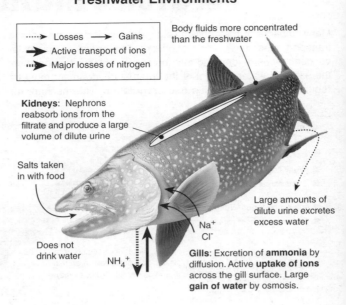

Kidneys: Nephrons reabsorb ions from the filtrate and produce a large volume of dilute urine

Salts taken in with food

Large amounts of dilute urine excretes excess water

Na^+
Cl^-

Does not drink water

NH_4^+

Gills: Excretion of **ammonia** by diffusion. Active **uptake of ions** across the gill surface. Large **gain of water** by osmosis.

Most marine invertebrates do not regulate salt and water balance; they are **osmoconformers** and their body fluids fluctuate with changes in the environment. Animals, such as fish and marine mammals, that regulate their salt and water fluxes are termed **osmoregulatory**. Bony fish lose water osmotically and counter the loss by drinking salt water and excreting the excess salt. Marine elasmobranchs generate osmotic concentrations in their body fluids similar to seawater by tolerating high urea levels. Excess salt from the diet is excreted via a salt gland in the rectum. Marine mammals produce a urine that is high in both salt and urea. Some intertidal animals tolerate frequent dilutions of normal seawater and may actively take up salts across the gill surfaces to compensate for water gain and salt loss.

Freshwater animals have body fluids that are osmotically more concentrated than the water they live in. Water tends to enter their tissues by osmosis and must be expelled to avoid flooding the body. Simple protozoans use contractile vacuoles to collect the excess water and expel it. Other invertebrates expel water and nitrogenous wastes using simple nephridial organs. Bony fish and aquatic arthropods produce dilute urine (containing ammonia) and actively take up salts across their gills (in aquatic insects these are often non-respiratory, anal gills). The kidneys of **freshwater bony fish** (above) also reabsorb salts from the filtrate through active transport mechanisms. These ion gains are important because some loss of valuable ions occurs constantly as a result of the high urine volume.

Sea anemones
Most marine invertebrates, like these sea anemones, are osmoconformers.

Stingray
Elasmobranchs maintain an osmotic concentration similar to seawater.

Amoeba
Contractile vacuoles in protozoans collect excess water and expel it.

Damselfly larva
In aquatic insect larvae, the gills actively take up salts from the water.

3. Describe how freshwater animals can compensate for salt losses that occur when they excrete large amounts of water:

_____

4. (a) Explain what is meant by an **osmoregulator** and give an example: _____

_____ Example: _____

(b) Explain what is meant by an **osmoconformer** and give an example: _____

_____ Example: _____

5. Freshwater and marine bony fish have contrasting excretion and osmoregulation problems. Explain why:

(a) Marine bony fish drink vast quantities of salt water: _____

(b) Freshwater bony fish do not drink water at all: _____

6. Describe the salt and water balance problems faced by migrating fish as they move from a marine environment to freshwater (as happens during spawning runs in salmon):

_____

_____

Waste Products in Humans

In humans and other mammals, a number of organs are involved in the excretion of the waste products of metabolism: mainly the kidneys, lungs, skin, and gut. The liver is a particularly important organ in the initial treatment of waste products, particularly the breakdown of haemoglobin and the formation of urea from ammonia. Excretion should not be confused with the elimination or egestion of undigested and unabsorbed food material from the gut. Note that the breakdown products of haemoglobin (blood pigment) are excreted in bile and pass out with the faeces, but they are not the result of digestion.

CO₂
Water

Lungs
Excretion of carbon dioxide (CO₂) with some loss of water.

Skin
Excretion of water, CO₂, hormones, salts and ions, and small amounts of urea as sweat.

Liver
Produces urea from ammonia in the urea cycle. Breakdown of haemoglobin in the liver produces the bile pigments e.g. bilirubin.

Gut
Excretion of bile pigments in the faeces. Also loses water, salts, and carbon dioxide.

Bladder
Storage of urine before it is expelled to the outside.

All cells
All the cells that make up the body carry out cellular respiration; they break down glucose to release energy and produce the waste products, carbon dioxide and water.

Excretion In Humans

In mammals, the kidney is the main organ of excretion, although the skin, gut, and lungs also play important roles. As well as ridding the body of nitrogenous wastes, the kidney is also able to excrete many unwanted poisons and drugs that are taken in from the environment. Usually these are ingested with food or drink, or inhaled. As long as these are not present in toxic amounts, they can usually be slowly eliminated from the body.

Kidney
Filtration of the blood to remove urea. Unwanted ions, particularly hydrogen (H^+) and potassium (K^+), and some hormones are also excreted by the kidneys. Some poisons and drugs (e.g. penicillin) are also excreted by active secretion into the urine. Water is lost in excreting these substances and extra water may be excreted if necessary.

Substance	Origin*	Organ(s) of excretion
Carbon dioxide		
Water		
Bile pigments		
Urea		
Ions (K+, H+)		
Hormones		
Poisons		
Drugs		

* Origin refers to from where in the body each substance originates

Homeostasis

1. In the diagram above, complete the table summarising the origin of excretory products and the main organ (or organs) of excretion involved for each substance.

2. Explain the role of the liver in excretion, even though it is not an excretory organ itself: _____

_____

_____

3. Tests for pregnancy are sensitive to an excreted substance in the urine. Suggest what this substance might be:

_____

4. People sometimes suffer renal (kidney) failure, where the kidneys cease to operate and can no longer produce urine. Given that the kidney rids the body of excessive ions and water, as well as nitrogenous wastes, predict the probable effects of kidney failure. You may wish to discuss this as a group. HINT: Consider the effects of salt and water retention, and the effect of high salt levels on blood pressure and on the heart.

_____

_____

_____

_____

Related activities: The Physiology of the Kidney, Kidney Dialysis

A 3

The Urinary System

The mammalian urinary system consists of the kidneys and bladder, and their associated blood vessels and ducts. The kidneys have a plentiful blood supply from the renal artery. The blood plasma is filtered by the kidneys to form urine. Urine is produced continuously, passing along the ureters to the bladder, a hollow muscular organ lined with smooth muscle and stretchable epithelium. Each day the kidneys filter about 180 dm³ of plasma.

Most of this is reabsorbed, leaving a daily urine output of about 1 dm³. By adjusting the composition of the fluid excreted, the kidneys help to maintain the body's internal chemical balance. All vertebrates have kidneys, but their efficiency in producing a concentrated urine varies considerably. Mammalian kidneys are very efficient, producing a urine that is concentrated to varying degrees depending on requirements.

The kidneys of **rats** (above), humans, and other mammals are distinctive, bean shaped organs that lie at the back of the abdominal cavity to either side of the spine. The kidneys lie outside the peritoneum of the abdominal cavity and are partly protected by the lower ribs. Each kidney is surrounded by three layers of tissue. The innermost renal capsule is a smooth fibrous membrane that acts as a barrier against trauma and infection. The two outer layers comprise fatty tissue and fibrous connective tissue. These act to protect the kidney and anchor it firmly in place.

The Human Urinary System

Vena cava returns blood to the heart.

Dorsal aorta supplies oxygenated blood to the body.

Adrenal glands are associated with, but not part of, the urinary system.

Renal vein returns the blood from the kidney to the venous circulation.

Renal artery carries blood from the aorta into the kidney.

Kidney produces urine (blood filtration, the removal of waste products, and the regulation of blood volume).

Ureter carries urine to the bladder.

Bladder (sectioned) stores the urine before it passes out of the body. It can expand to hold about 80% of the daily urine output.

Urethra conducts urine from the bladder to the outside. The urethra is regulated by a voluntary sphincter muscle.

The very precise alignment of the nephrons (the filtering elements of the kidney) and their associated blood vessels gives the kidney tissue a striated appearance, as seen in this cross section.

Transitional epithelium lines the bladder. This type of epithelium is layered, or stratified, and can be stretched without the outer cells breaking apart from each other.

1. Identify the components of the urinary system and describe their functions: _____

_____

_____

_____

2. Calculate the percentage of the plasma reabsorbed by the kidneys: _____

3. The kidney receives blood at a higher pressure than other organs. Suggest why this is the case: _____

_____

_____

4. Explain why the kidneys are surrounded by fatty connective tissue: _____

_____

Related activities: The Physiology of the Kidney, Control of Kidney Function

The Physiology of the Kidney

The functional unit of the kidney, the **nephron**, is a selective filter element, comprising a renal tubule and its associated blood vessels. Filtration, i.e. forcing fluid and dissolved substances through a membrane by pressure, occurs in the first part of the nephron, across the membranes of the capillaries and the glomerular capsule. The passage of water and solutes into the nephron and the formation of the glomerular filtrate depends on the pressure of the blood entering the afferent arteriole (below). If it increases, filtration rate increases; when it falls, glomerular filtration rate also falls. This process is so precisely regulated that, in spite of fluctuations in arteriolar pressure, glomerular filtration rate per day stays constant. After formation of the initial filtrate, the **urine** is modified through secretion and tubular reabsorption according to physiological needs at the time.

Nephrons are arranged with all the collecting ducts pointing towards the ureter.

Outer **cortex**

Inner **medulla**

The urine collects in a space near the ureter called the **renal pelvis** before flowing out of the kidney via the ureter.

Ureter

Urine flow

Internal Structure of the Human Kidney

Human kidneys are about 100-120 mm long and 25 mm thick. The functional unit of the kidney is the **nephron**. The other parts of the urinary system are primarily passageways and storage areas. The inner tissue of the kidney appears striated (striped), due to alignment of the nephrons and their surrounding blood vessels. It is the precise alignment of the nephrons in the kidney that makes it possible to fit in all the filtering units required. Each kidney contains more than 1 million nephrons. They are **selective filter elements**, which regulate blood composition and pH, and excrete wastes and toxins. The initial urine is formed by **filtration** in the glomerulus. Plasma is filtered through three layers: the capillary wall, and the basement membrane and epithelium of Bowman's capsule. The epithelium is made up of very specialised cells called **podocytes**. The filtrate is modified as it passes through the tubules of the nephron and the final urine passes out the ureter.

Homeostasis

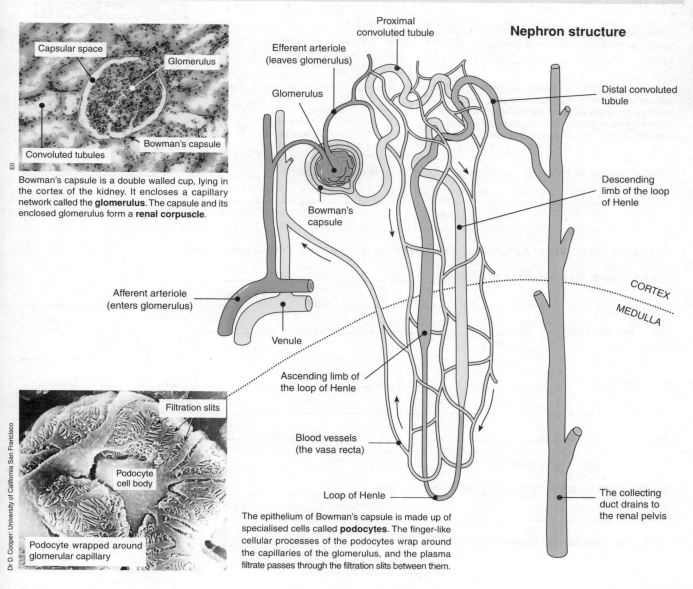

Capsular space

Glomerulus

Bowman's capsule

Convoluted tubules

Bowman's capsule is a double walled cup, lying in the cortex of the kidney. It encloses a capillary network called the **glomerulus**. The capsule and its enclosed glomerulus form a **renal corpuscle**.

Nephron structure

Proximal convoluted tubule

Efferent arteriole (leaves glomerulus)

Glomerulus

Distal convoluted tubule

Bowman's capsule

Descending limb of the loop of Henle

Afferent arteriole (enters glomerulus)

Venule

CORTEX

MEDULLA

Ascending limb of the loop of Henle

Blood vessels (the vasa recta)

Loop of Henle

The collecting duct drains to the renal pelvis

Filtration slits

Podocyte cell body

Podocyte wrapped around glomerular capillary

The epithelium of Bowman's capsule is made up of specialised cells called **podocytes**. The finger-like cellular processes of the podocytes wrap around the capillaries of the glomerulus, and the plasma filtrate passes through the filtration slits between them.

Related activities: The Urinary System, Control of Kidney Function

A 2

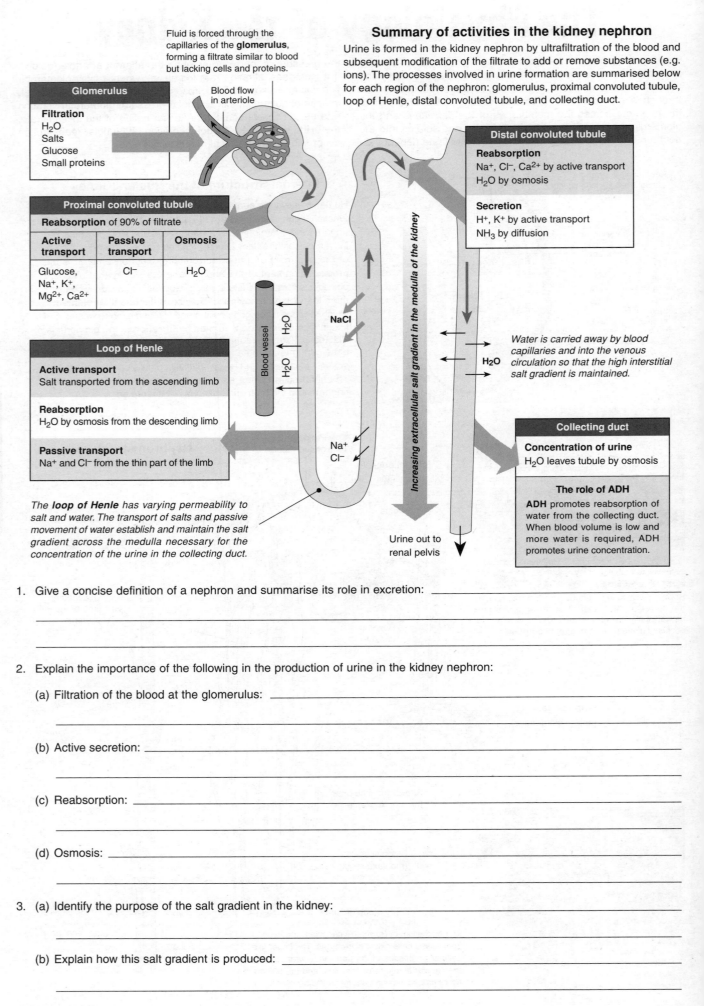

Fluid is forced through the capillaries of the **glomerulus**, forming a filtrate similar to blood but lacking cells and proteins.

Summary of activities in the kidney nephron

Urine is formed in the kidney nephron by ultrafiltration of the blood and subsequent modification of the filtrate to add or remove substances (e.g. ions). The processes involved in urine formation are summarised below for each region of the nephron: glomerulus, proximal convoluted tubule, loop of Henle, distal convoluted tubule, and collecting duct.

Glomerulus

Filtration
H_2O
Salts
Glucose
Small proteins

Blood flow in arteriole

Proximal convoluted tubule

Reabsorption of 90% of filtrate

Active transport	Passive transport	Osmosis
Glucose, Na^+, K^+, Mg^{2+}, Ca^{2+}	Cl^-	H_2O

Loop of Henle

Active transport
Salt transported from the ascending limb

Reabsorption
H_2O by osmosis from the descending limb

Passive transport
Na^+ and Cl^- from the thin part of the limb

*The **loop of Henle** has varying permeability to salt and water. The transport of salts and passive movement of water establish and maintain the salt gradient across the medulla necessary for the concentration of the urine in the collecting duct.*

Blood vessel

H_2O
H_2O
H_2O

NaCl

Na^+
Cl^-

Increasing extracellular salt gradient in the medulla of the kidney

Distal convoluted tubule

Reabsorption
Na^+, Cl^-, Ca^{2+} by active transport
H_2O by osmosis

Secretion
H^+, K^+ by active transport
NH_3 by diffusion

Water is carried away by blood capillaries and into the venous circulation so that the high interstitial salt gradient is maintained.

H_2O

Collecting duct

Concentration of urine
H_2O leaves tubule by osmosis

The role of ADH
ADH promotes reabsorption of water from the collecting duct. When blood volume is low and more water is required, ADH promotes urine concentration.

Urine out to renal pelvis

---

1. Give a concise definition of a nephron and summarise its role in excretion: _____

_____

_____

2. Explain the importance of the following in the production of urine in the kidney nephron:

(a) Filtration of the blood at the glomerulus: _____

_____

(b) Active secretion: _____

_____

(c) Reabsorption: _____

_____

(d) Osmosis: _____

_____

3. (a) Identify the purpose of the salt gradient in the kidney: _____

_____

(b) Explain how this salt gradient is produced: _____

_____

Control of Kidney Function

Variations in salt and water intake, and in the environmental conditions to which we are exposed, contribute to fluctuations in blood volume and composition. The primary role of the kidneys is to regulate blood volume and composition (including the removal of nitrogenous wastes), so that homeostasis is maintained. This is achieved through varying the volume and composition of the urine. Two hormones, antidiuretic hormone and aldosterone, are involved in the process.

Brain

ADH
ACTS ON KIDNEY

Variation in urine output

Control of Blood Volume

Osmoreceptors in the hypothalamus of the brain detect a fall in the concentration of water in the blood. They stimulate **neurosecretory cells in the hypothalamus** to synthesise and secrete the hormone ADH (antidiuretic hormone).

ADH passes from the hypothalamus to the posterior pituitary where it is released into the blood. ADH increases the permeability of the kidney collecting duct to water so that more water is reabsorbed and urine volume decreases.

Factors inhibiting ADH release
• High fluid intake
• High blood volume
• Low blood sodium levels
• Alcohol consumption

ADH levels decrease

Water reabsorption decreases. Urine output increases.

Factors causing ADH release
• Low fluid intake
• Low blood volume
• High blood sodium levels
• Nicotine and morphine

ADH levels increase

Water reabsorption increases. Urine output decreases.

Factors causing release of aldosterone
Low blood volumes also stimulate secretion of aldosterone from the adrenal cortex. This is mediated through a complex pathway involving the hormone renin from the kidney.

Aldosterone

Sodium reabsorption increases, water follows, blood volume restored.

Homeostasis

1. (a) *Diabetes insipidus* is a type of diabetes, caused by a lack of ADH. Based on what you know of the role of ADH in kidney function describe the symptoms of this disease:

_____

_____

(b) Suggest how this disorder might be treated: _____

_____

2. Explain why alcohol consumption (especially to excess) causes dehydration and thirst: _____

_____

_____

3. (a) State the effect of aldosterone on the kidney nephron: _____

(b) Explain the net result of this effect: _____

4. Explain how negative feedback mechanisms operate to regulate blood volume and urine output:

_____

_____

_____

Related activities: Water Budget in Mammals, The Urinary System, The Human Brain

RA 2

Kidney Dialysis

A dialysis machine is a machine designed to remove wastes from the blood. It is used when the kidneys fail, or when blood acidity, urea, or potassium levels increase much above normal. In kidney dialysis, blood flows through a system of tubes composed of semi-permeable membranes. Dialysis fluid (**dialysate**) has a composition similar to blood except that the concentration of wastes is low. It flows in the opposite direction to the blood on the outside of the dialysis tubes. Consequently, waste products like urea diffuse from the blood into the dialysis fluid, which is constantly replaced. The dialysis fluid flows at a rate of several 100 cm³ per minute over a large surface area. For some people dialysis is an ongoing procedure, but for others dialysis just allows the kidneys to rest and heal or recover.

A patient undergoing kidney dialysis at a hospital.

Principles of Kidney Dialysis

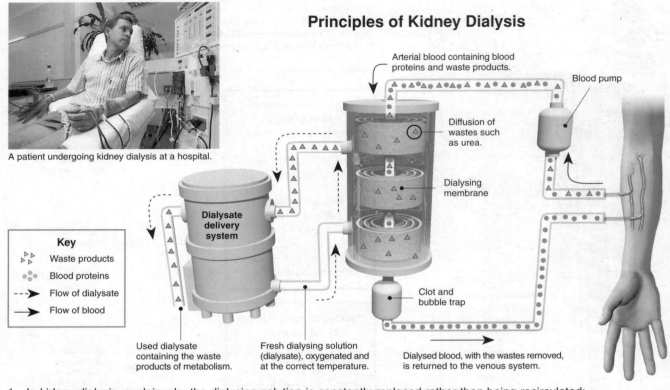

Arterial blood containing blood proteins and waste products.

Blood pump

Diffusion of wastes such as urea.

Dialysing membrane

Dialysate delivery system

Key
- ▷▷ Waste products
- ∘•∘ Blood proteins
- --▷ Flow of dialysate
- ──▶ Flow of blood

Clot and bubble trap

Used dialysate containing the waste products of metabolism.

Fresh dialysing solution (dialysate), oxygenated and at the correct temperature.

Dialysed blood, with the wastes removed, is returned to the venous system.

1. In kidney dialysis, explain why the dialysing solution is constantly replaced rather than being recirculated:

_____

_____

2. Explain why ions such as potassium and sodium, and small molecules such as glucose, do not diffuse rapidly from the blood into the dialysing solution along with the urea:

_____

_____

3. Explain why the urea passes from the blood into the dialysing solution: _____

_____

4. Describe the general transport process involved in dialysis: _____

_____

5. Give a reason why the dialysing solution flows in the opposite direction to the blood: _____

_____

_____

6. Explain why a clot and bubble trap is needed after the blood has been dialysed but before it re-enters the body:

_____

_____

_____

Water Budget in Mammals

Water loss is a major problem for most mammals. The degree to which urine can be concentrated (and water conserved) depends on the number of nephrons present in the kidney and the length of the loop of Henle. The highest urine concentrations are found in mammals from desert environments, such as kangaroo rats (below). Under normal conditions these animals will not drink water, obtaining most of their water from the metabolic break down of food instead.

Regulation of Water Balance in Humans

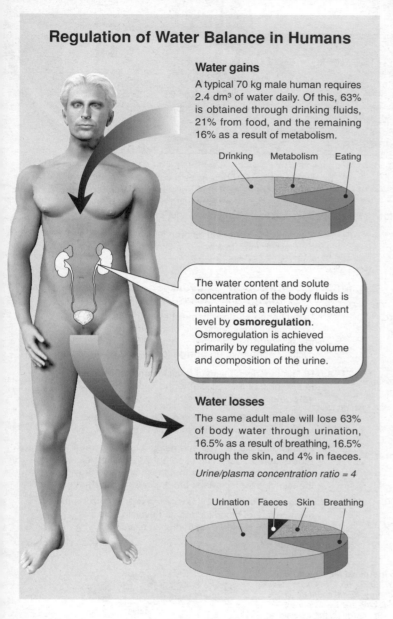

Water gains

A typical 70 kg male human requires 2.4 dm^3 of water daily. Of this, 63% is obtained through drinking fluids, 21% from food, and the remaining 16% as a result of metabolism.

Drinking   Metabolism   Eating

The water content and solute concentration of the body fluids is maintained at a relatively constant level by **osmoregulation**. Osmoregulation is achieved primarily by regulating the volume and composition of the urine.

Water losses

The same adult male will lose 63% of body water through urination, 16.5% as a result of breathing, 16.5% through the skin, and 4% in faeces.

Urine/plasma concentration ratio = 4

Urination   Faeces   Skin   Breathing

Adaptations of Arid Adapted Rodents

Most desert-dwelling mammals are adapted to tolerate a low water intake. Arid adapted rodents, such as jerboas and kangaroo rats, conserve water by reducing losses to the environment and obtain the balance of their water needs from the oxidation of dry foods (respiratory metabolism). The table below shows the water balance in a kangaroo rat after eating 100 g of dry pearl barley. Note the high urine to plasma concentration ratio relative to that of humans.

Water balance in a kangaroo rat
(*Dipodomys spectabilis*)

Water gains		Water losses	
Absorbed from food	6.0 cm^3	Breathing	43.9 cm^3
From metabolism	54.0 cm^3	Urination	13.5 cm^3
		Defaecation	2.6 cm^3

Urine/plasma concentration ratio = 17

Adaptations of kangaroo rats

Kangaroo rats, and other arid-adapted rodents, tolerate long periods without drinking, meeting their water requirements from the metabolism of dry foods. They dispose of nitrogenous wastes with very little output of water and they neither sweat nor pant to keep cool.

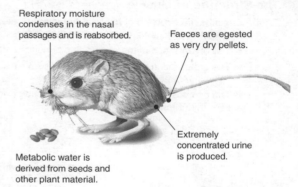

Respiratory moisture condenses in the nasal passages and is reabsorbed.

Faeces are egested as very dry pellets.

Extremely concentrated urine is produced.

Metabolic water is derived from seeds and other plant material.

1. Explain why most mammals need to drink regularly: _____

_____

2. Using the tabulated data for the kangaroo rat (above), graph the water gains and losses in the space provided below.

3. Describe three physiological adaptations of desert adapted rodents to low water availability:

(a) _____

_____

(b) _____

_____

(c) _____

_____

Water gains   Water losses

Related activities: Water Balance in Animals, The Physiology of the Kidney

DA 2

Homeostasis

Muscles and Movement

AQA-A	AQA-B	SNAB	Edexcel	OCR
Not applicable to the A2 course	Complete: 1-5, 7-8	Complete: 1-11	Complete the learning objectives for Option C (TRC)	Complete the learning objectives for Option 05 (TRC)

Learning Objectives

☐ 1. Compile your own glossary from the **KEY WORDS** displayed in **bold type** in the learning objectives below.

Muscle Action *(pages 267-268)*

☐ 2. Understand how joints, together with **antagonistic muscles**, permit particular movement of specific parts of the skeleton.

☐ 3. Explain examples of movement in terms of **antagonistic muscle action**. Include reference to the terms **extensor**, **flexor**, **protraction**, and **retraction**. If required, identify the role of reflex inhibition in the movement of antagonistic muscle pairs. Identify the functional role of **agonists**, **antagonists**, and **synergists** as appropriate to your described examples.

The Structure of Muscle *(pages 269-271)*

☐ 4. Describe the structure of **skeletal muscle**, as seen with light and electron microscopy. Include reference to:
- The organisation of the muscle into bundles of **fibres**.
- The organisation of the individual fibres (muscle cells) into bundles of smaller **myofibrils**.
- The composition and arrangement of the thick and thin (**myo**)**filaments** within each myofibril.

☐ 5. Recognise a **sarcomere** as a complete contractile unit. Describe the relationship between the structure of the sarcomere and the distribution of **actin** and **myosin** (thin and thick filaments) within the myofibril.

☐ 6. Describe the structural differences between fast and slow twitch muscle fibres. Explain the physiological basis of these differences.

Muscle Contraction *(pages 269-270, 272 and the TRC: Energy and Exercise, Effects of Training)*

☐ 7. Describe the **sliding-filament hypothesis** of muscle contraction. Identify the role of **calcium ions** (Ca^{2+}), the **troponin-tropomyosin** complex, and **ATP** in the cycle of **actin/myosin bridge** (cross-bridge) formation.

☐ 8. Relate the mechanism of muscle contraction (shortening of the muscle) to the appearance of a **sarcomere** in a contracted or relaxed state.

☐ 9. EXTENSION: Define VO_2 and VO_2max. Recognise the relationship between fitness and VO_2max.

☐ 10. Explain **muscle fatigue** and relate it to the increase in **blood lactate**, depletion of carbohydrate supplies, and decreased pH. Recognise that these changes provide the stimulus for increased breathing (and heart) rates.

☐ 11. Understand the concept of **oxygen debt**. Describe the ultimate fate of blood lactate and explain how the oxygen debt is repaid after intense exercise.

See pages 5-7 for additional details of these texts:
- Adds, J. *et al.* 2003. **Respiration and Coordination** (NelsonThornes), chpt. 12 in part.
- Clegg, C.J., 1998. **Mammals: Structure and Function** (John Murray), chpt. 9.
- Jones, M. & G. Jones, 2002. **Mammalian Physiology and Behaviour** (CUP), chpt. 3.
- Rowett, H.G.Q., 1999. **Basic Anatomy & Physiology** (John Murray), pp. 38-59.

See page 7 for details of publishers of periodicals:

STUDENT'S REFERENCE

■ **Skeletal Muscle: Is Bigger Always Better?** Biol. Sci. Rev., 11(2) Nov. 1998, pp. 36-39. *The structure of muscle and physiology of contraction.*

■ **Energy Saving in Animal Movement** Biol. Sci. Rev., 18(2) Nov. 2005, pp. 2-5. *Locomotion is costly in terms of energy; this account describes how structure, physiology, and behaviour are adapted to conserving energy during locomotion.*

■ **Lactic Acid** Biol. Sci. Rev., 18(2) Nov. 2005, pp. 6-9. *Lactic acid: its production as a result of anaerobic metabolism and its role in muscle.*

■ **Lactic Acid** Biol. Sci. Rev., 8(1) September 1995, pp. 20-25. *The role of lactic acid as an important source of energy for working muscles, even in the presence of oxygen. This article examines how lactic acid is produced and its role in exercise science and clinical medicine.*

■ **Roots of Fatigue** New Scientist, 21 May 1994, (Inside Science). *Muscle contraction and fatigue. Includes good diagrams of muscle contraction and discusses the role of ATP and Ca^{2+} in this process.*

■ **The Physiology of Muscle** Biol. Sci. Rev., 8(3) January 1996, pp. 2-6. *Contraction of smooth and striated muscle and the role of Ca^{2+} in this.*

■ **Running Ragged** New Scientist, 16 August, 1997, pp. 28-29. *Human exercise physiology and the role of the muscle proteins actin and myosin.*

■ **Muscles for the Job** New Scientist, 15 April 1989, pp. 50-53. *The type and activity of a muscle is tuned specifically to the tasks it has to do.*

■ **Bodyworks** New Scientist, 13 December 1997, pp. 38-41. *The physiology of muscles and microtubules: what makes our muscles move.*

See pages 10-11 for details of how to access **Bio Links** from our web site: **www.biozone.co.uk**. From Bio Links, access sites under the topics:

GENERAL BIOLOGY ONLINE RESOURCES

• Biology I interactive animations • Instructional multimedia, University of Alberta • How stuff works *... and others* > **Online Textbooks and Lecture Notes**: • Human biology help • Learn.co.uk • S-Cool! A level biology revision guide *... and others* > **Glossaries**: Animal anatomy glossary • Kimball's biology glossary... *and others*

ANIMAL BIOLOGY: • Anatomy and physiology • Human physiology lecture notes *... and others* > **Support & Movement**: • Bone and joint sources • Muscle spindle and stretch reflexes • Muscle structure and function • Muscles • Skeletal muscle structure and function

Software and video resources for this topic are provided on the Teacher Resource CD-ROM

The Mechanics of Locomotion

Bones are too rigid to bend without damage. To allow movement, the skeletal system consists of many bones held together at **joints** by flexible connective tissues called **ligaments**. All movements of the skeleton occur at joints: points of contact between bones, or between cartilage and bones. Joints may be classified according to the amount of movement they permit: none (sutures); slight movement (symphyses), and free movement in one or more planes (e.g. **synovial joints**). Bones are made to move about a joint by the force of muscles acting upon them. The muscles are attached to bone by **tendons**. Many muscles act in **antagonistic** pairs, one set causing the joint to move one way, the other set causing its return. The skeleton of an animal works as a system of levers. The joint acts as a fulcrum, the muscles exert the force, and the weight of the bone being moved represents the load. Contraction causes a muscle to shorten and this shortening moves attached bones. When only a few fibres in a muscle contract, the muscle will tighten but not produce movement. This partly contracted state is responsible for **muscle tone** and is important in maintaining **posture**. The amount of muscle contraction is monitored by sensory receptors in the muscle called **muscle spindle organs**. These provide the sensory information necessary to adjust movement as required.

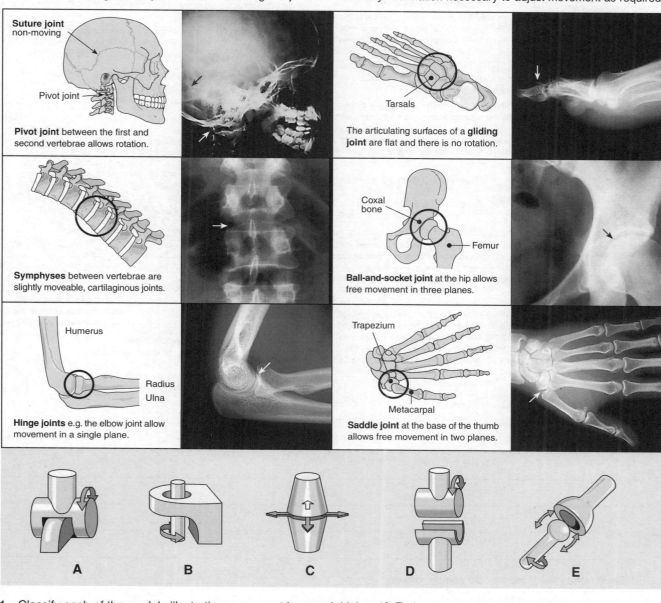

Suture joint non-moving

Pivot joint between the first and second vertebrae allows rotation.

Symphyses between vertebrae are slightly moveable, cartilaginous joints.

Humerus
Radius
Ulna

Hinge joints e.g. the elbow joint allow movement in a single plane.

Tarsals

The articulating surfaces of a **gliding joint** are flat and there is no rotation.

Coxal bone
Femur

Ball-and-socket joint at the hip allows free movement in three planes.

Trapezium
Metacarpal

Saddle joint at the base of the thumb allows free movement in two planes.

A   B   C   D   E

Muscles and Movement

1. Classify each of the models illustrating movement in synovial joints (**A-E** above), according to the labels given below:

   (a) Pivot: _____ (b) Hinge: _____ (c) Ball-and-socket: _____ (d) Saddle: _____ (e) Gliding: _____

2. Briefly state the role of each of the following in human locomotion:

   (a) Ligament: _____

   (b) Tendon: _____

   (c) Antagonistic muscles: _____

   (d) Bones: _____

   (e) Joints: _____

Related activities: Muscle Structure and Function

RA 2

The Action of Antagonistic Muscles

The flexion (bending) and extension (unbending) of limbs is caused by the action of **antagonistic muscles**; muscles that work in pairs and whose actions oppose each other. Every coordinated movement in the body requires the application of muscle force. This is accomplished by the action of agonists, antagonists, and synergists. The opposing action of agonists and antagonists also produces muscle tone. Note that either muscle in an antagonistic pair can act as the prime mover, depending on the movement (flexion or extension).

Biceps brachii

Agonists or prime movers: muscles that are primarily responsible for the movement and produce most of the force required.

Antagonists: muscles that oppose the prime mover. They may also play a protective role by preventing overstretching of the prime mover.

Synergists: muscles that assist the prime movers and may be involved in fine-tuning the direction of the movement.

During flexion of the forearm at the elbow, (left) the **biceps brachii** acts as the prime mover, and the antagonist, the **triceps brachii** at the back of the arm, is relaxing. During extension, their roles are reversed.

Quadriceps

Hamstrings

Movement of the leg is accomplished through the action of several large groups of muscles, collectively called the **quadriceps** and the **hamstrings**.

The hamstrings are actually a collection of three muscles, which act together to flex the leg. The quadriceps at the front of the thigh (a collection of four large muscles) opposes the motion of the hamstrings and extends the leg. When the prime mover contracts very forcefully, the antagonist also contracts very slightly. This prevents any overstretching and allows greater control over thigh movement.

The Role of the Muscle Spindle

Changes in length of a muscle are monitored by the **muscle spindle organ**, a stretch receptor located within skeletal muscle, parallel to the muscle fibres themselves. The muscle spindle is stimulated in response to sustained or sudden stretch on the central region of its specialised intrafusal fibres. Sensory information from the muscle spindle is relayed to the spinal cord. The motor response brings about adjustments to the degree of stretch in the muscle. These adjustments help in the coordination and efficiency of muscle contraction. Muscle spindles are important in the maintenance of muscle tone, postural reflexes, and movement control, and are concentrated in muscles that exert fine control over movement.

Sensory nerves monitor stretch in the non-contractile region of the spindle and send impulses to the spinal cord.

Motor nerves send impulses to adjust the degree of contraction in the intrafusal and extrafusal fibres.

Striated appearance of contractile elements

Nucleus of muscle fibre

The spindle is surrounded by the muscle fibres of the skeletal muscle.

The **muscle spindle organ** comprises special **intrafusal fibres** which lie parallel to the muscle fibres within a lymph-filled capsule. Only the regions near the end can contract.

3. Using appropriate terminology, explain how antagonistic muscles act together to raise and lower a limb:

_____

_____

_____

_____

_____

4. (a) Explain the role of the muscle spindle organ: _____

_____

(b) With reference to the following, describe how the structure of the muscle spindle organ is related to its function:

Intrafusal fibres lie parallel to the extrafusal fibres: _____

_____

Sensory neurones are located in the non-contractile region of the organ: _____

_____

Motor neurones synapse in the extrafusal fibres and the contractile region of the intrafusal fibres: _____

_____

Muscle Structure and Function

There are three kinds of muscle: **skeletal, cardiac,** and **smooth** muscle, each with a distinct structure. The muscles used for posture and locomotion are skeletal (striated) muscles. Their distinct striped appearance is the result of the regular arrangement of contractile elements within the muscle cells. Muscle fibres are innervated by motor neurones, each of which terminates in a specialised cholinergic synapse called the **motor end plate**. A motor neurone and all the fibres it innervates is called a **motor unit**.

Skeletal muscle
Also called striated or striped muscle. It has a banded appearance under high power microscopy. Sometimes called voluntary muscle because it is under conscious control. The cells are large with many nuclei at the edge of each cell.

Cardiac muscle
Specialised striated muscle that does not fatigue. Cells branch and connect with each other to assist the passage of nerve impulses through the muscle. Cardiac muscle is not under conscious control (it is involuntary).

Smooth muscle
Also called involuntary muscle because it is not under conscious control. Contractile filaments are irregularly arranged so the contraction is not in one direction as in skeletal muscle. Cells are spindle shaped with one central nucleus.

Structure of Skeletal Muscle

Skeletal muscle is organised into bundles of muscle cells or **fibres**. Each fibre is a single cell with many nuclei and each fibre is itself a bundle of smaller **myofibrils** arranged lengthwise. Each myofibril is in turn composed of two kinds of **myofilaments** (thick and thin), which overlap to form light and dark bands. It is the alternation of these light and dark bands which gives skeletal muscle its striated or striped appearance. The **sarcomere**, bounded by the dark Z lines, forms one complete contractile unit.

Longitudinal section of a sarcomere

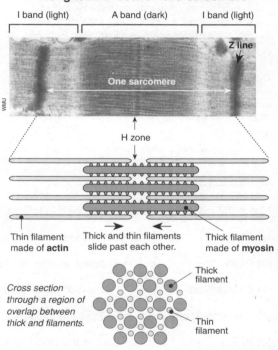

The photograph of a sarcomere (above) illustrates the distinctive banding pattern of skeletal muscle. The pattern, which arises because of the arrangement of thin and thick filaments, is shown in schematic form below the photograph. A representation of the region of overlap is also illustrated.

An action potential is conducted to all myofibrils of the muscle fibre.

The arrival of a nerve impulse at the **motor end plate** causes the release of acetylcholine, stimulating an action potential in the sarcolemma.

Motor end plate

Axon of motor neurone

Sarcolemma (=plasma membrane)

Sarcoplasmic reticulum: a network of membranous tubules containing a store of calcium ions.

Sarcoplasm (=cytoplasm)

Nucleus

One sarcomere

Filaments of a myofibril seen in cross section.

Mitochondrion

Several myofibrils of a muscle fibre

The Banding Pattern of Myofibrils

Within a myofibril, thin filaments, held together by the **Z lines**, project in both directions. The arrival of a nerve impulse sets in motion a series of events that cause the thick and thin filaments to slide past each other. This **contraction** results in shortening of the muscle and is accompanied by a visible change in the appearance of the myofibril: the I band and the sarcomere shorten and H zone shortens or disappears.

Related activities: Neurone Structure & Function, The Mechanics of Locomotion
Web links: Muscle Cell Contraction, Muscle Structure and Contraction

A 2

Muscles and Movement

The Sliding Filament Hypothesis

Muscle contraction requires calcium ions (Ca^{2+}) and energy (in the form of ATP) in order for the thick and thin filaments to slide past each other. The steps are:

1. The binding sites on the **actin** molecule (to which myosin 'heads' will locate) are blocked by a complex of two molecules: tropomyosin and troponin.

2. Prior to muscle contraction, ATP binds to the heads of the myosin molecules, priming them in an erect high energy state. Arrival of an action potential causes a release of Ca^{2+} from the sarcoplasmic reticulum. The Ca^{2+} binds to the troponin and causes the blocking molecules to move so that the myosin binding sites on the actin filament become exposed.

3. The heads of the cross-bridging myosin molecules attach to the binding sites on the actin filament. Release of energy from the hydrolysis of ATP accompanies the cross bridge formation.

4. The energy released from ATP hydrolysis causes a change in shape of the myosin **cross bridge**, resulting in a bending action (*the power stroke*). This causes the actin filaments to slide past the myosin filaments towards the centre of the sarcomere.

5. (Not illustrated). Fresh ATP attaches to the myosin molecules, releasing them from the binding sites and repriming them for a repeat movement. They become attached further along the actin chain as long as ATP and Ca^{2+} are available.

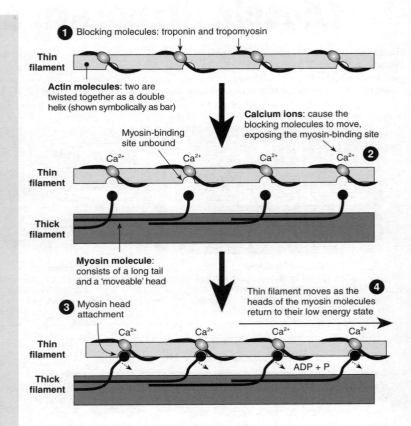

1. Describe the structural and physiological features of skeletal muscle that distinguish it from smooth and cardiac muscle:

_____

_____

_____

_____

2. Match the following chemicals with their functional role in muscle movement (draw a line between matching pairs):

(a) Myosin   • Bind to the actin molecule in a way that prevents myosin head from forming a cross bridge

(b) Actin   • Supplies energy for the flexing of the myosin 'head' (power stroke)

(c) Calcium ions   • Has a moveable head that provides a power stroke when activated

(d) Troponin-tropomyosin   • Two protein molecules twisted in a helix shape that form the thin filament of a myofibril

(e) ATP   • Bind to the blocking molecules, causing them to move and expose the myosin binding site

3. (a) Explain the cause of the banding pattern visible in striated muscle: _____

_____

_____

(b) Explain the change in appearance of a myofibril during contraction with reference to the following:

The I band: _____

The H zone: _____

The sarcomere: _____

4. **Rigor mortis** is a state of partial contraction of the muscle that occurs after death. During rigor mortis, the muscles are locked and the body is stiff. From what you understand of muscle contraction, suggest why the muscle enters this state:

_____

_____

Muscle Physiology and Performance

The overall effect of **aerobic training** on muscle is improved oxidative function and better endurance. Regardless of the type of training, some of our ability to perform different types of activity depends on our genetic make-up. This is particularly true of aspects of muscle physiology, such as the relative proportions of different fibre types in the skeletal muscles. Muscle fibres are primarily of two types: **fast twitch** (FT) or **slow twitch** (ST). Fast twitch fibres predominate during anaerobic, explosive activity, whereas slow twitch fibres predominate during endurance activity. In the table below, note the difference in the degree to which the two fibre types show fatigue (a decrease in the capacity to do work). Training can increase fibre size and, to some extent, the makeup of the fibre, but not the proportion of ST to FT, which is genetically determined.

The Effects of Aerobic Training on Muscle Physiology

Improved oxidation of glycogen. Training increases the capacity of skeletal muscle to generate ATP aerobically.

An increased capacity of the muscle to oxidise fats. This allows muscle and liver glycogen to be used at a slower rate. The body also becomes more efficient at mobilising free fatty acids from adipose tissue for use as fuel.

Increased myoglobin content. Myoglobin stores oxygen in the muscle cells and aids oxygen delivery to the mitochondria. Endurance training increases muscle myoglobin by 75%-80%.

Increase in lean muscle mass and decrease in body fat. Trained endurance athletes typically have body fat levels of 15-19% (women) or 6-18% (men), compared with 26% (women) and 15% (men) for non-athletes.

The size of **slow twitch fibres** increases. This change in size is associated with increased aerobic capacity.

An increase in the size and density of mitochondria in the skeletal muscles and an increase in the activity and concentration of Krebs cycle enzymes.

An increase in the number of capillaries surrounding each muscle fibre. Endurance trained men have 5%-10% more capillaries in their muscles than sedentary men.

Fast vs Slow Twitch Muscle

Feature	Fast twitch	Slow twitch
Colour	White	Red
Diameter	Large	Small
Contraction rate	Fast	Slow
ATP production	Fast	Slow
Metabolism	Anaerobic	Aerobic
Rate of fatigue	Fast	Slow
Power	High	Low

Skeletal muscle fibres

There are two basic types of muscle fibres. Slow twitch (type I) muscle and fast twitch (type II) muscle fibre. Human muscles contain a genetically determined mixture of both slow and fast fibre types. On average, we have about 50% slow and 50% fast fibres in most of the muscles used for movement.

The slow twitch fibres contain more mitochondria and myoglobin than fast twitch fibres, which makes them more efficient at using oxygen to generate ATP without lactate acid build up. In this way, they can fuel repeated and extended muscle contractions such as those required for endurance events like a marathon.

Both fibre types generally produce the same force per contraction, but fast twitch fibres produce that force at a higher rate, so fast fibres are important when there is a limited time in which to generate maximal force (as in a sprint)

Muscles and Movement

1. Explain three ways in which aerobic (endurance) training improves the oxidative function of muscle:

(a) _____

_____

(b) _____

_____

(c) _____

_____

2. Contrast the properties of fast and slow twitch skeletal muscle fibres, identifying how these properties contribute to their performance in different conditions:

_____

_____

_____

Related activities: Muscle Structure and Function

RA 2

Muscle Fatigue

Muscle fatigue refers to the decline in a muscle's ability to maintain force in a prolonged or repeated contraction. It is a normal result of vigorous exercise but the reasons for it are complex. During strenuous activity, the oxygen demand of the working muscle exceeds supply and the muscle metabolises anaerobically. This leads to the metabolic changes (including lactic acid accumulation) that result in fatigue. The lactic acid (lactate) produced during muscle contraction diffuses into the bloodstream and is carried to the liver where it is reconverted to glucose via the **Cori cycle**. This requires energy from aerobic respiration and accounts for the **oxygen debt** which causes the continued heavy breathing after exercise has stopped. During moderate intensity exercise, the rate at which lactate enters the blood from the muscles is equal to its rate of removal by the liver. This steady-state situation is suitable for endurance training, because the lactic acid doesn't accumulate in the muscle tissue.

At Rest

- Muscles produce a surplus of ATP
- This extra energy is stored in CP (creatine phosphate) and glycogen

During Moderate Activity

- ATP requirements are met by the aerobic metabolism of glycogen and lipids

During Peak Activity

- Effort is limited by ATP. ATP production is ultimately limited by availability of oxygen.
- During short-term, intense activity, ATP is increased through anaerobic metabolism of glycogen (glycolysis).
- A by-product of this is lactic acid, which lowers tissue pH and affects cellular activity.
- When **fatigued**, the muscle can no longer contract fully.

Lactic Acid and Muscle Fatigue

Lactic acid is a by-product of ATP production through anaerobic metabolism when oxygen demand exceeds supply. Lactic acid accumulation in the muscle causes a fall in tissue pH and inhibits the activity of the key enzymes involved in ATP production. This decline in ATP supply limits muscular performance during peak activity (see graph below). Together with the effects of ATP and creatine phosphate breakdown (accumulating phosphate (Pi) for example), lactic acid buildup also acts to slow the release of calcium into the T tubules and affects the ion pumps which move calcium ions back into the sarcoplasmic reticulum. This contributes to fatigue because calcium is an key component in muscle contraction.

Effect of pH on muscle tension

Relative velocity — Relative tension

Normal pH

Low pH

Increased lactate
Decline in pH
Elevated Pi

Decline in ATP
Fall in Ca^{2+} release

Fatigue

Short term maximal exertion (sprint)
- Lactic acid build-up lowers pH ● Depletion of creatine phosphate ● Buildup of phosphate (P_i) affects the sensitivity of the muscle to Ca^{2+}

Mixed aerobic and anaerobic (5 km race)
- Lactate accumulation in the muscle ● Build-up of ADP and P_i ● Decline in Ca^{2+} release affects the ability of the muscle to contract

Extended sub-maximal effort (marathon)
- Depletion of all energy stores (glycogen, lipids, amino acids) leads to a failure of Ca^{2+} release
- Repetitive overuse damages muscle fibres.

1. Explain the mechanism by which lactic acid accumulation leads to muscle fatigue: _____

_____

2. Identify the two physiological changes in the muscle that ultimately result in a decline in muscle performance:

(a) _____   (b) _____

3. Suggest why the reasons for fatigue in a long distance race are different to those in a 100 m sprint: _____

_____

_____

Related activities: Muscle Performance and Physiology
Web links: Energy and Exercise, Effects of Training

Responses and Coordination

Learning Objectives

☐ 1. Compile your own glossary from the **KEY WORDS** displayed in **bold type** in the learning objectives below.

Nervous Systems *(pages 239-240, 275-280)*

☐ 2. Appreciate the structural and functional differences between nervous and hormonal coordination.

☐ 3. With respect to structure and function, distinguish between the two primary divisions of the mammalian nervous system: the **central nervous system** (CNS) and the **peripheral nervous system** (PNS).

☐ 4. Recognise main regions of the **brain** and their primary functions: brain stem (**medulla oblongata**, pons, and midbrain), **cerebellum**, thalamus, **hypothalamus**, **cerebral hemispheres** (cerebrum).

☐ 5. Describe the location and principal functions of the cerebral hemispheres with reference to the following:
- **Sensory** and **motor areas**.
- Association areas, e.g. the **visual association area**.
- The areas associated with speech.

Relate the size of a cerebral region to the complexity of its innervation and describe how each side of the body is controlled by the opposite hemisphere.

☐ 6. EXTENSION: Describe an example of brain malfunction, e.g. as occurs in **Alzheimer's disease**.

☐ 7. Explain the basic function of the two divisions of the PNS: **sensory division** and **motor division**.

☐ 8. Appreciate that the motor division of the PNS is further divided into the **autonomic** and **somatic nervous systems** and outline the role of each.

☐ 9. Recognise the components of the autonomic nervous system (**ANS**): the **parasympathetic** and **sympathetic nervous systems**. Describe their roles and recognise that they have generally opposite effects.

☐ 10. Describe examples of the effects of the autonomic nervous system with reference to the following:
(a) Control of pupil diameter and tear production.
(b) Control of bladder emptying.
(c) Control of heart rate and force of contraction.
(d) Control of digestive secretions and gut motility.

Understand that some reflex autonomic activity can be modified by conscious control, and provide an example.

Neurone Structure and Function *(pages 281-283)*

☐ 11. Understand the term **neurone**. Recognise the structural and functional features that distinguish different types of neurones: **sensory**, **effector** (**motor**), and **relay neurones** (**interneurones**).

☐ 12. Draw a labelled diagram to illustrate the structure of a **myelinated motor neurone**. Appreciate the role of the **Schwann cells** in the **myelination** of a motor neurone.

☐ 13. Explain what is meant by a **reflex**. Using a diagram, describe the functioning of a simple spinal reflex arc involving three neurones and identify the neurone types involved. Appreciate the adaptive value of reflexes.

☐ 14. Explain what is meant by the **resting potential** of a neurone and explain how it is established. Include reference to **ion pumps** and the movement of Na^+ and K^+, the differential permeability of the membrane, and the generation of an **electrochemical gradient**.

☐ 15. Describe the generation of the **action potential (nerve impulse)** with reference to the change in membrane permeability of the nerve leading to **depolarisation** and the **all-or-nothing** nature of the impulse.

☐ 16. Describe how an action potential is propagated along a myelinated nerve by **saltatory conduction**, identifying the roles of **myelin** and the **nodes of Ranvier**.

☐ 17. Contrast the nature of the nerve impulse in **myelinated** and **non-myelinated** fibres and explain reasons for the difference. Describe how myelination, axon diameter, and temperature affect speed of conductance.

☐ 18. Explain the nature and importance of the **refractory period** in producing discrete impulses.

Synapses *(pages 284-286)*

☐ 19. Identify the role of synapses in mammals. Describe the structure of a **cholinergic synapse** (as seen using electron microscopy). Recognise the **neuromuscular junction** as a specialised cholinergic synapse.

☐ 20. Outline the sequence of events involved in the action of a **cholinergic synapse**. Include reference to the arrival of the **action potential** at the **presynaptic terminal**, the role of Ca^{2+} and **acetylcholine**, and the generation of the action potential in the **postsynaptic neurone**.

☐ 21. Understand that transmission at chemical synapses involves **neurotransmitters** and understand how these differ functionally from hormones. Identify the roles of **acetylcholine** and **noradrenaline** at different synapses.

☐ 22. Describe the effects of drugs at synapses as shown by:
(a) Amplification at the synapse by mimicking the action of natural transmitters, e.g. **nicotine**, caffeine.
(b) Amplification at the synapse by preventing neuro-transmitter breakdown, e.g. strychnine, nerve gas.
(c) Inhibition at the synapse by blocking transmission across the synapse, e.g. atropine, curare.

☐ 23. Understand the terms: **agonist** and **antagonist** with respect to the actions of drugs at synapses.

☐ 24. Appreciate the role of synapses in:
(a) **Unidirectionality**: impulses travel in one direction.
(b) Allowing the interconnection of nerve pathways.
(c) **Integration** through **summation** and **inhibition**.

Sensory Reception in Mammals *(pages 287-288)*

☐ 25. Outline the need for communication systems to respond to environmental **stimuli**. Distinguish between internal and external stimuli and identify some of the types of **sense organs** that respond to these.

☐ 26. Appreciate the role of sensory receptors as **biological transducers**. With reference to specific examples, understand the following features of sensory receptors:
   (a) They respond to specific stimuli by producing generator potentials.
   (b) The strength of receptor response is proportional to the stimulus strength.
   (c) They show **sensory adaptation**.

☐ 27. Using an example, describe the structure of a sensory receptor. Identify the stimulus for response in the receptor, and explain how the response is achieved.

The mammalian eye *(pages 290-292)*

☐ 28. Describe the basic structure and properties of the mammalian **eye** with respect to the transmission and refraction of light, and the focusing of an image on the **retina**. Include reference to the following:
   (a) The structure and function of the iris.
   (b) The role of the cornea, lens, ciliary muscles, and suspensory ligaments (including **accommodation**).
   (c) The significance of the central fovea and blind spot.

☐ 29. Describe the structure and function of the photo-receptor cells (**rods** and **cones**) in the retina, including:
   (a) The role of the photosensitive pigments (e.g. **rhodopsin**) and their response to absorbed light.
   (b) The basis of monochromatic and trichromatic vision.
   (c) The creation of generator potentials in the rod or cone and action potentials in the optic nerve.
   (d) The reasons for differences in sensitivity and acuity.

☐ 30. In simple terms, explain how the nerve impulses from the eye are interpreted.

The mammalian ear *(page 289)*

☐ 31. Understand the nature of sound as a stimulus. Describe the basic structure of the mammalian **ear**, and outline the function of each of its main regions. Describe how sound is received by the ear and how the sense cells produce a nerve impulse.

Animal Behaviour *(pages 293-300 and the TRC: Animal Behaviour Record)*

☐ 32. Describe examples of simple responses as illustrated by **taxes** and **kineses**. Identify the adaptive role of these and other simple behaviours (e.g. reflexes).

☐ 33. Distinguish between **innate** and **learned behaviour**. Describe examples of learned behaviours, identifying their adaptive role in each case. Include reference to **habituation**, **imprinting**, **classical coditioning**, and **operant conditioning**

☐ 34. Recognise that many behaviours have both an inherited (innate) and a learned component and describe an example, e.g. song development in birds.

☐ 35. Describe each of the following as they relate to successful breeding: **courtship behaviour**, including the role of hormones or **pheromones**, **species recognition**, and **pair bonding**. Appreciate the stereotyped nature of courtship displays, including the role of sign stimuli and innate releasing mechanisms.

Responses in Plants *(pages 301-304, and TRC: Applications of Plant Hormones, Plant Defences)*

☐ 36. Appreciate the need for communication systems in flowering plants to respond to environmental stimuli. Describe responses in plants and their physiological basis, including, where appropriate, the commercial significance of these. Include reference to the following:
   (a) The role of **auxins**, **gibberellins**, and **abscisic acid** on plant growth and development.
   (b) The role of **cytokinins** and **ethene** on plant growth.
   (c) The role of **phytochrome pigments** in the detection and response to daylength.
   (d) The nature of **tropisms**, especially phototropism.
   (e) The nature of plant defence mechanisms.

See the 'Textbook Reference Grid' on pages 8-9 for textbook page references relating to material in this topic.

Supplementary Texts

See pages 5-7 for additional details of these texts:

■ Adds, J. *et al.* 2003. **Respiration and Coordination** (NelsonThornes), chpt. 2-3 as reqd.

■ Clegg, C.J., 1998. **Mammals: Structure and Function** (John Murray), pp. 58-69.

■ Dockerty, M. & M. Reiss, 1999. **Behaviour** (CUP), entire text as required.

■ Jones, M. & G. Jones, 2002. **Mammalian Physiology and Behaviour** (CUP), pp. 48-91.

■ Murray, P. and N. Owens, 2001. **Behaviour and Populations** (Collins), pp. 6-27.

■ Rowett, H.G.Q., 1999. **Basic Anatomy & Physiology** (John Murray), pp. 60-72.

See page 7 for details of publishers of periodicals:

STUDENT'S REFERENCE

■ **Serotonin - The Brain's Mood Modulator** Biol. Sci. Rev., 12(1), Sept. 1999, pp. 28-31. *The role of serotonin as a neurotransmitter.*

■ **Making the Connection** Biol. Sci. Rev.,13(3) January 2001, pp. 10-13. *The central nervous system, neurotransmitters, and synapses.*

■ **Synapses** Biol. Sci. Rev., 8(4) March 1996, pp. 26-29. *Types and functions of synapses including synaptic plasticity and the neural basis of learning.*

■ **What's Your Poison** Bio. Sci. Rev. 16(2) Nov. 2003, pp. 33-37. *The action of naturally derived poisons on synaptic transmission. This account includes an account of toxins and their actions.*

■ **A Receptor** Biol. Sci. Rev., 12(2) Nov. 1999, pp. 6-7. *The nature of sensory receptors.*

■ **A Pacinian Corpuscle** Biol. Sci. Rev., 12(3) Jan. 2000, pp. 33-34. *An account of the structure and operation of a common pressure receptor.*

■ **Remodelling the Eye** Biol. Sci. Rev., 17(2) Nov. 2004, pp. 9-12. *An account of the structure of the human eye and disorders of vision.*

■ **Before Your Very Eyes** New Scientist 15 March 1997 (Inside Science). *Excellent account of eye structure, & perception & processing of visual info.*

■ **Colour Vision and Colour Blindness** Biol. Sci. Rev., 19(3) Feb. 2007, pp. 28-32. *Colour is a quality our brains give to light based on the different sensitivities of three types of cones in the retina to different wavelengths of light.*

■ **In the Realm of Your Senses** New Scientist, 31 Jan. 2004, pp. 40-43. *There is a genetic basis to our highly individualised sensory responses.*

■ **How Plants Know Their Place** Biol. Sci. Rev., 17(3) Feb. 2005, pp. 33-36. *Plant responses to light and the action of phytochromes.*

■ **Sending Plants around the Bend** Biol. Sci. Rev., 12(4) March 2000, pp. 14-17. *How plants perceive and respond to stimuli around them.*

■ **The Autonomic Nervous System** Biol. Sci. Rev., 18(3) Feb. 2006, pp. 21-25. *Description of the structure and roles of the ANS.*

■ **Beyond the Brain** National Geographic March 2005, .pp. 2-31. *Brain development in babies, musicians, autistic people, meditating monks etc.*

See pages 10-11 for details of how to access **Bio Links** from our web site: **www.biozone.co.uk**. From Bio Links, access sites under the topics:

GENERAL BIOLOGY ONLINE RESOURCES > Online Textbooks and Lecture Notes: • S-Cool! A level biology revision guide *... and others*

ANIMAL BEHAVIOUR: • Animal behaviour • Innate behaviour • Pavlovian conditioning • Sign stimuli and motivation... *and others*

ANIMAL BIOLOGY > Neuroscience: • Basic neural processes • Cow's eye dissection • Nervous system • Seeing, hearing, smelling the world • The human eye *... and others*

PLANT BIOLOGY: • Kimball's plant biology lecture notes *... and others* > **Hormones & Responses:** • Botany online: Plant hormones • Plant hormones • Plant hormones and growth regulators • Plant tropisms • Plants in motion

Software and video resources for this topic are provided on the Teacher Resource CD-ROM

The Mammalian Nervous System

The **nervous system** is the body's control and communication centre. It has three broad functions: detecting stimuli, interpreting them, and initiating appropriate responses. Its basic structure is outlined below. Further detail is provided in the following pages.

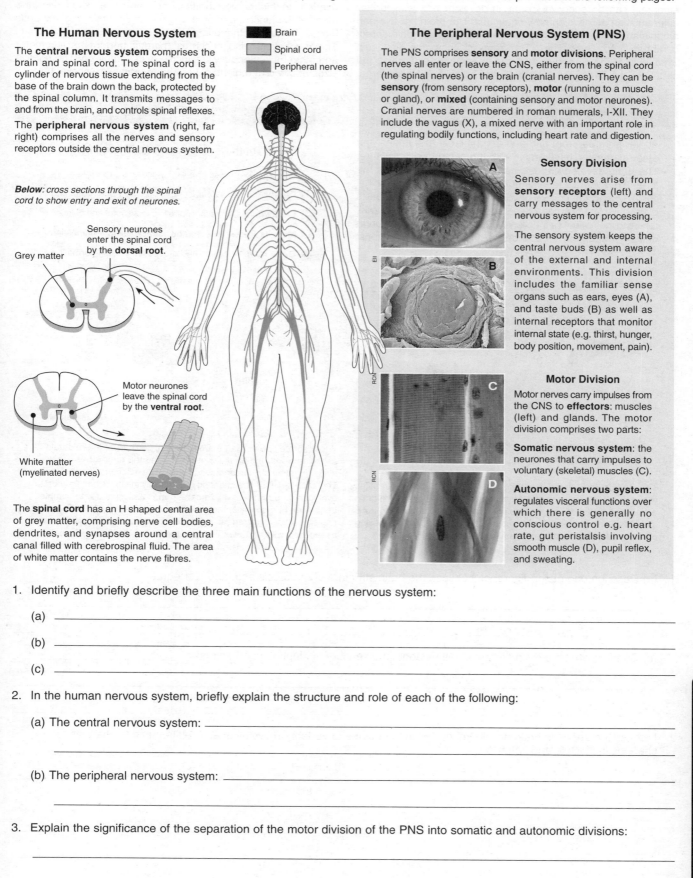

The Human Nervous System

The **central nervous system** comprises the brain and spinal cord. The spinal cord is a cylinder of nervous tissue extending from the base of the brain down the back, protected by the spinal column. It transmits messages to and from the brain, and controls spinal reflexes.

The **peripheral nervous system** (right, far right) comprises all the nerves and sensory receptors outside the central nervous system.

Below: cross sections through the spinal cord to show entry and exit of neurones.

Grey matter

Sensory neurones enter the spinal cord by the **dorsal root**.

Motor neurones leave the spinal cord by the **ventral root**.

White matter (myelinated nerves)

The **spinal cord** has an H shaped central area of grey matter, comprising nerve cell bodies, dendrites, and synapses around a central canal filled with cerebrospinal fluid. The area of white matter contains the nerve fibres.

■ Brain
▨ Spinal cord
▨ Peripheral nerves

The Peripheral Nervous System (PNS)

The PNS comprises **sensory** and **motor divisions**. Peripheral nerves all enter or leave the CNS, either from the spinal cord (the spinal nerves) or the brain (cranial nerves). They can be **sensory** (from sensory receptors), **motor** (running to a muscle or gland), or **mixed** (containing sensory and motor neurones). Cranial nerves are numbered in roman numerals, I-XII. They include the vagus (X), a mixed nerve with an important role in regulating bodily functions, including heart rate and digestion.

Sensory Division

Sensory nerves arise from **sensory receptors** (left) and carry messages to the central nervous system for processing.

The sensory system keeps the central nervous system aware of the external and internal environments. This division includes the familiar sense organs such as ears, eyes (A), and taste buds (B) as well as internal receptors that monitor internal state (e.g. thirst, hunger, body position, movement, pain).

Motor Division

Motor nerves carry impulses from the CNS to **effectors**: muscles (left) and glands. The motor division comprises two parts:

Somatic nervous system: the neurones that carry impulses to voluntary (skeletal) muscles (C).

Autonomic nervous system: regulates visceral functions over which there is generally no conscious control e.g. heart rate, gut peristalsis involving smooth muscle (D), pupil reflex, and sweating.

1. Identify and briefly describe the three main functions of the nervous system:

    (a) _____

    (b) _____

    (c) _____

2. In the human nervous system, briefly explain the structure and role of each of the following:

    (a) The central nervous system: _____

    _____

    (b) The peripheral nervous system: _____

    _____

3. Explain the significance of the separation of the motor division of the PNS into somatic and autonomic divisions:

    _____

    _____

    _____

    _____

Responses and Coordination

Related activities: Nervous Regulatory Systems, The Vertebrate Brain, The Human Brain, The Autonomic Nervous System

A 2

The Vertebrate Brain

The vertebrate brain develops as an expansion of the anterior end of the neural tube in embryos. The forebrain, midbrain, and hindbrain can be seen very early in development, with further differentiation as development continues. The brains of fish and amphibians are relatively unspecialised, with a rudimentary cerebrum and cerebellum. The reptiles show the first real expansion of the cerebrum, with the grey matter external in the cortex. In the birds and mammals, the brain is relatively large, with well developed cerebral and cerebellar regions. In primitive vertebrates, the cerebral regions act primarily as olfactory centres. In higher vertebrates, the cerebrum takes over the many of the functions of other regions of the brain (e.g. the optic lobes), becoming the primary integration centre of the brain. The cerebellum also becomes more important as locomotor and other muscular activities increase in complexity. The relative sizes of different regions of vertebrate brains are shown below.

Vertebrate Brains

All vertebrates, from fish and amphibians, to humans and other mammals, have brains with the same basic structure. The brain develops from a hollow tube and comprises the forebrain, midbrain, and hindbrain (which runs into the spinal cord). During the course of vertebrate evolution some parts of the brain (e.g. the medulla) have remained largely unchanged, retaining their primitive functions. Other parts (e.g. the cerebrum of the forebrain) have expanded and taken on new functions.

Key to Brain Regions and their Functions

- **Olfactory bulb**: Receives and processes olfactory signals.
- **Cerebrum**: Behaviour, complex thought and reasoning.
- **Cerebellum**: Centre for controlling movement and balance.
- **Medulla**: Reflex functions and relay for sensory information.
- **Optic lobe**: Receives and processes visual information.

Reptile

Fish

Amphibian

Bird

Mammal

The medulla (part of the hindbrain) in fish, amphibians, and reptiles is prominent. It relays sensory information to or from parts of the brain associated with sensory processing. In these groups both the optic lobe (visual processing) and the centres associated with processing olfactory information (in the thalamus) are also large.

In vertebrates other than mammals, a major part of the midbrain is associated with the analysis of vision, and the optic lobes are very large. In mammals, the analysis of vision is a function of the forebrain. The forebrain itself has changed dramatically in size during vertebrate evolution. In birds most of the cerebrum is associated with complex behaviour. In mammals, there has been an progressive increase in the size and importance of the cerebrum; particularly the parts associated with complex thought, reasoning, and communication.

1. (a) Describe one major difference between the brain structure of mammals and other vertebrates:

_____

_____

(b) Suggest what brain structure can tell us about the sensory perception of an animal: _____

_____

_____

2. Discuss the trends in brain development during the course of vertebrate evolution and relate these to changes to changes in lifestyle and behaviour:

_____

_____

_____

_____

_____

_____

The Human Brain

The brain is one the largest organs in the body. It is protected by the skull, the meninges (membranous coverings), and the cerebrospinal fluid (CSF). The brain is the control centre for the body. It receives a constant flow of information from the senses, but responds only to what is important at the time. Some responses are very simple (e.g. cranial reflexes), whilst others require many levels of processing. The human brain is noted for its large, well developed cerebral region, and the region responsible for complex thought and reasoning. Each cerebral hemisphere is divided into four lobes by deep sulci or fissures. These lobes: temporal, frontal, occipital, and parietal, correspond to the bones of the skull under which they lie.

Primary Structural Regions of the Brain

Cerebrum: the cerebrum is divided into two cerebral hemispheres and has many, complex roles. It contains sensory, motor, and association areas, and is involved in memory, emotion, language, reasoning, and sensory processing.

Ventricles: Cavities containing the CSF, which absorbs shocks and delivers nutritive substances.

Thalamus is the main relay centre for all sensory messages that enter the brain, before they are transmitted to the cerebrum.

Hypothalamus controls the autonomic nervous system and links nervous and endocrine systems. Regulates appetite, thirst, body temperature, and sleep.

Cerebellum coordinates body movements, posture, and balance.

Midbrain
Pons
Medulla

Brainstem: Relay centre for impulses between the rest of the brain and the spinal cord. Controls breathing, heartbeat, and the coughing and vomiting reflexes.

MRI scan of the brain viewed from above. The visual pathway has been superimposed on the image. Note the crossing of sensory neurones to the opposite hemisphere and the fluid filled ventricles (V) in the centre.

Sensory and Motor Regions in the Cerebrum

General sensory area receives sensations from receptors in the skin, muscles and viscera. Sensory information from receptors on one side of the body crosses over to the opposite side of the cerebral cortex where conscious sensations are produced. The size of the sensory region for different body parts depends on the number of receptors in that particular body part.

Visual areas within the occipital lobe receive, interpret, and evaluate visual stimuli. In vision, each eye views both sides of the visual field but the brain receives impulses from left and right visual fields separately (see photo caption above). The visual cortex combines the images into a single impression or **perception** of the image.

Primary motor area controls muscle movement. Stimulation of a point one side of the motor area results in muscular contraction on the opposite side of the body.

Primary gustatory area interprets sensations related to taste.

Language areas: The motor speech area (Broca's area) is concerned with speech production. The sensory speech area (Wernicke's area) is concerned with speech recognition and coherence.

Auditory areas interpret the basic characteristics and meaning of sounds.

Parietal lobe

Occipital lobe

Frontal lobe

Temporal lobe

1. For each of the following bodily functions, identify the region(s) of the brain involved in its control:

(a) Breathing and heartbeat: _____

(b) Memory and emotion: _____

(c) Posture and balance: _____

(d) Autonomic functions: _____

(e) Visual processing: _____

(f) Body temperature: _____

(g) Language: _____

Related activities: The Vertebrate Brain

A 2

Responses and Coordination

The Malfunctioning Brain: The Effects of Alzheimer's Disease

Alzheimer's disease is a disabling neurological disorder affecting about 5% of the population over 65. Its causes are largely unknown, its effects are irreversible, and it has no cure. Sufferers of Alzheimer's have trouble remembering recent events and they become confused and forgetful. In the later stages of the disease, people with Alzheimer's become very disorientated, lose past memories, and may become paranoid and moody. Dementia and loss of reason occur at the end stages of the disease.

Upper Brain	Lower Brain
Normal	

Cerebral cortex: Conscious thought, reasoning, and language. Alzheimer's sufferers show considerable loss of function from this region.

Early Alzheimer's

Hippocampus: A swelling in the floor of the lateral ventricle. It contains complex foldings of the cortical tissue and is involved in the establishment of memory patterns. In Alzheimer's sufferers, it is one of the first regions to show loss of neurones and accumulation of amyloid.

Late Alzheimer's

It is not uncommon for Alzheimer's sufferers to wander and become lost and disorientated.

Alzheimer's is associated with accelerated loss of neurones, particularly in regions of the brain that are important for memory and intellectual processing, such as the cerebral cortex and hippocampus. The disease has been linked to abnormal accumulations of protein-rich **amyloid** plaques, which invade the brain tissue and interfere with synaptic transmission. The brain scans above show diminishing brain function in certain areas of the brain in Alzheimer's sufferers. Note, particularly in the two lower scans, how much the brain has shrunk (original size indicated by the dotted line). Light areas indicate brain activity.

(h) Muscular contraction: _____

(i) Sensory processing related to taste: _____

(j) Sensory processing related to sound: _____

2. Describe the likely effect of a loss of function (through injury) to the primary motor area in the left hemisphere:

_____

_____

3. Describe the role of the ventricles of the brain: _____

_____

_____

4. Some loss of neuronal function occurs normally as a result of ageing. Identify the features distinguishing Alzheimer's disease from normal age related loss of neuronal function:

_____

_____

_____

The Autonomic Nervous System

The **autonomic nervous system** (ANS) regulates involuntary visceral functions by means of **reflexes**. Although most autonomic nervous system activity is beyond our conscious control, voluntary control over some basic reflexes (such as bladder emptying) can be learned. Most visceral effectors have dual innervation, receiving fibres from both branches of the ANS. These two branches, the **parasympathetic** and **sympathetic** divisions, have broadly opposing actions on the organs they control (excitatory or inhibitory). Nerves in the parasympathetic division release acetylcholine. This neurotransmitter is rapidly deactivated at the synapse and its effects are short lived and localised. Most sympathetic postganglionic nerves release noradrenaline, which enters the bloodstream and is deactivated slowly. Hence, sympathetic stimulation tends to have more widespread and long lasting effects than parasympathetic stimulation. Aspects of autonomic nervous system structure and function are illustrated below. The arrows indicate nerves to organs or ganglia (concentrations of nerve cell bodies).

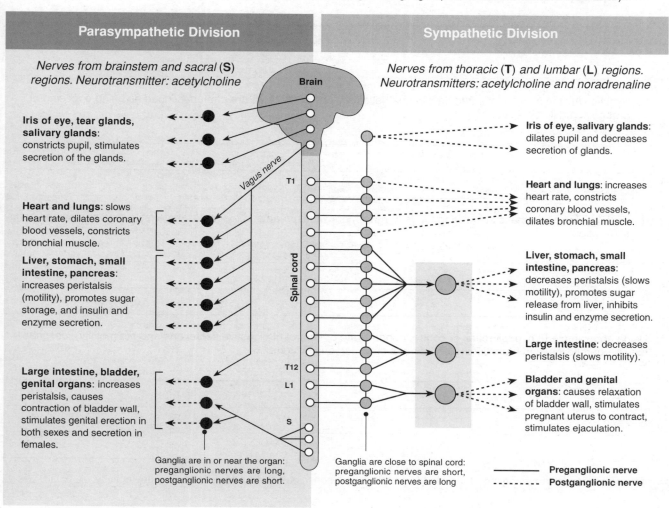

Parasympathetic Division

Nerves from brainstem and sacral (**S**) regions. Neurotransmitter: acetylcholine

Iris of eye, tear glands, salivary glands: constricts pupil, stimulates secretion of the glands.

Heart and lungs: slows heart rate, dilates coronary blood vessels, constricts bronchial muscle.

Liver, stomach, small intestine, pancreas: increases peristalsis (motility), promotes sugar storage, and insulin and enzyme secretion.

Large intestine, bladder, genital organs: increases peristalsis, causes contraction of bladder wall, stimulates genital erection in both sexes and secretion in females.

Ganglia are in or near the organ: preganglionic nerves are long, postganglionic nerves are short.

Sympathetic Division

Nerves from thoracic (**T**) and lumbar (**L**) regions. Neurotransmitters: acetylcholine and noradrenaline

Iris of eye, salivary glands: dilates pupil and decreases secretion of glands.

Heart and lungs: increases heart rate, constricts coronary blood vessels, dilates bronchial muscle.

Liver, stomach, small intestine, pancreas: decreases peristalsis (slows motility), promotes sugar release from liver, inhibits insulin and enzyme secretion.

Large intestine: decreases peristalsis (slows motility).

Bladder and genital organs: causes relaxation of bladder wall, stimulates pregnant uterus to contract, stimulates ejaculation.

Ganglia are close to spinal cord: preganglionic nerves are short, postganglionic nerves are long

——————— Preganglionic nerve
- - - - - - - - Postganglionic nerve

(Brain, Vagus nerve, Spinal cord, T1, T12, L1, S)

The Effects of the Autonomic Nervous System on the Body

When a person is fearful, their pupils enlarge as a consequence of sympathetic nervous system activity (the fight or flight response). The same response occurs during sexual arousal. Sympathetic stimulation also decreases secretion of the tear (lacrimal) glands.

Parasympathetic stimulation is responsible for bladder emptying through contraction of the bladder wall and relaxation of the urethral sphincter. This reflex activity can be inhibited by conscious control, but this ability does not develop until 2-4 years of age.

As part of the fight or flight response, the sympathetic nervous system dilates arteries and increases the rate and force at which the heart contracts. Heart rate is increased when increased blood flow to the heart causes reflex stimulation of the **accelerator** centre.

The **enteric nervous system** (ENS) is a recently recognised, interdependent part of the ANS. The ENS regulates itself but is influenced by sympathetic and parasympathetic nerves which are connected to it. The ENS innervates the gut, pancreas and gall bladder.

Related activities: Chemical Synapses, Drugs at Synapses

RA 2

Responses and Coordination

1. Explain the structure and role of each of the following divisions of the autonomic nervous system:

(a) The **sympathetic** nervous system: _____

_____

_____

_____

(b) The **parasympathetic** nervous system: _____

_____

_____

_____

2. (a) Explain why the sympathetic and parasympathetic divisions of the ANS are often described as being opposing or **antagonistic** in function:

_____

_____

_____

(b) Explain why sympathetic stimulation tends to have more widespread and longer lasting effects than parasympathetic stimulation:

_____

_____

_____

_____

3. Using an example (e.g. pupil reflex or control of heart rate or control of digestive secretions and gut motility), describe the role of reflexes in the functioning of the autonomic nervous system:

_____

_____

_____

_____

4. With reference to the emptying of the bladder, explain how conscious control can modify a reflex activity:

_____

_____

_____

_____

5. Predict how the sympathetic reflexes controlling blood vessels would respond to a sudden decrease in blood pressure:

_____

_____

_____

6. Asthma can be treated by inhaling a drug that mimics the action of noradrenaline on the sympathetic NS. Describe how this drug could be used to target the patient's respiratory system:

_____

_____

_____

Neurone Structure and Function

The nervous and endocrine systems are the body's regulatory and coordinating systems. Homeostasis depends on the ability of the nervous system to detect, interpret, and respond to, internal and external conditions. Sensory receptors relay information to the central nervous system (CNS) where it is interpreted and responses are coordinated. The information is transmitted along nerve cells (**neurones**) as electrical impulses. The speed of

impulse conduction depends primarily on the axon diameter and whether or not the axon is myelinated (see below). Within the tolerable physiological range, an increase in temperature also increases the speed of impulse conduction: in cool environments, impulses travel faster in endothermic than in ectothermic vertebrates. Neurones typically consist of a cell body, dendrites, and an axon. Basic types are described below.

The Structure of Neurones

Sensory neurone
Transmits impulses from sensory receptors to other neurones.

Dendron

Sense organ (pressure receptor) in the skin.

Cell body containing the organelles to keep the neurone alive and functioning.

Axon: A long extension of the cell transmits the nerve impulse to another neurone or to an **effector** (e.g. muscle). Axons may be very long and in the peripheral nervous system, many are myelinated.

Axon branches: Extensively branching with tiny knobs at each end. These release neurotransmitter chemicals which transmit the message between neurones or between a neurone and a muscle cell.

Relay neurone
Also called association or interneurones. Located in the CNS and carry impulses from sensory to motor neurones (as in reflexes).

Dendrites: Bushy extensions of the cell body, specialised to receive stimuli.

Axon branches

Axon

Axon surrounded by myelin sheath.

Axon branches

Cell body

Dendrites

Impulse direction

Myelin sheath

Area enlarged below

Motor neurone
Transmits impulses from the CNS to muscles or glands.

The Myelin Sheath
Many nerves outside the CNS are **myelinated**. Specialised **Schwann cells** form a tightly wrapped **myelin sheath** around the axon of the nerve. Myelin is rich in lipid and forms an electrical insulation layer around the axon. This increases the speed of impulse conduction.

At intervals along the axon, there are gaps between the sheath called *nodes of Ranvier*. The sheath prevents ion flow across the neurone membrane and forces the current to flow from node to node. In this way, impulses "jump" along the axon.

Axon diameter also affects the speed of impulse conduction; larger diameter axons conduct impulses more rapidly than axons with a smaller diameter. However, increasing speed must be balanced against the cost of increasing space, As a compromise, larger axons tend to be myelinated and conduct impulses very rapidly. Axons smaller than about 1 µm diameter tend to be non-myelinated and operate where conduction speed is less important.

Responses and Coordination

Related activities: The Mammalian Nervous System, Nervous Regulatory Systems
Web links: Nerve Action Potential

RA 2

A reflex is an automatic response to a stimulus involving a small number of neurones and a central nervous system (CNS) processing point (usually the spinal cord, but sometimes the brain stem). This type of circuit is often called a **reflex arc**. Reflexes permit rapid responses to stimuli. They are classified according to the number of CNS synapses involved: *monosynaptic reflexes* involve only one CNS synapse (e.g. knee jerk), *polysynaptic reflexes* involve two or more (e.g. pain withdrawal reflex). Both are spinal reflexes. The pupil reflex (opening and closure of the pupil) is an example of a cranial reflex.

A Polysynaptic Reflex Arc: Pain Withdrawal

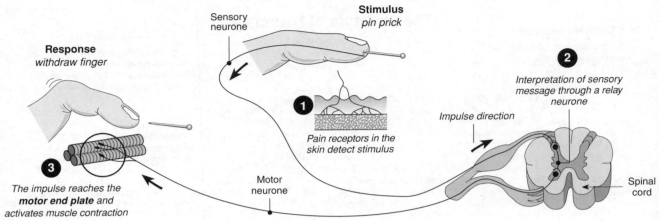

1. (a) Describe a structural difference between a motor and a sensory neurone: _____

_____

(b) Describe a functional difference between a motor and a relay neurone: _____

_____

2. (a) Predict what would happen to the part of an axon if it is cut so that it is no longer connected to its nerve cell body:

_____

(b) Explain your prediction: _____

_____

3. (a) Describe one way (other than insulation of the axon) in which impulse conduction speed could be increased:

_____

(b) Name one animal that uses this method: _____

(c) Describe the adaptive advantage of faster conduction of nerve impulses: _____

_____

4. (a) Briefly, describe the cause of the disorder, multiple sclerosis (MS): _____

_____

(b) Explain why MS impairs nervous system function even though axons are undamaged: _____

_____

5. (a) Explain why higher reasoning or conscious thought are not necessary or desirable features of reflexes:

_____

(b) Explain when it might be adaptive for conscious thought to intervene and modify a reflex action: _____

_____

6. Distinguish between a spinal and a cranial reflex and give an example of each: _____

_____

Transmission of Nerve Impulses

Neurones, like all cells, contain ions or charged atoms. Those of special importance include sodium (Na^+), potassium (K^+), and negatively charged proteins. Neurones are **electrically excitable** cells: a property that results from the separation of ion charge either side of the neurone membrane. They may exist in either a resting or stimulated state. When stimulated, neurones produce electrical impulses that are transmitted along the axon. These impulses are transmitted between neurones across junctions called **synapses**. Synapses enable the transmission of impulses rapidly all around the body.

Axon

Impulse travels in this direction

Na^+ Na^+

Area of impulse

Next area to be stimulated

Area returning to resting state

The Resting Neurone

When a neurone is not transmitting an impulse, the inside of the cell is negatively charged compared with the outside of the cell. The cell is said to be electrically polarised, because the inside and the outside of the cell are oppositely charged. The potential difference (voltage) across the membrane is called the resting potential and for most nerve cells is about -70 mV. Nerve transmission is possible because this membrane potential exists.

The Nerve Impulse

When a neurone is stimulated, the distribution of charges on each side of the membrane changes. For a millisecond, the charges reverse. This process, called **depolarisation**, causes a burst of electrical activity to pass along the axon of the neurone. As the charge reversal reaches one region, local currents depolarise the next region. In this way the impulse spreads along the axon. An impulse that spreads this way is called an **action potential**.

The Action Potential

The depolarisation described above can be illustrated as a change in membrane potential (in millivolts). In order for an action potential to be generated, the stimulation must be strong enough to reach the **threshold** potential; the potential (voltage) at which the depolarisation of the membrane becomes "unstoppable" and the action potential is generated. The action potential is **all or none** in its generation. Either the **threshold** is reached and the action potential is generated or the nerve does not fire. The resting potential is restored by the movement of potassium ions (K^+) out of the cell. During this **refractory period**, the nerve cannot respond.

Depolarisation: Na^+ influx speeds up causing rapid depolarisation of the membrane (called the all or none response).

Repolarisation: K^+ channels open and K^+ flows out.

Direction of impulse travel

Refractory period Na^+ channels inactivated and neurone cannot respond

Threshold potential

Resting potential

When stimulated, Na^+ channels open and Na^+ begins to flow into the cell.

Hyperpolarisation: Overshoot caused by a delay in closing the K^+ channels.

Membrane potential: +50 mV, 0, -50 mV, -70 mV

Elapsed time in milliseconds: 0, 1, 2, 3, 4, 5

1. Explain how an action potential is able to pass along a nerve: _____

_____

2. Explain how the refractory period influences the direction in which an impulse will travel: _____

_____

3. Action potentials themselves are indistinguishable from each other. Explain how the nervous system is able to interpret the impulses correctly and bring about an appropriate response:

_____

_____

Responses and Coordination

Related activities: Chemical Synapses

RA 2

Chemical Synapses

Action potentials are transmitted between neurones across synapses: junctions between the end of one axon and the dendrite or cell body of a receiving neurone. **Chemical synapses** are the most widespread type of synapse in nervous systems. The axon terminal is a swollen knob, and a small gap separates it from the receiving neurone. The synaptic knobs are filled with tiny packets of chemicals called **neurotransmitters**. Transmission involves the diffusion of the neurotransmitter across the gap,

where it interacts with the receiving membrane and causes an electrical response. The response of a receiving cell to the arrival of a neurotransmitter depends on the nature of the cell itself, on its location in the nervous system, and on the neurotransmitter involved. Synapses that release acetylcholine (ACh) are termed **cholinergic**. In the example below, ACh causes membrane depolarisation and the generation of an action potential (termed excitation or an excitatory response).

The Structure of a Cholinergic Synapse

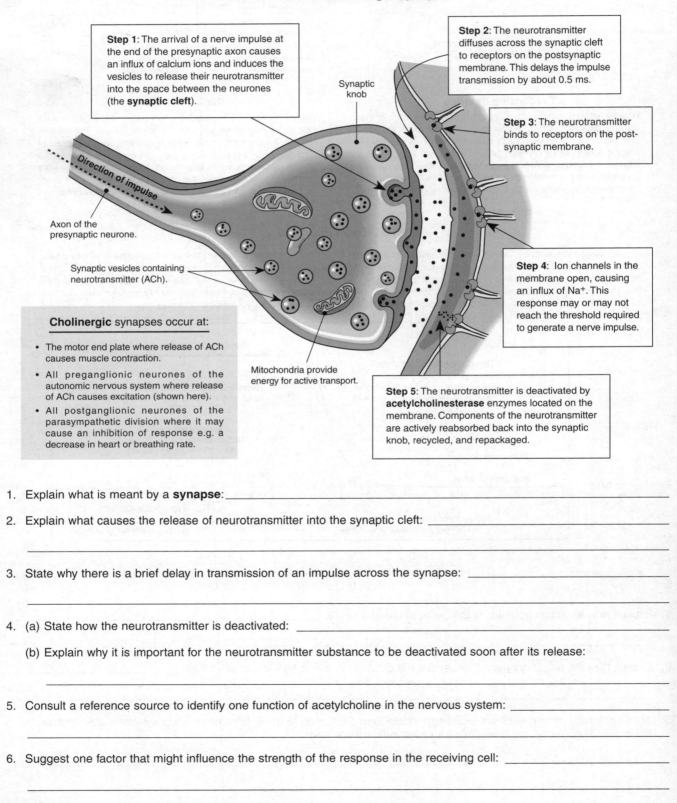

Step 1: The arrival of a nerve impulse at the end of the presynaptic axon causes an influx of calcium ions and induces the vesicles to release their neurotransmitter into the space between the neurones (the **synaptic cleft**).

Step 2: The neurotransmitter diffuses across the synaptic cleft to receptors on the postsynaptic membrane. This delays the impulse transmission by about 0.5 ms.

Step 3: The neurotransmitter binds to receptors on the post-synaptic membrane.

Step 4: Ion channels in the membrane open, causing an influx of Na^+. This response may or may not reach the threshold required to generate a nerve impulse.

Step 5: The neurotransmitter is deactivated by **acetylcholinesterase** enzymes located on the membrane. Components of the neurotransmitter are actively reabsorbed back into the synaptic knob, recycled, and repackaged.

Synaptic knob

Direction of impulse

Axon of the presynaptic neurone.

Synaptic vesicles containing neurotransmitter (ACh).

Mitochondria provide energy for active transport.

Cholinergic synapses occur at:

- The motor end plate where release of ACh causes muscle contraction.
- All preganglionic neurones of the autonomic nervous system where release of ACh causes excitation (shown here).
- All postganglionic neurones of the parasympathetic division where it may cause an inhibition of response e.g. a decrease in heart or breathing rate.

1. Explain what is meant by a **synapse**: _____

2. Explain what causes the release of neurotransmitter into the synaptic cleft: _____
   _____

3. State why there is a brief delay in transmission of an impulse across the synapse: _____
   _____

4. (a) State how the neurotransmitter is deactivated: _____

   (b) Explain why it is important for the neurotransmitter substance to be deactivated soon after its release:
   _____

5. Consult a reference source to identify one function of acetylcholine in the nervous system: _____
   _____

6. Suggest one factor that might influence the strength of the response in the receiving cell: _____
   _____

Related activities: The Autonomic Nervous System, Transmission of Nerve Impulses, Chemical Synapses, Integration at Synapses, Drugs At Synapses

Integration at Synapses

Synapses play a pivotal role in the ability of the nervous system to respond appropriately to stimulation and to adapt to change. The nature of synaptic transmission allows the **integration** (interpretation and coordination) of inputs from many sources. These inputs need not be just excitatory (causing depolarisation). Inhibition results when the neurotransmitter released causes negative chloride ions (rather than sodium ions) to enter the postsynaptic neurone. The postsynaptic neurone then becomes more negative inside (hyperpolarised) and an action potential is less likely to be generated. At synapses, it is the sum of **all** inputs (excitatory and inhibitory) that leads to the final response in a postsynaptic cell. Integration at synapses makes possible the various responses we have to stimuli. It is also the most probable mechanism by which learning and memory are achieved.

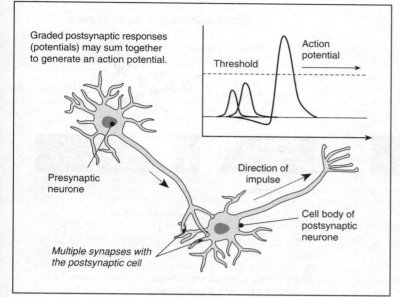

Graded postsynaptic responses (potentials) may sum together to generate an action potential.

Threshold

Action potential

Presynaptic neurone

Direction of impulse

Cell body of postsynaptic neurone

Multiple synapses with the postsynaptic cell

Synapses and Summation

Nerve transmission across chemical synapses has several advantages, despite the delay caused by neurotransmitter diffusion. Chemical synapses transmit impulses in one direction to a precise location and, because they rely on a limited supply of neurotransmitter, they are subject to fatigue (inability to respond to repeated stimulation). This protects the system against overstimulation.

Synapses also act as centres for the **integration** of inputs from many sources. The response of a postsynaptic cell is often graded; it is not strong enough on its own to generate an action potential. However, because the strength of the response is related to the amount of neurotransmitter released, subthreshold responses can sum to produce a response in the post-synaptic cell. This additive effect is termed **summation**. Summation can be **temporal** or **spatial** (below left and centre). A neuromuscular junction (photo below) is a specialised form of synapse between a motor neurone and a skeletal muscle fibre. Functionally, it is similar to any excitatory cholinergic synapse.

❶ Temporal summation

Presynaptic neurone

Action potential

Postsynaptic cell

Several impulses may arrive at the synapse in quick succession from a single axon. The individual responses are so close together in time that they sum to reach threshold and produce an action potential in the postsynaptic neurone.

❷ Spatial summation

Presynaptic neurones

Neurotransmitter

Individual impulses from spatially separated axon terminals may arrive **simultaneously** at different regions of the same postsynaptic neurone. The responses from the different places sum to reach threshold and produce an action potential.

❸ Neuromuscular junction

Axons

Motor end plate

Muscle fibre (cell)

The arrival of an impulse at the neuromuscular junction causes the release of acetylcholine from the synaptic knobs. This causes the muscle cell membrane (sarcolemma) to depolarise, and an action potential is generated in the muscle cell.

1. Explain the purpose of nervous system integration: _____

_____

2. (a) Explain what is meant by **summation**: _____

_____

(b) In simple terms, distinguish between temporal and spatial summation: _____

_____

_____

3. Describe two ways in which a neuromuscular junction is similar to any excitatory cholinergic synapse:

(a) _____

(b) _____

Related activities: Transmission of Nerve Impulses, Chemical Synapses

RA 3

Responses and Coordination

Drugs at Synapses

Synapses in the peripheral nervous system are classified according to the neurotransmitter they release; **cholinergic** synapses release acetylcholine (Ach) while **adrenergic** synapses release adrenaline or noradrenaline (NA). The effect produced by these neurotransmitters depends, in turn, on the type of receptors present on the postsynaptic membrane. Ach receptors are classified as nicotinic or muscarinic according of their response to nicotine or muscarine (a fungal toxin).

Adrenergic receptors are also of two types, alpha (α) or beta (β), classified according to their particular responses to specific chemicals. **Drugs** exert their effects on the nervous system by mimicking (**agonists**) or blocking (**antagonists**) the action of neurotransmitters at synapses. Because of the small amounts of chemicals involved in synaptic transmission, drugs that affect the activity of neurotransmitters, or their binding sites, can have powerful effects even in small doses.

Drugs at cholinergic synapses

RESULT: Depolarisation and impulse generation

Nicotine acts as a **direct agonist** at nicotinic synapses. Nicotine binds to and activates acetylcholine (Ach) receptors on the postsynaptic membrane. This opens sodium gates, leading to a sodium influx and membrane depolarisation. Some agonists work indirectly at the synapse by preventing Ach breakdown. Such drugs are used to treat elderly patients with Alzheimer's disease.

RESULT: No depolarisation

Atropine and **curare** act as antagonists at some cholinergic synapses. These molecules compete with Ach for binding sites on the postsynaptic membrane, and block sodium influx so that impulses are not generated. If the postsynaptic cell is a muscle cell, muscle contraction is prevented. In the case of curare, this causes death by flaccid paralysis.

Drugs at adrenergic synapses

RESULT: Neurone keeps firing

Under normal circumstances, the continued activity of the neurotransmitter noradrenaline (NA) at the synapse is prevented by reuptake of NA by the presynaptic neurone. **Cocaine** and **amphetamine** drugs act indirectly as agonists by preventing this reuptake. This action allows NA to linger at the synapse and continue to exert its effects.

RESULT: Response to noradrenaline is blocked or reduced.

Therapeutic drugs called **beta** (β) **blockers** act as direct antagonists at adrenergic synapses (sympathetic nervous system). They compete for the adrenergic β receptors on the postsynaptic membrane and block impulse transmission. Beta blockers are prescribed primarily to treat hypertension and heart disorders because they slow heart rate and reduce the force of contraction.

1. Providing an example of each, outline two ways in which drugs can act at a cholinergic synapse:

   (a) _____

   (b) _____

2. Providing an example, outline one way in which drugs can operate at adrenergic synapses: _____

   _____

3. Explain why atropine and curare are described as direct antagonists: _____

   _____

4. Suggest why curare (carefully administered) is used during abdominal surgery: _____

   _____

Detecting Changing States

A stimulus is any physical or chemical change in the environment capable of provoking a response in an organism. Animals respond to stimuli in order to survive. This response is adaptive; it acts to maintain the organism's state of homeostasis. Stimuli may be either external (outside the organism) or internal (within its body). Some of the stimuli to which humans and other mammals respond are described below, together with the sense organs that detect and respond to these stimuli. Note that sensory receptors respond only to specific stimuli. The sense organs an animal possesses therefore determine how it perceives the world.

Hair cells in the vestibule of the inner ear respond to **gravity** by detecting the rate of change and direction of the head and body. Other hair cells in the cochlea of the inner ear detect **sound** waves. The sound is directed and amplified by specialised regions of the outer and middle ear (pinna, canal, middle ear bones).

Photoreceptor cells in the eyes detect colour, intensity, and movement of **light**.

Olfactory receptors in the nose detect airborne **chemicals**. The human nose has about 5 million of these receptors, a bloodhound nose has more than 200 million. The taste buds of the tongue detect dissolved chemicals (gustation). Tastes are combinations of five basic sensations: sweet, salt, sour, bitter, and savoury (umami receptor).

Chemoreceptors in certain blood vessels, e.g. carotid arteries, monitor carbon dioxide levels (and therefore pH) of the blood. Breathing and heart rate increase or decrease (as appropriate) to adjust blood composition.

Baroreceptors in the walls of some arteries, e.g. aorta, monitor blood pressure. Heart rate and blood vessel diameter are adjusted accordingly.

Proprioreceptors (stretch receptors) in the muscles, tendons, and joints monitor limb position, **stretch**, and **tension**. The muscle spindle is a stretch receptor that monitors the state of muscle contraction and enables muscle to maintain its length.

Pressure deforms the skin surface and stimulates sensory receptors in the dermis. These receptors are especially abundant on the lips and fingertips.

Pain and temperature are detected by simple nerve endings in the skin. Deep tissue injury is sometimes felt on the skin as referred pain.

Humans rely heavily on their hearing when learning to communicate; without it, speech and language development are more difficult.

Breathing and heart rates are regulated in response to sensory input from chemoreceptors.

Baroreceptors and osmoreceptors act together to keep blood pressure and volume within narrow limits.

1. Provide a concise definition of a **stimulus**: _____

_____

2. Using humans as an example, discuss the need for communication systems to respond to changes in the environment:

_____

_____

_____

3. (a) Name one internal stimulus and its sensory receptor: _____

(b) Describe the role of this sensory receptor in contributing to homeostasis: _____

_____

_____

Related activities: The Basis of Sensory Perception
Web links: Sensory Systems

A 2

Responses and Coordination

The Basis of Sensory Perception

Sensory receptors are specialised to detect stimuli and respond by producing an electrical discharge. In this way they act as **biological transducers**, converting the energy from a stimulus into an electrochemical signal. Stimulation of a sensory receptor cell results in an electrical impulse with specific properties. The frequency of impulses produced by the receptor cell encodes information about the strength of the stimulus; a stronger stimulus produces more frequent impulses. Sensory receptors also show **sensory adaptation** and will cease responding to a stimulus of the same intensity. The simplest sensory receptors consist of a single sensory neuron (e.g. free nerve endings). More complex sense cells form synapses with their sensory neurons (e.g. taste buds). Sensory receptors are classified according to the stimuli to which they respond (for example, photoreceptors respond to light). The response of a simple **mechanoreceptor**, the Pacinian corpuscle, to a stimulus (pressure) is described below.

The Pacinian Corpuscle

Pacinian corpuscles are pressure receptors occurring in deep subcutaneous tissues all over the body. They are relatively large and simple in structure, consisting of a sensory nerve ending (dendrite) surrounded by a capsule of layered connective tissue. Pressure deforms the capsule, stretching the nerve ending and leading to a localized depolarization. Once a **threshold** value is reached, an **action potential** propagates along the axon.

Deforming the corpuscle leads to an increase in the permeability of the nerve to sodium. Na^+ diffuses into the nerve ending creating a localized depolarization. This depolarization is called a **generator potential**.

Pacinian corpuscle (above, left), illustrating the distinctive layers of connective tissue. The photograph on the right shows corpuscles grouped together in the pancreas. Pacinian corpuscles are rapidly adapting receptors; they fire at the beginning and end of a stimulus, but do not respond to unchanging pressure.

A volley of **action potentials** is triggered once the generator potential reaches or exceeds a **threshold value**. These action potentials are conducted along the sensory axon. A strong stimulus results in a high frequency of impulses.

1. Explain why sensory receptors are termed 'biological transducers': _____

_____

_____

2. Explain the significance of linking the magnitude of a sensory response to stimulus intensity: _____

_____

_____

3. Explain the physiological importance of sensory adaptation: _____

_____

4. (a) Describe the properties of a generator potential: _____

_____

(b) Suggest why a simple mechanoreceptor, such as the Pacinian corpuscle, does not fire action potentials unless a stimulus of threshold value is reached:

_____

_____

_____

Related activities: Transmission of Nerve Impulses, The Physiology of Vision
Web links: Neuron Information Coding and Transfer, Sensory Systems

Hearing

Most animals respond to sound and so have receptors for the detection of sound waves. In mammals, these receptors are organised into hearing organs called ears. Sound is produced by the vibration of particles in a medium and it travels in waves that can pass through solids, liquids, or gases. The distance between wave 'crests' determines the frequency (pitch) of the sound. The absolute size (amplitude) of the waves determines the intensity or loudness of the sound. Sound reception in mammals is the role of **mechanoreceptors**: tiny hair cells in the cochlea of the inner ear. The hair cells are very sensitive and are easily damaged by prolonged exposure to high intensity sounds. Gradual hearing loss with age is often caused by the cumulative loss of sensory hair cell function, especially at the higher frequencies. Such hearing loss is termed perceptive deafness.

The Human Ear

In mammals, sound waves are converted to pressure waves in the inner ear. The ears of mammals use mechanoreceptors (sensory hair cells) to change the pressure waves into nerve impulses. The mammalian ear contains not only the organ of hearing, the **cochlea**, but all the specialised structures associated with gathering, directing, and amplifying the sound. The cochlea is a tapered, coiled tube, divided lengthwise into **three fluid filled canals**. The cochlea is shown below, unrolled to indicate the way in which sound waves are transmitted through the canals to the sensory cells. The mechanisms involved in hearing are outlined in a simplified series of steps. In mammals, the inner ear is also associated with the organ for detecting balance and position (the vestibular apparatus), although this region is not involved in hearing.

2 Ear ossicles transmit the sound waves from the ear drum to the smaller oval window.

3 The sound waves are amplified at the **oval window**. They enter the fluid filled inner ear as pressure waves.

1 Ear drum vibrates in response to sound waves.

4 Pressure waves are transmitted through the fluid of the inner ear canals and across the membrane of the middle canal. The fluid is moved by the pressure waves.

7 At the end of the lower canal, the **round window** membrane bulges out, absorbing the shock of the fluid movement.

6 Each hair cell has a nerve fibre attached. The fibres form the auditory nerve sending nerve impulses to the brain.

5 The **membrane** bearing the sensory hair cells flexes in response to the fluid movement. When the membrane flexes, the hair cells are stimulated.

1. In a short sentence, outline the role of each of the following in the reception and response to sound:

(a) The ear drum: _____

(b) The ear ossicles: _____

(c) The oval window: _____

(d) The sensory hair cells: _____

(e) The auditory nerve: _____

2. Explain the significance of the inner ear being fluid filled: _____

_____

_____

Responses and Coordination

Related activities: The Basis of Sensory Perception

A 2

The Structure of the Eye

The eye is a complex and highly sophisticated sense organ specialised to detect light. The adult eyeball is about 25 mm in diameter. Only the anterior one-sixth of its total surface area is exposed; the rest lies recessed and protected by the **orbit** into which it fits. The eyeball is protected and given shape by a fibrous tunic. The posterior part of this structure is the **sclera** (the white of the eye), while the anterior transparent portion is the **cornea**, which covers the coloured iris.

The Structure and Function of the Mammalian Eye

The human eye is essentially a three layered structure comprising an outer fibrous layer (the sclera and cornea), a middle vascular layer (the choroid, ciliary body, and iris), and inner **retina** (neurones and **photoreceptor cells**). The shape of the eye is maintained by the fluid filled cavities (aqueous and vitreous humours), which also assist in light refraction. Eye colour is provided by the pigmented iris. The iris also regulates the entry of light into the eye through the contraction of circular and radial muscles.

Forming a Visual Image

Before light can reach the photoreceptor cells of the retina, it must pass through the cornea, aqueous humour, pupil, lens, and vitreous humour. For vision to occur, light reaching the photoreceptor cells must form an image on the retina. This requires **refraction** of the incoming light, **accommodation** of the lens, and **constriction** of the pupil.

The anterior of the eye is concerned mainly with **refracting** (bending) the incoming light rays so that they focus on the retina (below left). Most refraction occurs at the cornea. The lens adjusts the degree of refraction to produce a sharp image. **Accommodation** adjusts the eye for near or far objects. Constriction of the pupil narrows the diameter of the hole through which light enters the eye, preventing light rays entering from the periphery.

The point at which the nerve fibres leave the eye as the optic nerve, is the **blind spot** (the point at which there are no photoreceptor cells). Nerve impulses travel along the optic nerves to the visual processing areas in the cerebral cortex. Images on the retina are inverted and reversed by the lens but the brain interprets the information it receives to correct for this image reversal.

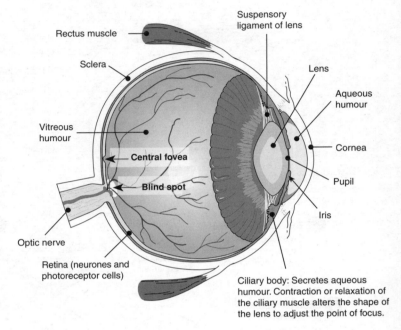

Ciliary body: Secretes aqueous humour. Contraction or relaxation of the ciliary muscle alters the shape of the lens to adjust the point of focus.

Normal vision	Accommodation for near and distant vision

In normal vision, light rays from an object are bent sufficiently by the cornea and lens, and converge on the central fovea. A clear image is formed. Images are focused upside down and mirror reversed on the retina. The brain automatically interprets the image as right way up.

The degree of refraction occurring at each surface of the eye is precise. The light rays reflected from an object 6 m or more away are nearly parallel to one another. Those reflected from near objects are divergent. The light rays must be refracted differently in each case so that they fall exactly on the central fovea. This is achieved through adjustment of the shape of the lens (**accommodation**). Accommodation from distant to close objects occurs by rounding the lens to shorten its focal length, since the image distance to the object is essentially fixed.

1. Identify the function of each of the structures of the eye listed below:

   (a) Cornea: _____

   (b) Ciliary body: _____

   (c) Retina: _____

   (d) Iris: _____

Related activities: The Physiology of Vision
Web links: Anatomy of the Eye

The Physiology of Vision

Vision involves essentially two stages: formation of the image on the retina (see previous activity), and generation and conduction of nerve impulses. When light reaches the retina, it is absorbed by the photosensitive pigments associated with the membranes of the photoreceptor cells (the rods and cones). The pigment molecules are altered by the absorption of light in such a way as to lead to the generation of nerve impulses. It is these impulses that are conducted via nerve fibres to the visual processing centre of the cerebral cortex (see the activity *The Human Brain* for the location of this region).

The Structure and Arrangement of Photoreceptor Cells in the Retina

Arrangement of photoreceptors and neurones in the retina

Structure of a rod photoreceptor cell

The photoreceptor cells of the mammalian retina are the **rods** and **cones**. Rods are specialised for vision in dim light, whereas cones are specialised for colour vision and high visual acuity. Cone density and visual acuity are greatest in the **central fovea** (rods are absent here). After an image is formed on the retina, light impulses must be converted into nerve impulses. The first step is the development of **generator potentials** by the rods and cones. Light induces structural changes in the **photochemical pigments** (or photopigments) of the rod and cone membranes. The generator potential that develops from the pigment breakdown in the rods and cones is different from the generator potentials that occur in other types of sensory receptors because stimulation results in a **hyperpolarisation** rather than a depolarisation (in other words, there is a net loss of Na^+ from the photoreceptor cell). Once generator potentials have developed, the graded changes in membrane conductance spread through the photoreceptor cell. Each photoreceptor makes synaptic connection with a bipolar neurone, which transmits the potentials to the **ganglion cells**. The ganglion cells become **depolarised** and initiate nerve impulses which pass through the optic chiasma and eventually to the visual areas of the cerebral cortex. The frequency and pattern of impulses in the optic nerve conveys information about the changing visual field.

1. (a) The first stage of vision involves forming an image on the retina. Explain what this involves (see the previous activity):

_____

_____

(b) Explain how accommodation is achieved (see the previous activity for help): _____

_____

_____

_____

2. Contrast the structure of the blind spot and the central fovea: _____

_____

_____

_____

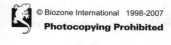
Related activities: The Structure of the Eye, The Human Brain
Web links: Eye Structure and Function

RA 3

Responses and Coordination

The Basis of Trichromatic Vision

·There are three classes of **cones**, each with a maximal response in either short (blue), intermediate (green) or long (yellow-green) wavelength light (coded B,G, and R below). The yellow-green cone is also sensitive to the red part of the spectrum and is often called the red cone (R). The differential responses of the cones to light of different wavelengths provides the basis of trichromatic colour vision.

Cone response to light wavelengths

Synaptic connection   Nucleus   Mitochondrion

Each **cone** synapses with only one bipolar cell giving high acuity.

Membranes containing bound **iodopsin** pigment molecules.

3. Complete the table below, comparing the features of rod and cone cells:

Feature	Rod cells	Cone cells
Visual pigment(s):		
Visual acuity:		
Overall function:		

4. Account for the differences in acuity and sensitivity between rod and cone cells: _____

_____

_____

_____

5. (a) Explain clearly what is meant by the term photochemical pigment (photopigment): _____

_____

_____

(b) Identify two photopigments and their location: _____

_____

6. In your own words, explain how light is able to produce a nerve impulse in the ganglion cells: _____

_____

_____

_____

_____

_____

7. Explain the physiological basis for colour vision in humans: _____

_____

_____

_____

_____

The Components of Behaviour

Behaviour in animals can be attributed to two components; **innate behaviour** that has a genetic basis, and **learned behaviour**, which results from the experiences of the animal. Together they combine to produce the total behaviour exhibited by the animal. It should also be noted that experience can modify innate behaviour. Animals behave in 'automatic ways' in many situations. The innate behaviour follows a classical pathway called a **fixed-action pattern** (FAP) where an innate behavioural programme is activated by a stimulus or **releaser** to direct some kind of behavioural response. Innate behaviours are generally adaptive and are performed for a variety of reasons. Learning, which involves the modification of behaviour by experience, occurs in various ways.

Innate Behaviours

Reflex behaviour
Simplest type of animal behaviour. A sudden stimulus induces an automatic involuntary and stereotyped response. Many reflexes are protective.

Kinesis
Random movement of an animal in which the rate of movement is related to the intensity of the stimulus, but not to its direction.

Taxis
A movement in response to the direction of a stimulus. Movement towards a stimulus are positive while those away from a stimulus are negative.

Stereotyped behaviour
Occurs when the same response is given to the same stimulus on different occasions. This behaviour shows fixed patterns of coordinated movements called fixed action patterns.

The complex behaviour patterns exhibited by an animal

Learned Behaviours

Classical conditioning
Animals may associate one stimulus with another.

Habituation
Response to a stimulus wanes when it is repeated with no apparent effect.

Insight behaviour
Correct behaviour on the first attempt where the animal has no prior experience.

Imprinting behaviour
During a critical period, an animal can adopt a behaviour by latching on to its first stimulus.

Operant conditioning
Also called trial and error learning; an animal is rewarded or punished after chance behaviour.

Fixed Action Pattern

A **releaser** (sign stimulus) triggers the operation of an **innate behavioural programme** in the brain that results in a **fixed-action pattern** (FAP), i.e. a predictable behavioural response.

Releaser → Innate behavioural programme → FAP

---

1. Distinguish between innate and learned behaviours: _____

   _____

2. (a) Explain the role of releasers in innate behaviours: _____

   _____

   (b) Name a releaser for a fixed action pattern and the animal involved, and describe the behaviour elicited:

   _____

   _____

Responses and Coordination

Related activities: Simple Behaviours, Learned Behaviour
Web links: Animal Behaviour Record

RA 3

Simple Behaviours

Taxes and kineses are examples of **orientation behaviours**. Such behaviours describe the way in which motile organisms (or gametes) **position** themselves and move in response to external cues (stimuli). Common stimuli are gravity, light, chemicals, and temperature. Some animals and many protozoa respond to certain stimuli simply by changing their rate of movement or by randomly turning without actually orientating to the stimulus. These movements are called **kineses**. In contrast, **taxes** involve orientation and movement directly to or away from one or more stimuli, such as temperature. Taxes often involve moving the head (which carries the sensory receptors) from side to side until the sensory input from both sides is equal (a klinotaxic response). Note that many taxic responses are complicated by a simultaneous response to more than one stimulus. For example, fish orientate dorsal side up by responding to both light and gravity. Male moths orientate positively to pheromones but use the wind to judge the direction of the odour source (the female moth).

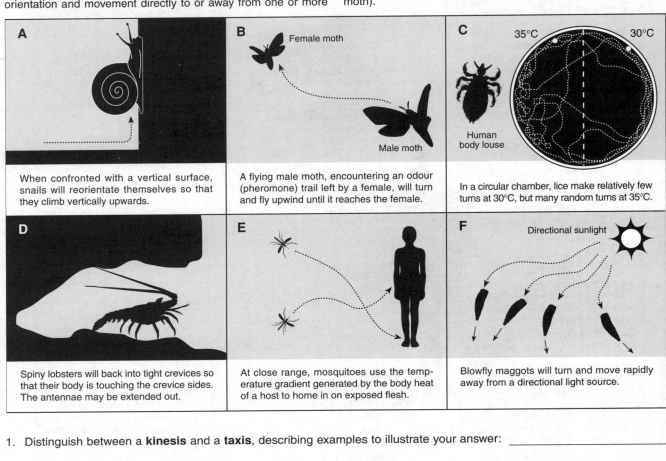

A When confronted with a vertical surface, snails will reorientate themselves so that they climb vertically upwards.

B Female moth / Male moth
A flying male moth, encountering an odour (pheromone) trail left by a female, will turn and fly upwind until it reaches the female.

C 35°C 30°C
Human body louse
In a circular chamber, lice make relatively few turns at 30°C, but many random turns at 35°C.

D Spiny lobsters will back into tight crevices so that their body is touching the crevice sides. The antennae may be extended out.

E At close range, mosquitoes use the temperature gradient generated by the body heat of a host to home in on exposed flesh.

F Directional sunlight
Blowfly maggots will turn and move rapidly away from a directional light source.

1. Distinguish between a **kinesis** and a **taxis**, describing examples to illustrate your answer: _____
   _____
   _____
   _____

2. Giving an example, describe the adaptive value of simple orientation behaviours such as kineses: _____
   _____
   _____

3. Name the physical stimulus for each of the following **prefixes** used in naming orientation responses:

   (a) Gravi- _____   (b) Hydro- _____   (c) Thigmo- _____

   (d) Photo- _____   (e) Chemo- _____   (f) Thermo- _____

4. For each of the above examples (A-F), describe the orientation response. Indicate whether the response is positive or negative (e.g. positive phototaxis):

   (a) **A:** _____   (b) **B:** _____

   (c) **C:** _____   (d) **D:** _____

   (e) **E:** _____   (f) **F:** _____

5. Suggest what temperature body lice "prefer", given their response in the chamber (in C): _____

Related activities: The Components of Behaviour, Pheromones
Web links: Assessing Habitat Preference

Learned Behaviour

Imprinting occurs when an animal learns to make a particular response only to one type of animal or object. Imprinting differs from most other kinds of learned behaviour in that it normally can occur only at a specific time during an animal's life. This **critical period** is usually shortly after hatching (about 12 hours) and can last for several days. While a critical period and the resulting imprinted behaviour are normally irreversible, they are not considered rigidly fixed. There are examples of animals that have had abnormal imprinted behaviours revert to the 'wild type'. There are two main types of imprinting using visual and auditory stimuli: filial and sexual imprinting. Breeding ground imprinting uses olfactory (smell) stimuli.

Filial (Parent) Imprinting

Filial imprinting is the process by which animals develop a social attachment. It differs from most other kinds of learning (including other types of imprinting), in that it normally can occur only at a specific time during an animal's life. This **critical period** is usually shortly after hatching (about 12 hours) and may last for several days. Ducks and geese have no innate ability to recognise *mother* or even their own species. They simply respond to, and identify as mother, the first object they encounter that has certain characteristics.

Breeding Ground Imprinting

Salmon undertake long migrations in the open ocean where they feed, grow and mature after hatching in freshwater streams. Some species remain at sea for several years, after which each fish returns to its exact home stream to spawn. Research has shown that this ability is based on **olfactory imprinting** where the fish recognise the chemical odours of their specific stream and swim towards its source. Other animals (bears, humans, etc.) have learned to exploit this behaviour.

Sexual Identity Imprinting

Individuals learn to direct their sexual behaviour at some stimulus objects, but not at others. Termed **sexual imprinting**, it may serve as a species identifying and species isolating mechanism. The mate preferences of birds have been shown to be imprinted according to the stimulus they were exposed to (other birds) during early rearing. Sexual imprinting generally involves longer periods of exposure to the stimulus than filial imprinting (*see left*).

Habituation

Habituation is a very simple type of learning involving a loss of a response to a repeated stimulus when it fails to provide any form of reinforcement (reward or punishment). Habituation is different to fatigue, which involves loss of efficiency in a repeated activity, and arises as a result of the nature of sensory reception itself. An example of habituation is the waning response of a snail attempting to cross a platform that is being tapped at regular time intervals. At first, the snail retreats into its shell for a considerable period after each tap. As the tapping continues, the snail stays in its shell for a shorter duration, before resuming its travel.

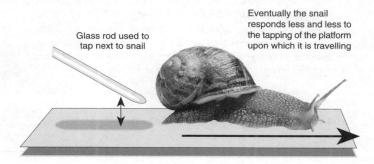

Glass rod used to tap next to snail

Eventually the snail responds less and less to the tapping of the platform upon which it is travelling

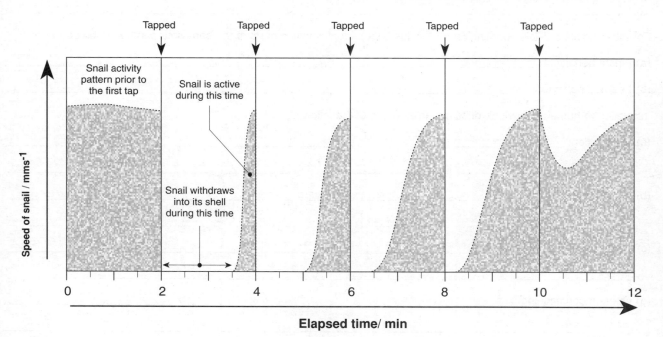

Tapped   Tapped   Tapped   Tapped   Tapped

Snail activity pattern prior to the first tap

Snail is active during this time

Snail withdraws into its shell during this time

Speed of snail / mm s⁻¹

0   2   4   6   8   10   12

Elapsed time/ min

Responses and Coordination

Related activities: The Components of Behaviour, Learning to Sing
Web links: Optimal Foraging

A 2

Classical (Pavlovian) Conditioning

Classical conditioning, founded by **Ivan Pavlov**, describes a type of **associative learning** in which behaviour that is normally triggered by a certain stimulus comes to be triggered by a substitute stimulus that previously had no effect on the behaviour. Between 1890 and 1900, Pavlov noticed that the dogs he was studying would salivate when they knew they were to be fed. It was determined that the dogs were alerted by a bell that rung every time the door into the lab was opened. Through experimentation, Pavlov discovered that the ringing of the bell initially brought about no salivation, but the dogs could be **conditioned** to relate the ringing of the bell to the presentation of food. Eventually the ringing of the bell elicited the same salivation response as the presentation of food, indicating that the dog was conditioned to associate the two stimuli.

1 A bell is rung immediately prior to feeding. The bell alone produces no salivary response in the dog.

2 Food is introduced after the bell has rung. Steps one are two are repeated a number of times (association of bell and food).

3 Eventually the dog becomes conditioned to salivate whenever the bell is rung, even when no food is presented.

Operant Conditioning

Operant conditioning is used to describe a situation where an animal learns to associate a particular behavioural act with a **reward** (as opposed to a stimulus in classical conditioning). This behaviour determines whether or not the reward appears. **Burrhus Skinner** studied operant conditioning using an apparatus he invented called a **Skinner box** *(see right)*. Skinner designed the box so that when an animal (usually a pigeon or rat) pushed a particular button it was rewarded with food. The animals learned to associate the pushing of the button with obtaining food (the reward). The behavioural act that leads the animal to push the button in the first place is thought to be generated spontaneously (by accident or curiosity). This type of learning is also called **instrumental learning** because the spontaneous behaviour is instrumental in obtaining the reward. Operant conditioning is the predominant learning process found in animals.

A Skinner box typically contains one or more buttons, which can be pressed to obtain a reward.

Food is dispensed when the correct button is pushed.

1. Explain what is meant by **filial imprinting**: _____

2. For the example of filial imprinting (see previous page), identify which parts of the behaviour can be attributed to:

   (a) Innate behaviour: _____

   (b) Learned behaviour: _____

3. In relation to human behaviour, describe an example of the following:

   (a) Habituation: _____

   (b) Imprinting: _____

   (c) Classical conditioning: _____

   (d) Operant conditioning: _____

Learning to Sing

Song birds use vocalisations (songs and calls) as a way to communicate, establish territories, and attract mates. The characteristics of a song may also be an indicator of fitness, as it has been shown that parasites and disease affect the song produced. While singing is instinctive, learning to sing the correct song is a learned behaviour, and without it, a song bird is unlikely to gain a territory or a mate. Analyses of many bird species show that there are at least two major strategies for song development:

(1) imitation of other birds, particularly adults of the same species, and (2) invention or improvisation. These strategies overlie the genetic template for the song learning process. The window during which a song can be learned varies between species. In some species, the inherited song pattern can be modified by learning only during early life. In others, the song is modified according to experience for at least another year, and some (e.g. blackbirds) modify their songs throughout life.

Learning to Sing

The songs of different bird species vary but are generally characteristic of the species. The structure of bird song is studied using a technique called **acoustic spectroscopy**, a technique which produces a graphical representation of the sounds being made. This enables song patterns to be compared between individuals and has led to experimental work to establish how birds learn song and how much of the song is genetically determined.

Blackbirds modify their songs throughout their life

The sound spectrographs of **chaffinch** song (below) illustrate how the final song that is produced can be altered by exposure to the songs of same-species individuals during the first three months when the song is learned. The upper trace shows the song of a normal male, while the lower trace is the song of a male reared in isolation from the nest. The isolated male's song is the right pitch and relatively normal in length, but it is simpler and lacks the acoustic 'flourish' at the end (arrow), which is typical of the chaffinch's song.

All data modified after A. Manning: An Introduction to Animal Behaviour (1979)

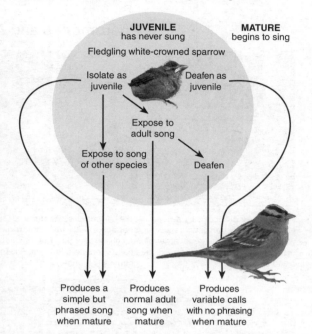

The diagram above summarises experiments investigating song development in **white-crowned sparrows**. This small finch has a wide range on the Pacific Coast of America and birds from different regions have different **song dialects**. The experiments found:

▶ Isolated males will eventually sing similar and simplified versions of the normal song, regardless of which region they come from.

▶ Isolated males can be trained to sing their own dialect by playing them tape recorded songs of birds from their region.

▶ After 4 months of age, birds are unreceptive to further learning.

▶ A bird needs to be able to hear itself in order to produce the normal song. It requires auditory feedback to adjust the notes.

▶ Once birds have learned their normal adult song, they can continue to sing normally even if deafened.

1. From the studies of song development in white-crowned sparrows, describe the evidence for the following statements:

(a) The inherited song pattern can be modified by learning only in the early stages of life: _____

_____

(b) Young birds need to hear themselves sing to produce a normal song as adults: _____

_____

(c) The song produced by an adult bird is determined by a genetic predisposition and what it hears when it is a juvenile:

_____

_____

2. Discuss the possible evolutionary significance of modifying a basic (genetically determined) song pattern by learning:

_____

_____

Responses and Coordination

Related activities: Learned Behaviour

A 2

Pheromones

A **pheromone** is a chemical produced by an animal and released into the external environment where it has an effect on the the physiology or behaviour of members of the same species. Hundreds of pheromones, some of which are sex attractants, are known. They are especially common amongst insects and mammals, and commonly relate to reproductive behaviour. Many mammals, including canids and all members of the cat family, commonly use scent marking to mark territorial boundaries and to advertise their sexual receptivity to potential mates. Other mammals, including rabbits, release a mammary pheromone that triggers nursing behaviour in the young. Phermones are also used as signalling molecules in social insects such as bees, wasps, and ants. They may be used to mark a scent trail to a food source or to signal alarm. Species specific odour cues are now widely used as bait in traps when controlling insect or mammalian pests or to capture animals for study.

Pheromones and Animal Communication

Pheromones produced by a honey bee queen and her daughters, the workers, maintain the social order of the colony. The pheromone is a blend of unsaturated fatty acids.

Like mammals, reptiles detect chemicals with the vomeronasal (Jacobson's) organ. The flicking of a snake's tongue allows the snake to chemically sample its environment.

In mammals, pheromones are used to signal sexual receptivity and territorial presence, or to synchronise group behaviour. Pheromone detection relies on the vomeronasal organ (VNO), an area of receptor tissue in the nasal cavity. Mammals use a flehmen response, in which the upper lip is curled up, to better expose the VNO to the chemicals of interest.

Photo courtesy of Cereal Research Centre, AAFC

Communication in ants and other social insects occurs through detection of pheromones. Foraging ants will leave a trail along the ground which other ants will follow and reinforce until the food source is depleted. Ants also release alarm substances, which will send other ants in the vicinity into an attack frenzy. These signals dissipate rapidly if not reinforced.

The feathery antennae of male moths are stereochemically specialised to detect the pheromone released by a female moth. Male moths can detect concentrations as low as 2ppm and will fly upwind toward the source to mate with the female. This sex attractant property of pheromones is exploited in pheromone traps, which are widely used to trap insect pests in orchards.

1. (a) Distinguish between hormones and **pheromones**: _____

   _____

   _____

   (b) Explain the significance of pheromones being species specific: _____

   _____

2. Giving examples, briefly describe the role of pheromones in three aspects of animals behaviour:

   (a) _____

   (b) _____

   (c) _____

3. From what you know of pheromone activity, suggest how a pheromone trap would operate to control an insect pest:

   _____

   _____

Related activities: Simple Behaviours, The Endocrine System
Web links: Cichlid Response to Pheromones

Breeding Behaviour

Many of the behaviours observed in animals are associated with reproduction, reflecting the importance of this event in an individual's life cycle. Many types of behaviour are aimed at facilitating successful reproduction. These include **courtship** behaviours, which may involve attracting a mate to a particular breeding site (often associated with high availability of resources such as food or nesting sites). Courtship behaviours are aimed at reducing conflict between the sexes and are often **stereotyped** or ritualistic. They rely on **sign stimuli** to elicit specific responses in potential mates. Other reproductive behaviours are associated with assessing the receptivity of a mate, defending mates against others, and rearing the young.

Courtship gift
The male gives the female a meal (an insect wrapped up in a cocoon) to keep her occupied while he mates.

FEMALE

MALE

Male hangs on
The male empid fly grips on to a twig with its front legs during mating. It uses the other four legs to grip on to the female.

Lock and key
The empid flies lock the tips of their abdomens together, so the male's sperm cells can enter the female's body. If the sperm cells were exposed to the air, they would dry out and die.

Mating Between Empid Flies

Female **empid flies** are aggressive hunters, so males have to be careful about how they approach them. Ritualised courtship behaviour by the male facilitates acceptance by the female during attempts to mate.

Courtship behaviour occurs as a prelude to mating and is most common in vertebrates and insects. One function of courtship is to synchronise the behaviours of the male and female so that mating can occur, and to override attack or escape behaviour. Here, a male greater frigatebird calls, spreads its wings, and inflates its throat pouch to court a female.

It is common in some birds (and many arthropods) for the male to provide an offering, such as food or nesting material, to the female. These **rituals** reduce aggression in the male and promote appeasement behaviour by the female. For some **monogamous** species, e.g. the blue-footed boobies mating above, the pairing begins a long term breeding partnership.

Many marine birds, such as the emperor penguins above, form breeding colonies where large numbers of birds come together to lay their eggs and raise their young. The adults feed at sea, leaving the young unattended for varying lengths of time. Penguin chicks congregate in large, densely packed groups to conserve body heat while their parents are away.

1. (a) Explain what is meant by **courtship behaviour**: _____

_____

(b) Suggest a reason why courtship behaviour may be necessary prior to mating: _____

_____

(c) Explain why courtship behaviour is often ritualised and involves stereotyped displays: _____

_____

_____

Responses and Coordination

Related activities: Sexual Selection

RA 2

Territorial Behaviour and Courtship Displays

A territory is any area that is defended against members of the same species. Territories are usually defended by clear acts of aggression or ritualised signals (e.g. vocal, visual or chemical signals).

Birds often engage in complex courtship behaviour. Males display in some way and females select their mates. Courtship displays are usually species specific and include ritualised acts such as dancing, feeding, and nest-building.

Resource availability often determines territory size and the population becomes spread out accordingly. Gannet territories are relatively small, with hens defending only the area they can reach while sitting on their nest.

(d) Describe one example of a courtship behaviour: _____

_____

2. Using an example, describe an advantage of cooperation during breeding: _____

_____

3. (a) Explain the nature and purpose of the courtship display in empid flies: _____

_____

(b) Identify the **sign stimulus** in this behaviour: _____

_____

4. (a) Explain what is meant by the formation of a **pair-bond**: _____

_____

(b) Using an example, describe an advantage of pair-bonding behaviour: _____

_____

_____

5. (a) Explain what is meant by a **territory**: _____

_____

(b) Explain why a male will defend a particular breeding territory against other males: _____

_____

(c) Suggest how a female in this situation would select a mate: _____

_____

6. Discuss the advantages and disadvantages of maintaining a territory: _____

_____

_____

_____

_____

_____

Plant Responses

Even though most plants are firmly rooted in the ground, they are still capable of responding and making adjustments to changes in their external environment. This ability is manifested chiefly in changing patterns of growth. These responses may involve relatively sudden physiological changes, as occurs in flowering, or a steady growth response, such as a tropism. Other responses made by plants include nastic movements, circadian rhythms, photoperiodism, dormancy, and vernalisation.

TROPISMS
Tropisms are growth responses made by plants to directional external stimuli, where the direction of the stimulus determines the direction of the growth response. A tropism may be positive (towards the stimulus), or negative (away from the stimulus). Common stimuli for plants include light, gravity, touch, and chemicals.

LIFE CYCLE RESPONSES
Plants use seasonal changes in the environment as cues for the commencement or ending of particular life cycle stages. Such changes are mediated by **plant growth factors**, such as phytochrome and gibberellin. Examples include flowering and other **photoperiodic responses**, dormancy and germination, and leaf fall.

RAPID RESPONSES TO ENVIRONMENTAL STIMULI
Plants are capable of quite rapid responses. Examples include the closing of **stomata** in response to water loss, opening and closing of flowers in response to temperature (photo, below right), and **nastic responses** (photos, below left). These responses often follow a circadian rhythm.

PLANT COMPETITION AND ALLELOPATHY
Although plants are rooted in the ground, they can still compete with other plants to gain access to resources. Some plants produce chemicals that inhibit the growth of neighbouring plants. Such chemical inhibition is called **allelopathy**. Plants also compete for light and may grow aggressively to shade out slower growing competitors.

PLANT RESPONSES TO HERBIVORY
Many plant species have responded to grazing or browsing pressure with evolutionary adaptations enabling them to survive constant cropping. Examples include rapid growth to counteract the constant loss of biomass (grasses), sharp spines or thorns to deter browsers (acacias, cacti), or toxins in the leaf tissues (eucalyptus).

Shoots are positively phototropic and grow toward the light.

Roots are positively gravitropic and grow towards the Earth's gravitational pull.

Some plants, such as *Mimosa* (above), are capable of **nastic responses**. These are relatively rapid, reversible movements, such as leaf closure in response to touch. Unlike tropisms, nastic responses are independent of stimulus direction.

The growth of tendrils around a support is a response to a mechanical stimulus, and is called thigmomorphogenesis.

The opening and closing of this tulip flower is temperature dependent. The flowers close when it is cool at night.

1. Identify the stimuli to which plants typically respond: _____

2. Explain how plants benefit by responding appropriately to the environment: _____

_____

_____

Responses and Coordination

Related activities: Auxins, Gibberellins, and ABA, Photoperiodism in Plants
Web links: Plant Defences, Plants in Motion

A 2

Auxins, Gibberellins, and ABA

Auxin was the first substance to be identified as a plant hormone. Charles Darwin and his son Francis were first to recognise its role in stimulating cell elongation, while Frits W. Went isolated this growth-regulating substance, which he called auxin. **Indole-acetic acid** (IAA) is the only known naturally occurring auxin. Shortly after its discovery it was found to have a role in suppressing the growth of lateral buds. This inhibitory influence of a shoot tip or apical bud on the lateral buds is called **apical dominance**. Two Japanese scientists isolated **gibberellin** in 1934, eight years after the isolation of auxin, and more than 78 gibberellins have now been identified. Gibberellins are involved in stem and leaf elongation, as well as breaking dormancy in seeds. Specifically, they stimulate cell division and cell elongation, allowing stems to 'bolt' and the root to penetrate the testa in germination. During the 1960s, Frederick T. Addicott discovered a substance apparently capable of accelerating **abscission** in leaves and fruit (which he called abscisin), and which is now called **abscisic acid** (ABA). Although it now seems that ABA has very little to do with leaf abscission, it is a growth inhibitor and also stimulates the closing of stomata in most plant species. It is also involved in preventing premature germination and development of seeds.

Auxins, Gibberellins, and Abscisic Acid (ABA) and Plant Responses

ABA stimulates the closing of stomata in most plant species. Its synthesis is stimulated by water deficiency (water stress).

Cytokinins promote the development of leaf buds in calluses (above). In contrast, increased auxin levels will promote the development of roots.

Gibberellins are responsible for breaking dormancy in seeds and promote the growth of the embryo and emergence of the seedling.

ABA promotes seed dormancy. It is concentrated in senescent leaves, but it is probably not involved in leaf abscission except in a few species.

Gibberellins cause stem and leaf elongation by stimulating cell division and cell elongation. They are responsible for **bolting** in brassicas.

Gibberellins are used to hasten seed germination and ensure germination uniformity in the production of barley malt in brewing.

ABA is produced in ripe fruit and induces fruit fall. The effects of ABA are generally opposite to those of cytokinins.

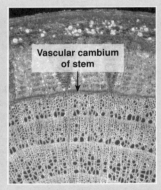

Auxins promote the activity of the vascular cambium (above), stem length, differentiation of tissues, and **apical dominance**.

1. Describe the role of **auxins** in apical dominance: _____

_____

_____

2. Describe the role of **gibberellins** in stem elongation and in the germination of grasses such as barley:

_____

_____

_____

3. Describe the role of abscisic acid in closure of stomata: _____

_____

Related activities: Plant Responses

Photoperiodism in Plants

Photoperiodism is the response of a plant to the relative lengths of daylight and darkness. Flowering is a photoperiodic activity; individuals of a single species will all flower at much the same time, even though their germination and maturation dates may vary. The exact onset of flowering varies depending on whether the plant is a short-day or long-day type (see next page). Photoperiodic activities are controlled through the action of a pigment called **phytochrome**. Phytochrome acts as a signal for some biological clocks in plants and is also involved in other light initiated responses, such as germination, shoot growth, and chlorophyll synthesis. Plants do not grow at the same rate all of the time. In temperate regions, many perennial and biennial plants begin to shut down growth as autumn approaches. During unfavourable seasons, they limit their growth or cease to grow altogether. This condition of arrested growth is called **dormancy**, and it enables plants to survive periods of water scarcity or low temperature. The plant's buds will not resume growth until there is a convergence of precise environmental cues in early spring. Short days and long, cold nights (as well as dry, nitrogen deficient soils) are strong cues for dormancy. Temperature and daylength change seasonally in most parts of the world, so changes in these variables also influence many plant responses, including germination and flowering. In many plants, flowering is triggered only after a specific period of exposure to low winter temperatures. As described in the previous activity, this low-temperature stimulation of flowering is called **vernalisation**.

Photoperiodism

Photoperiodism is based on a system that monitors the day/night cycle. The photoreceptor involved in this, and a number of other light-initiated plant responses, is a blue-green pigment called **phytochrome**. Phytochrome is universal in vascular plants and has two forms: active and inactive. On absorbing light, it readily converts from the inactive form (P_r) to the active form (P_{fr}). P_{fr} predominates in daylight, but reverts spontaneously back to the inactive form in the dark. The plant measures daylength (or rather night length) by the amount of phytochrome in each form.

Summary of phytochrome related activities in plants

Process	Effect of daylight	Effect of darkness
Conversion of phytochrome	Promotes $P_r \rightarrow P_{fr}$	Promotes $P_{fr} \rightarrow P_r$
Seed germination	Promotes	Inhibits
Leaf growth	Promotes	Inhibits
Flowering: long-day plants	Promotes	Inhibits
Flowering: short-day plants	Inhibits	Promotes
Chlorophyll synthesis	Promotes	Inhibits

Inactive phytochrome — In natural light, P_r converts rapidly to P_{fr}. In the dark, P_{fr} reverts slowly back to P_r. **Active phytochrome** — P_{fr} may trigger the synthesis of specific enzymes in specific cells (see table above) → **Response**

Day length and life cycle in plants (Northern Hemisphere)

The cycle of active growth and dormancy shown by temperate plants is correlated with the number of daylight hours each day (right). In the southern hemisphere, the pattern is similar, but is six months out of phase. The duration of the periods may also vary on islands and in coastal regions because of the moderating effect of nearby oceans.

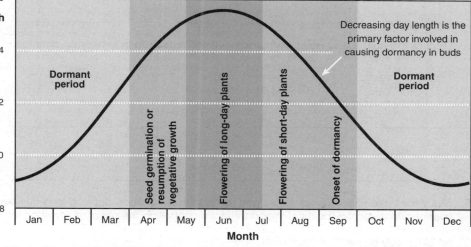

Decreasing day length is the primary factor involved in causing dormancy in buds

1. Describe two plant responses, initiated by exposure to light, which are thought to involve the action of phytochrome:

   (a) _____

   (b) _____

2. Discuss the role of phytochrome in a plant's ability to measure daylength: _____

   _____

   _____

   _____

   _____

Responses and Coordination

Related activities: Plant Responses

A 2

Long-day plants

When subjected to the light regimes on the right, the 'long-day' plants below flowered as indicated:

Flowering

No flowering

Flowering

Examples: *lettuce, clover, delphinium, gladiolus, beetscorn, coreopsis*

Photoperiodism in Plants

An experiment was carried out to determine the environmental cue that triggers flowering in 'long-day' and 'short-day' plants. The diagram below shows 3 different light regimes to which a variety of long-day and short-day plants were exposed.

Long night interrupted by a short period exposed to light

Short-day plants

When subjected to the light regimes on the left, the 'short-day' plants below flowered as indicated:

No flowering

Flowering

No flowering

Examples: *potatoes, asters, dahlias, cosmos, chrysanthemums, pointsettias*

3. (a) Identify the environmental cue that synchronises flowering in plants: _____

   (b) Describe one biological advantage of this synchronisation to the plants: _____

   _____

   _____

4. Discuss the role of environmental cues in triggering and breaking **dormancy** in plants:

   _____

   _____

   _____

5. Discuss the adaptive value of **dormancy** and **vernalisation** in temperate climates: _____

   _____

   _____

   _____

   _____

6. Study the three light regimes above and the responses of short-day and long-day flowering plants to that light. From this observation, describe the most important factor controlling the onset of flowering in:

   (a) Short-day plants: _____

   (b) Long-day plants: _____

7. Using information from the experiment described above, discuss the evidence for the statement "*Short-day plants are really better described as long-night plants.*"

   _____

   _____

   _____

   _____

   _____

Index

Advanced Biology AS

This companion title to Advanced Biology A2 provides coverage of the following topics for AQA-A, AQA-B, CIE, Edexcel, OCR & SNAB:

- *Skills in Biology*
- *Biological Molecules*
- *Cell Structure*
- *Cell Membranes and Transport*
- *The Genetic Code*
- *Cell Division and Cloning*
- *Gene Technology*
- *Gas Exchange*
- *Transport Systems*
- *Defence and the Immune System*

- *Aspects of Human Health*
- *Infectious Disease*
- *Energy and Ecosystems*
- *Sexual Reproduction*

Price: £10.95 for student purchase.
This price represents a 35% discount off the RRP.
NOTE: ADD £3.00 for Postage & Packaging.

Model Answers

*Model Answers books are available for **both** Advanced Biology AS and Advanced Biology A2. Each provides suggested answers to nearly all of the activities in the workbook.*

Price: £4.95 each
NOTE: ADD £1.00 for Postage & Packaging

How to Order

1. **Clearly list all titles you want to purchase.**
2. **Enclose a cheque, including postage and packaging.**
3. **Include your name, address, and the name of your school.**
4. **Send to address below:**

BIOZONE Learning Media
P.O. Box 23698
Edinburgh, EH5 2WX, UK
Phone: 0131 557 5060
Email: sales@biozone.co.uk

Please do not send cash. Sorry, we do not accept credit cards, nor will we send books without first receiving payment. Your order will be dispatched the same day your payment is received.